A Model for the Flow of Students
through the Swiss University
System

Europäische Hochschulschriften

Publications Universitaires Européennes
European University Papers

Reihe XXII

Soziologie

Série XXII Series XXII

Sociologie
Sociology

Bd./Vol. 21

PETER LANG
Berne · Frankfurt/M. · Las Vegas

Hanspeter Kriesi

A Model for the Flow of Students through the Swiss University System

PETER LANG
Berne · Frankfurt/M. · Las Vegas

Printed from Manuscript

ISBN 3-261-02126-8
Auflage 300 Ex.

Printed by Fotokop Wilhelm Weihert KG, Darmstadt

For Rachel

PREFACE

In writing this dissertation, I have tried to satisfy two quite conflicting principles: I have attempted to write a text readily comprehensible for the educated layman and, at the same time, I have tried to stay as close to the figures as possible.

According to the first principle, the model outlined is not intended for scientific use only, but the policy makers and the administrators in the educational system should be able to benefit from it, too. I believe sociology to inherently be an applied science, designed to provide the tools to cope with the problems of present day society (1).

Today's discussion on higher education in Switzerland, however, seems to focus on two issues (2): The restriction of the input into the system ('numerus clausus') and the possible overabundance of graduates from the system. Although these issues are highly related, they are not always seen as such. By linking the input into the university system with the output therefrom in a systematic fashion, the present model provides an instrument for treating the two issues in a unified perspective.

As to the second principle, I have been so much preoccupied with methodological questions and computation techniques, because I wanted to give the interested reader all the instruments he needs to himself use the model. Secondly, I have stressed methodological considerations, because I also believe that

---

1) Prof. Joseph Ben-David from the University of Chicago has instilled this view in me.

2) For a recent point of view on these issues expressed by a Swiss cantonal government, see: Botschaft des Regierungsrats des Kantons Aargau an den Grossen Rat: Hochschulgesetz, Aarau, den 19. Januar 1976.

sociological results cannot be understood correctly without having a good grasp of the methodology involved in deriving them.

I hope to have done some justice to both of these principles. In writing this dissertation, I have incurred the debt of a number of persons. To begin with, I am grateful to Dr. Heinz Ries from the Preparatory Stage of the University for Educational Sciences at Aarau for having left the data to me for the purposes of this model. These data have been collected in the course of the research project on the cohort 1965 of Swiss students headed by Dr. Ries. It is only due to the far-sighted planning of Dr. Ries that they have been made available in the great detail required by the model.

I was very lucky to have had the opportunity to discuss the outline of this dissertation with Prof. David D. McFarland and Prof. James S. Coleman from the University of Chicago. I have greatly benefitted from the advice and the criticism of these two renowned scholars in the field of mathematical sociology. In particular, Prof. McFarland has helped me with respect to Markovian models and it was Prof. Coleman, who suggested the methodological innovation developed in Part II.

The people from the computation center of the University of ZUERICH and Wolfgang Otto from the Institute of Social Psychology at the same university have time and again been helpful in debugging my programs and in the data management.

Above all, however, this dissertation owes to the most generous assistance and the trust I have received from Prof. Dr. Peter Heintz, who has been its supervisor at the University of ZUERICH. Without his help it would not have been written.

Zürich, 26th of April, 1976

Table of Contents

PART I:

A MODEL FOR THE FLOW OF STUDENTS THROUGH THE SWISS UNIVERSITY SYSTEM

1. INTRODUCTION

1.1. The Purpose of Part I

The purpose of Part I is to construct on an exmpirical basis a mathematical model for the flow of Swiss students through the Swiss University System. This model should allow a condensed description of the flow of students, the prediction of future numbers of students and graduates and the simulation of hypothetical developments.

The flow of students aimed at has two major aspects:

- The movement between the system and its environment: Students enter and leave the system at different points in time and space: They enter a specific subject at a specific university in a specific semester. They leave temporarily, when they interrupt their studies or study at foreign universities for a couple of semesters; or they leave definitely having received their final degree.
- The movements within the system: Students change subjects and universities, and they move up the academic hierarchy by acquiring degrees.

For both types of movements data have been collected by a research team at the preparatory stage of the University for Educational Sciences in Aarau, Switzerland.

1.2. The Data

Following the methodology of several studies conducted earlier in Germany and Switzerland (KATH u.a. 1966, HITPASS 1967, KRAEHE 1970, SCHALLBERGER 1971), the team at Aarau collected longitudinal data retrospectively. The total cohort of Swiss students entering the ten Swiss universities in 1965 was interviewed by a mailed questionnaire. The questionnaires were sent out in May 1973, i.e. they cover the period of 16 or 17 semesters for the cohort 1965 (depending on when the individual student started in the year 1965).

The data are subject to three major limitations:

- They pertain only to students of Swiss origin plus a small fraction of foreign students of the cohort 1965 who were still living in Switzerland at the time of the interviews and could thus be considered as having settled in this country. (These foreign students make up approximately 3% of the total number of students included in the study.) Since only a small group of the total number of foreign students (which make up about 25% of the total student population in Switzerland) was included, the data do not give an overall picture of the flow of students through the Swiss University System.
- The response rate of the mailed questionnaire was 81.4%. This is equivalent to 4302 students who answered the questionnaire. The respondents were shown to represent the total population rather well, with one notable exception: Dropouts are largely underrepresented. They make up 12.2% of the respondents, although it was possible to estimate from additional sources that they amount to at least 15.1% of the total cohort (RIES 1974).
- The data do not give a complete picture of the development of the studies of the cohort 1965, because they only cover the period up to the summer 1973. At this point, 27.4% of the cohort were still enrolled in the system (RIES/KRIESI, 1974c).

Whereas the first limitation places no restrictions on the model, if it is supposed to be constructed for Swiss students only, the other two limitations have serious consequences and have to be met by additional assumptions. These will be discussed in 2.4. below.

1.3. Elements of the Swiss University System: The States for the Model

The Swiss University System, first of all, consists of ten universities (see Table 1). There are three Swiss German speaking universities (BASEL, BERN, ZUERICH), three French speaking universities (GENEVE, LAUSANNE, NEUCHATEL), two technical schools (ETHL, ETHZ), a university for Economics and Business (HSG) at St. Gallen and the university of FRIBOURG. FRIBOURG is special in many respects: Its language is both, French and German, it is predominantly Catholic and, like the technical schools, but unlike the other universites which have primarily regional importance, it recruits students

Table 1

The States for the Model: Swiss Universities and Groups of Subjects

Universities	Groups of Subjects
Basel	Theology
Bern	Medicine
ETH Lausanne (ETHL)	Law
ETH Zürich (ETHZ)	Economics/Business
Fribourg	Social Sciences
HSG St. Gallen	Humanities
Genève	Historical Sciences
Lausanne	Natural Sciences I
Neuchâtel	Natural Sciences II
Zürich	Engineering
Foreign Universities	Architecture
	Others

from all parts of the country. (For a more detailed description of the universities, see RIES/KRIESI 1974b).

Internally, the universities are organized into faculties ('Fakultät') or departments (at the technical schools). The term 'faculty' however, denotes very diverse structures indeed (DEPPELER 1969). Some of these faculties resemble the American Department (e.g. the Veterinarian Faculties at the universities of Bern and Zürich). Others originally showed similarities with the American Professional Schools (e.g. the Medical Faculty or the Faculty of Law at all the universities), but often have been enlarged by the addition of several related branches of science. Some of these faculties have developed into 'artificial conglomerates', sheer administrative structures with no clear-cut correlate in the disciplines of science (e.g. to Philosophical Faculty I and II).

It is, therefore, inadvisable to use the entity 'faculty' as state in the model. Instead, I have regrouped the 60 odd different subjects that are currently studied at Swiss universities into 12 fields of study (see Table 1). Some of these fields are quite heterogenous. Thus, the group of medicine not only consists of the usual branches of medicine, but also of dentistry, veterinarian sciences and pharmacy. Among the social sciences are included: Anthropology, education, philosophy, political science, psychology and sociology. The field called humanities is restricted to language studies, which primarily consist of the study of German, French, English, Italian, Spanish and the Classics in that order. Historical sciences include: Archeology, history proper, history of art and of music. The natural sciences are divided into two groups. Mathematics and physics make up the bulk of group I; chemistry, biology, zoology, geography, biochemistry and related subjects belong to group II. 'Others', finally, is a residual category which is included to make everything add up nicely, but which is not of any special interest. It includes sports sciences, interpreting and special courses in French. The constitution of the rest of the groups is self-explanatory.

It would, of course, be desirable to construct a model that takes into account both, the movements of students between different types of subjects and those between different universities. In such a model, the particular group of subjects at a specific university would be the level of analysis. Such a model is, however, not possible, because the 4300 cases of the cohort 1965 are too few to introduce such a multiplication of different states between which moves have to be estimated. Instead, there will have to be two distinct models, one for the movements between the universities, and one for the movements between the different groups of subjects.

Students not only move in space and between fields of study, they also make progress within their especialties. Now, the most striking aspect of the development of studies at Swiss universities is its diversity. The organization of studies varies not only from subject to subject, but also within subjects between different universities. There is a multiplicity of available degrees and a variety of ways to obtain them. For example, for students of the cohort 1965 it was possible to get a doctorate without ever having obtained any prior degree. Another way to get to the doctorate is to get a license ('Lizentiat') or a diploma first, and then to continue for the highest degree.

Although the Swiss hierarchy of degrees is far from being as clearly structured as the American one, it is possible to group the Swiss degrees into three major levels corresponding roughly to the three levels of the B.A., M.A. and Ph.D.: Degrees of the low level primarily include degrees for teaching at the lower secondary schools and for teaching sports. Medium level degrees range from professional degrees in medicine, pharmacy, law, theology, engineering and architecture as well as business to the rather unspecific diplomas and licenses which allow the student to continue for a Ph.D., if he passes at a certain level and is willing to go on. Degrees for teaching at higher secondary schools, the so called 'Gymnasium', are also included among the medium level ones. The high level degree, finally, is equivalent to the Ph.D.

Since the students do not always leave the system after having received a certain level, but go on studying for another degree on the same level or on a higher one, it would be desirable to differentiate the students in a specific state (a certain university or a certain group of subjects) according to the level they have already attained.

Although such a differentiation would be very interesting, the same argument that was used above in the case of the combination of universities and groups of subjects applies here, too: It would introduce a multiplication of states that could not be handled by the empirical basis.

There is a way around the dilemma of the multiplication of the states (see McFARLAND 1970, 1975; SPILERMAN 1972) that will be dealt with in the second part of this study.

At this point, the three levels of degrees plus dropping out will only be taken into consideration as four levels of leaving the system once and for all. If some student leaves the system after having received a degree, but then returns again to study for an additional degree, he will be treated as having interrupted his studies.

The four levels of definitely leaving the system plus interruption will be added as states to the 11 states of the university model, on the one hand, and to the 12 states of the subject model, on the other.

1.4. The Measurement of Time

Before turning to the description of the two models, one additional point has to be clarified: The movements through the university system are mapped into time. Now, the problem is how time is measured. This is by no means a trivial one. As James S. Coleman points out, there are three ways of measuring elapsed time: It can be measured, first, by calendar time, secondly, by the time one has to spend in a certain activity (e.g. as a student in a certain subject), and, finally, by the number of events that occur over time. The third measure, which has been called 'psychological time' (SØRENSEN 1972, 1974), can be disregarded for the problem presently considered. The difference between the first two types of measures, however, is well documented by Table 2 (next page). In this table, the amount of time spent in the university system by the cohort 1965 (duration status) is cross-tabulated with calendar time. The table shows that the later the point in time, the greater the variance of the distribution of the remaining students along the dimension of the duration status (1).

1) The upper limits of the distributions of the different columns are given by the definition of the variables, but the lower limits are, to some extent, artifacts of the construction of the table: In constructing the table, the duration status was allowed to differ from calendar time by at most 8 semesters. Even with this restriction, the construction used up a tremendous amount of computer time. The limitation resulting from this restriction, however, does not seem to have hampered the results at all.

Table 2
Duration Status (Number of Semesters passed in the System) vs. Calendar Time: Percentage Distribution of the Total Cohort

Duration Status	Calendar Time																
	S65 1	W65/66 2	S66 3	W66/67 4	S67 5	W67/68 6	S68 7	W68/69 8	S69 9	W69/70 10	S70 11	W70/71 12	S71 13	W71/72 14	S72 15	W72/73 16	S73 17
1	100	85.0	5.0	1.8	.4	.2	.2	.1	.1	.1							
2		15.0	80.9	7.7	4.9	.7	.4	.2	.2	.2	.2						
3			14.1	76.8	6.8	5.9	.8	.7	.2	.2	.2	-					
4				13.8	74.9	9.2	8.7	1.2	.9	.3	.2	.2	.1				
5					13.0	71.5	8.9	10.2	1.7	1.0	.4	.3	.2	-			
6						12.5	69.3	10.5	11.9	2.3	1.7	.8	.7	.3	.1		
7							11.6	66.0	10.0	12.7	2.9	2.0	.9	.8	.4	.1	
8								11.1	64.3	11.9	14.2	4.4	2.8	1.3	1.1	.7	.1
9									10.6	60.9	12.6	13.9	4.9	3.2	1.8	1.3	1.0
10										10.4	57.7	12.0	11.1	4.4	2.7	1.9	1.5
11											9.9	56.5	11.9	10.6	4.3	3.0	1.8
12												10.0	57.1	11.3	9.7	5.0	3.5
13													10.3	57.1	12.0	10.6	5.7
14														10.9	57.4	12.8	11.4
15															10.5	55.1	12.2
16																9.5	53.8
17																	9.1
Total	100	100	100	100	100	100	100	100	100	100	100	100	100	100	100	100	100
n	673	4189	4140	4078	4013	3885	3796	3621	3479	3246	2915	2604	2320	2069	1860	1512	1238

Table 3

Duration Status at Subsequent Calendar Times: Mean and Standard Deviation for Total Cohort

Calendar Time	Mean	S.D.
W65/66	1.150	.357
S66	2.091	.428
W66/67	3.025	.535
S67	3.925	.657
W67/68	4.885	.734
S68	5.801	.839
W68/69	6.721	.928
S69	7.650	1.018
W69/70	8.573	1.100
S70	9.475	1.194
W70/71	10.418	1.254
S71	11.397	1.345
W71/72	12.385	1.404
S72	13.367	1.450
W72/73	14.259	1.535
S73	15.178	1.607

Table 3 summarizes the information of Table 2 giving the mean and standard deviation of duration status at subsequent calendar times.

There are two reasons for this disparity between the two measures of time: Students of the cohort 1965 entered the system at two points in time. One sixth of them started in spring and five sixth in fall. That is, the first semester for the fall-starters is equivalent to the second semester of the spring-starters and so on all the way until the end of the period covered. The main diagonal in Table 2 is, therefore, only populated by spring-starters. But even if we consider the two subcohorts - spring-starters and fall-starters - separately (as is done in the Tables 4 and 5 respectively), we still find no congruence between the two measures of time. Responsible for the remaining disparity are the interruptions. Roughly 30% of the cohort have interrupted their studies at one point or another, 10.4% for one semester, 10.7% for two semesters, 3.2% for three and 4.6% for more semesters (see RIES/KRIESI 1974c).

This lack of congruence between the two measures of time has been overlooked by the Swiss public, which has been very critical about the large fraction of students still enrolled in the university system after 17 semesters. Only about one half of the spring-starters, a minority of the cohort to start out with, remaining in the system after 17 semesters actually had a duration status of 17 semesters. Among the fall-starters still enroled, less than two-thirds had completed a maximum of 16 semesters. Large fractions of the students still enroled had been in the system for a considerably shorter period of time.

What is left to do is to choose between these two measures of time. Calendar time is preferred, because it is the one that corresponds to the notion usually employed by administrators and politicians, people who might be interested in the results of the model.

Table 4

Duration Status (Number of Semesters Passed in the System) vs. Calendar Time: Percentage Distribution of the Spring Starters

Duration Status	Calendar Time S65 1	W65/66 2	S66 3	W66/67 4	S67 5	W67/68 6	S68 7	W68/69 8	S69 9	W69/70 10	S70 11	W70/71 12	S71 13	W71/72 14	S72 15	W72/73 16	S73 17
1	100	6.0	2.8	.9	.5	.2	.2	.2	.2	.2							
2		94.0	6.9	3.0	1.3	1.0	.5	.4	.4	.4	.5						
3			90.3	7.9	3.7	1.4	.9	.6	.2	.2	.2	-					
4				88.2	10.0	4.8	2.3	1.3	1.0	.2	.2	.5	.3				
5					84.5	10.5	5.7	3.6	2.4	1.3	.2	.3	.3	-			
6						82.2	11.6	5.4	3.2	2.6	2.4	.3	.5	.3	-		
7							78.8	11.7	5.6	3.2	2.4	2.6	.3	.6	.3	-	
8								76.8	12.5	6.6	3.8	2.6	2.7	.6	.3	.4	-
9									74.4	13.2	8.7	4.6	3.3	2.9	1.3	.4	.5
10										72.0	13.4	9.2	4.3	3.2	2.6	1.6	.5
11											68.2	14.0	9.8	5.7	3.9	2.8	1.5
12												66.1	13.6	9.2	5.8	4.5	4.0
13													65.0	12.9	10.0	7.7	5.0
14														64.8	12.3	10.1	7.4
15															63.4	14.2	9.9
16																58.3	15.3
17																	55.9
Total	100	100	100	100	100	100	100	100	100	100	100	100	100	100	100	100	100
n	673	668	648	636	619	589	561	522	497	468	424	392	369	349	309	247	

Table 5
Duration Status (Number of Semesters Passed in the System) vs. Calendar Time: Percentage Distribution of the Fall Starters

Duration Status	Calendar Time																
	S65 1	W65/66 2	S66 3	W66/67 4	S67 5	W67/68 6	S68 7	W68/69 8	S69 9	W69/70 10	S70 11	W70/71 12	S71 13	W71/72 14	S72 15	W72/73 16	S73 17
1	-	100	5.4	2.0	.4	.2	.2	.1	.1	-							
2			94.6	8.5	5.5	.6	.4	.2	.2	.1	.1						
3				89.5	7.3	6.8	.8	.7	.2	.3	.2	-					
4					86.7	10.0	9.9	1.2	.8	.3	.2	.2	.1				
5						82.4	9.5	11.3	1.6	1.0	.4	.3	.2	-			
6							79.2	11.4	13.4	2.3	1.6	.9	.7	.3	.1		
7								75.2	10.7	14.3	3.0	1.9	1.0	.9	.4	.2	
8									72.9	12.8	15.9	4.7	2.8	1.5	1.3	.7	.1
9										68.9	13.3	15.5	5.2	3.3	1.9	1.4	1.1
10											65.3	12.5	12.4	4.7	2.7	1.9	1.7
11												64.0	12.4	11.6	4.4	3.0	1.8
12													65.3	11.7	10.5	5.1	3.4
13														66.0	12.4	11.2	5.8
14															66.4	13.4	12.2
15																63.1	12.6
16																	61.3
17																	-
Total	100	100	100	100	100	100	100	100	100	100	100	100	100	100	100	100	100
n		3521	3492	3442	3394	3296	3235	3099	2982	2978	2491	2211	1951	1720	1551	1265	1036

2. THE MODEL

2.1. Description of the Model

There will be two models, one for the movements between universities and one for the flow between groups of subjects. The two will be identical except for the definition of part of the states. In both models two types of states can be distinguished: Transitory states and absorbing states. Transitory states are those, which, once entered, can also be left again. Absorbing states cannot be left anymore. In the case of the model for the universities, the transitory states consist of the ten universities, the foreign universities lumped together to form a single state, and a state for interruptions. As for the groups of subjects, there are 12 of these plus a state for interruptions. The absorbing states are the same for both models, consisting of the three levels of leaving the system permanently and the state for the dropouts.

Now, each model is basically constituted by a set of transition matrices. The elements of these matrices are transition rates expressing the rate of flow between any pair of states. These transition rates are defined as follows:

$\frac{n_{ij}(t,t+n)}{n_{i\cdot}(t)}$	=	probability of moving from state i at time t to state j at time t+n, where n stands for a number of semesters measured in calendar time
$n_{ij}(t,t+n)$	=	the number of students moving from state i to state j in the period between t and t+n
$n_{i\cdot}(t)$	=	the total number of students in state i at time t

The origin states i can only be transitory states, but both, transitory states and absorbing ones, are included among the destinations j.

Every transition matrix, say M(t,t+n), has two submatrices: A submatrix T(t,t+n) for the flow within the system and the temporary leaving of the system, and a submatrix Z(t,t+n) for the flow to the absorbing states:

$$M(t,t+n) = \left[\begin{array}{c|c} T(t,t+n) & Z(t,t+n) \end{array}\right]$$

Depending on the time period covered by the rates of flow in a matrix M, we can further distinguish two types of matrices:

- matrices consisting of transition rates between adjacent time periods, e.g. M(t,t+1), or, more generally, M(t+n-1,t+n)
- matrices consisting of transition rates between non-adjacent time periods, e.g. M(t,t+2), or, more generally, M(t,t+n)

Matrices between adjacent time periods will subsequently be referred to as matrices of type 1, the others will be called matrices of type 2.

Figure 1 gives an overview over the possible set of transition matrices for the university model (16 states). The hatched matrices are those of type 1, the rest is of type 2. In each matrix, the T submatrix is distinguished from the Z submatrix.

2.2. The Working of the Model

The working of the model closely follows the System of Accounts proposed by COLEMAN (197?) for a model of occupational mobility. In addition to the transition matrices just described, it requires one more type of information: The stocks of elements in the given states at a particular point in time have to be known. The stocks at time t, in the case of this model, are represented by the distribution of students over the different transitory states. The stocks are thus given by row vectors of 12 and 13 elements respectively, 12 for the

Figure 1

The Possible Set of Transition Matrices (for the University Model)

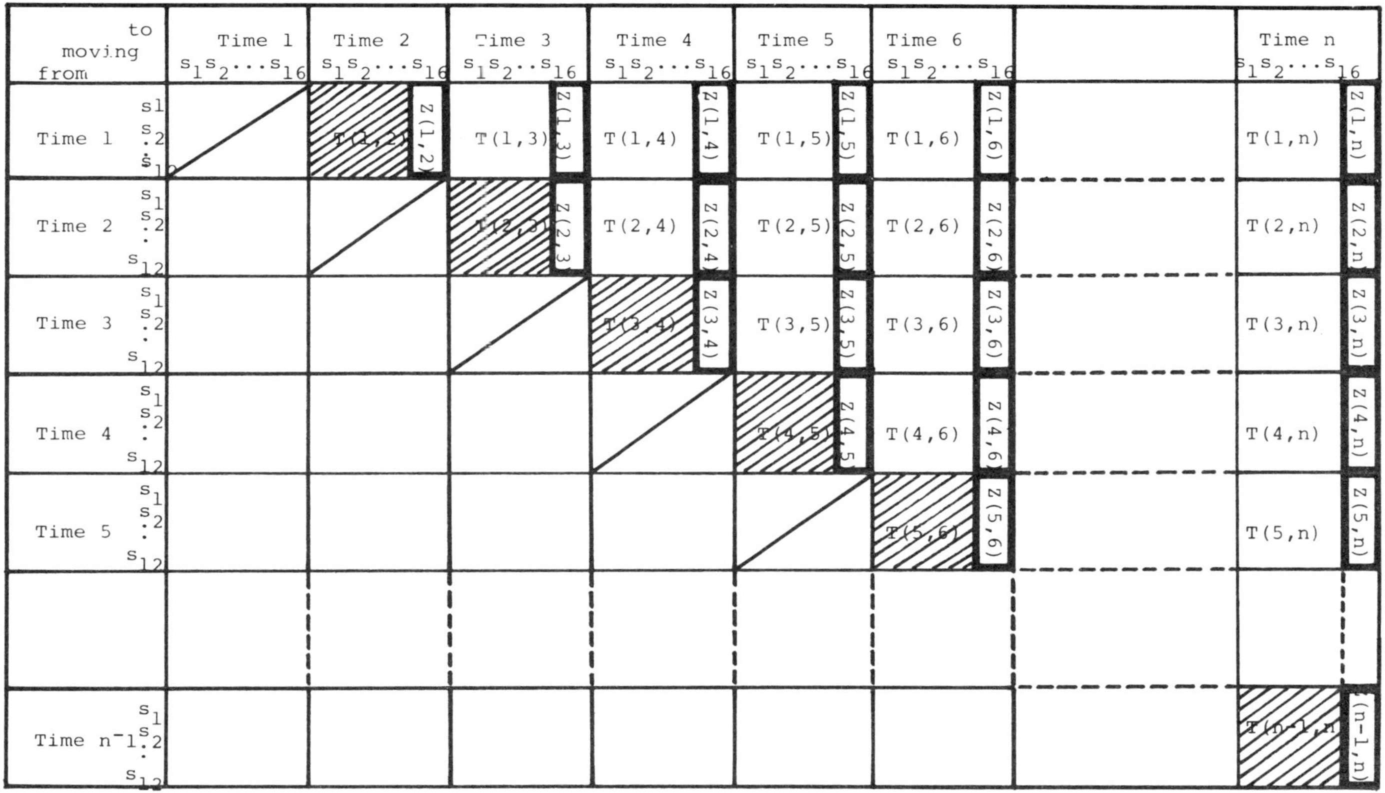

university model and 13 for the subject model.

By multiplication of these row vectors with the corresponding transition matrices, the stocks at time t can be transformed into stocks at subsequent points in time. Thus, by using type 1 matrices it is possible to compute a successive series of stocks, if the initial stocks are known.

Calling stocks at time t S(t), this series can be written as follows:

$$\begin{aligned} S(t+1) &= S(t) \cdot T(t,t+1) \\ S(t+2) &= S(t+1) \cdot T(t+1,t+2) \\ &\vdots \\ S(t+n) &= S(t+n-1) \cdot T(t+n-1,t+n) \end{aligned}$$

The number of students leaving the system between adjacent periods of time, call them L(t+n, t+n+1), follows a similar successive pattern:

$$\begin{aligned} L(t,t+1) &= S(t) \cdot Z(t,t+1) \\ L(t+1,t+2) &= S(t+1) \cdot Z(t+1,t+2) \\ &\vdots \\ L(t+n-1,t+n) &= S(t+n-1) \cdot Z(t+n-1,t+n) \end{aligned}$$

This successive scheme is, of course, not the only way to obtain stocks and leavers at subsequent points in time. By using type 2 matrices, it is possible to get the stocks at time t+n directly from the stocks at time t:

$$S(t+n) = S(t) \cdot T(t,t+n)$$

Similarly, the number of those leaving in the different absorbing states between time t and time t+n are directly given by the product

$$L(t,t+n) = S(t) \cdot Z(t,t+n)$$

Stocks of any two points in time can be related by a transition matrix of type 2.

2.3. The Construction of the Model

The construction of the model is based on the movements of the cohort 1965. The data collected for this cohort include:

- semester of entrance
- initial subject and initial university
- changes of subject and of university
- semester of occurrence of these changes
- interruptions and the semester of their occurrence
- date of graduation and degree received
- date of dropout
- current status (after 16 or 17 semesters)
- (for those still enroled) expected date of graduation and expected degree and degree already received

The first step in constructing the transition matrices is to obtain the stocks in the different semesters. A student is defined as belonging to state i of the stocks in semester n, if he is still in the system at that point in time and if he has already been in that state in the previous semester n-1, or, if he has moved to this state in semester n-1. In defining the stocks, the problem is that the data do not specify the exact date of the moves between universities and subjects, but only give the semester they were taken in. Thus, one cannot be sure, if the student moved to the new state already before or at the start of the semester, or, if he moved in the course or at the end of the term. The above definition assumes that students primarily move during the semester or at the end of it. The error due to this definition is, however, not very serious. Finally, if a student interrupts his studies in semester n, he belongs to the interruption state for that semester. APPENDIX B1 shows, how this definition is translated into SPSS statements in the case of the universities.

Once the stocks are obtained in this manner for any two semesters, we can crosstabulate them to get the number of moves between any two of their elements, i.e. we can get the n_{ij}'s.

For the moves to the absorbing states, the procedure to get to the n_{ij}'s is somewhat different. First, a variable is created that records the level of leaving for every case in the cohort 1965. Then, this variable is crosstabulated with the series of stocks, whereby the semester of leaving is controlled.

In order to compute the transition rates, we not only need to know the n_{ij}'s, but also the $n_{i.}$'s. But the derivation of these figures is very simple, because they correspond to the ith elements of the row vector for the origin stock.

If both, the n_{ij}'s and the $n_{i.}$'s are given, all that is left to do to arrive at the transition rates is to carry out the following multiplication:

$$\begin{bmatrix} \frac{1}{n_{1.}} & & & & \\ & \frac{1}{n_{2.}} & & & \\ & & \frac{1}{n_{3.}} & & \\ & & & \ddots & \\ & & & & \frac{1}{n_{12.}} \end{bmatrix} \cdot \begin{bmatrix} n_{ij} \end{bmatrix} = \begin{bmatrix} \frac{n_{ij}}{n_{i.}} \end{bmatrix}$$

The program in APPENDIX B2 performs this task. At the same time, this program punches the transition matrices and prints them.

2.4. Completing the Model: Adjusting for the Limitations of the Data

2.4.1. The Case of the Students still Enrolled

As was pointed out in the introduction, the data do not give a complete picture of the development of the studies of the cohort 1965, but cover only 16 or 17 semesters for fall- and spring-starters respectively. One way to complete the picture would be to set the type 1 transition matrices of the following semesters equal to the one of the last semester covered by the data. This would lead to a somewhat longer tail for the dying out of the process than could be expected from real data.

It is, however, not necessary to proceed in this fashion, because the data allow us to make some more realistic assmuptions. The 27.4% of the students still enrolled in the summer-semester 1973 (SS73) were asked, if they already had received a degree, if they expected to graduate and when and with what kind of degree. 60% of these students were already holding a degree and working for another one, and almost all of those not yet in possession of a degree (93.3%) indicated that they were expecting to get one. Only 33 students said they were not getting a degree or did not know yet, if they were ever getting one. For the students expecting a degree, the expected graduation dates show the following distribution over time:

Table 6

Expected Date of Graduation: Distribution of Total Number of Students still Enrolled in S73, who Expect to Graduate

Semester	raw	%
WS73/74	480	42.7
SS74	173	15.4
WS74/75	183	16.3
SS75	67	6.0
WS75/76	89	7.9
SS76	17	1.5
WS76/77	33	2.9
SS77	4	.4
WS77/78	4	.4
SS78	2	.2
WS78/79	8	.7
SS79	1	.1
WS79/80	1	.1
no date	63	5.6
Total	1125	100 %

If it is assumed that students who expect to graduate in the 'not too distant future' do this in realistically assessing their current progress, we can use dates and degrees they indicate as information for the estimation of the transition matrices for the following semesters. Students who mention a date too far away, on the other hand, probably cannot be taken seriously. They will most likely never graduate, but dropout some time before that date.

The decision about what is going to be 'too distant' necessarily is an arbitrary one. I decide that the cutting point is to lie after the conclusion of the SS76. This means that I assume that after 11 years or 22 semesters (23 for the minority of the spring-starters), the whole cohort 1965 will have left the university system.

This additional information allows an estimation of the moves to the absorbing states, but not for those between transitory states. Now, there is reason to assume that there will be practically no flows between the latter states in the semesters considered: As an inspection of the matrices of the last semesters for which there is detailed information shows, the flows between the transitory states are disappearing for all practical purposes.

Thus, the last transition matrices of the models can be estimated on the basis of these three assumptions (1):

- that informations about expected degrees are valid, if they specify a graduation date within three years from the completion of the questionnaire
- that expectations with a longer time horizon are unrealistic
- that there will be no more moves between transitory states after the 17th semester.

2.4.2 Adjusting for Dropout Errors

The second limitation of the data to be overcome is the underrepresentation of the dropouts among the respondents. As indicated, it was possible to show on the basis of additional information from university archives that at least 15.1% of the cohort have already dropped out before the SS73, as compared to the 12.1% among the respondents.

To get an overall dropout rate for the whole cohort, RIES (1975) has assumed

---

1) Excluded from this estimation are interruptions.

that all those still enroled in SS73 who were not yet in possession of a degree were potential dropouts. On the basis of this assumption he arrived at an overall dropout rate of 23%. SCHALLBERGER (1975) has critisized this assumption and has replaced it by one which posits that the proportion of dropouts among those still enrolled will be the same as among those who have already left the system. In doing so he obtains a dropout rate of 17% for the data used in this study and of 16% for his own data, which pertain to the cohort of 1961 at the University of Zürich and the ETHZ.

Both assumptions can be questioned: As has been seen in 2.4.1., 93.3% of those not yet in possession of a degree are expecting to get one and most of these students also mention a realistic date for graduation. The reasons for dropping out after a period of 17 and more semesters are, on the other hand, rather different from those of earlier dropouts. The late dropouts are a select group which does not necessarily reach similar proportions as the earlier ones. If we assume, as was done in 2.4.1., that those who do not mention a realistic graduation date are going to drop out, and if we add to this group the 33 students, who do not even know yet, if they are going to get a degree at all, we get a group of 149 late dropouts, i.e. 3.5% of the total cohort. Adding these 3.5% to the 15.1% yields an overall dropout rate of 18.6%, which is in fairly close agreement with SCHALLBERGER's estimate.

The transition matrices constructed according to the description in 2.3. do not take into consideration that the dropouts are underrepresented in the data available. They have to be adjusted to yield a more realistic dropout rate. The most adequate rate would be the 18.6% arrived at above. But, since it would have been rather difficult to retrieve the information necessary to adjust the rates of the different universities and groups of subjects correspondingly, I gave up on this ideal rate and chose the 15.1% rate instead. Thus, the dropouts are, most likely, underestimated by the models.

For the adjustment, a very simple procedure was chosen. First, an adjust-

Table 7

Adjusting the Drop out Rates: Drop out Rates Resulting from Additional Information, Original Rates and Correction Factor for Universities and Groups of Subjects

University/ Group of Subject	Rate using additional information (1)	Original Rate (2)	Correction Factor (1)/(2)
Basel	14.2	11.1	1.273
Bern	10.3	7.0	1.471
ETHL	18.3	5.2	3.519
ETHZ	12.7	9.5	1.337
Fribourg	12.9	9.9	1.303
HSG	10.5	7.2	1.458
Genève	22.9	15.1	1.517
Lausanne	19.0	15.1	1.258
Neuchâtel	11.7	9.8	1.194
Zürich	16.6	12.5	1.328
Theology	13.3	7.2	1.847
Medicine	9.7	6.6	1.470
Law	6.9	5.0	1.380
Econ/Bus	17.0	13.3	1.278
Social Sci	18.4	14.8	1.243
Humanities	21.0	15.1	1.391
Historical Sci	22.3	18.6	1.199
Natural Sci I	17.1	13.7	1.248
Natural Sci II	12.1	8.3	1.458
Engineering	14.7	10.6	1.387
Architecture	17.4	10.2	1.706
Others	22.7	12.3	1.846
Total	15.1	12.1	1.248

ment factor is computed for each university and group of subjects by dividing the dropout rate according to the complete information by the rate due to the questionnaire information alone. Table 7 shows the two series of dropout rates and the correction factors for each university and group of subjects.

Then, the transition rates in the column for dropouts in the different matrices are multiplied with the correction factor and the difference between the new transition rate and the original one is subtracted from the entry in the main diagonal of the matrix. In the case of the type 2 matrices, this procedure assumes that all the additional dropouts have come from among the stayers in the origin states. For type 2 matrices, this procedure, moreover, has the drawback that it can lead to negative numbers in the main diagonal. This happens, if the time period covered by the matrix is close to 22 semesters. This simple procedure also precludes adjustments for foreign universities and interruptions.

The second part of the program in APPENDIX B3 does the adjusting job. In the first part of this program, the stocks are recomputed, using the newly constructed transition matrices. Thus, a check is provided for the accuracy of the work done.

2.5. Description of the Transition Matrices

Four sets of transition matrices are provided in APPENDIX A, two sets for the university model and two sets for the subject model. The two sets in each model can be further distinguished according to their type:

	university model	subject model
Type 1	1.1. - 1.22. (p. 307-317)	3.1. - 3.22 (p. 329-339)
Type 2	2.1. - 2.21. (p. 318-328)	4.1. - 4.21. (p. 340-350)

The matrices provided are all unadjusted. The adjusted figures can, however, easily be derived from them as indicated above. Among the great number of possible type 2 matrices, one special series has been chosen, the one corresponding to the second row in Figure 1. It consists of the flows between the winter semester 1965 (WS65) and each of the following semesters.

The states are usually numbered in these matrices. The reader may find the following lists helpful for purposes of reference:

University Model	Subject Model
1 Basel	1 Theology
2 Bern	2 Medicine
3 ETHL	3 Law
4 ETHZ	4 Economics/Business
5 Fribourg	5 Social Sciences
6 HSG	6 Humanities
7 Genève	7 Historical Sciences
8 Lausanne	8 Natural Sciences I
9 Neuchâtel	9 Natural Sciences II
10 Zürich	10 Engineering
11 Foreign Universities	11 Architecture
12 Interruption	12 Others
13 Dropout	13 Interruptions
14 Low Level Degree	14 Dropout
15 Medium Level Degree	15 Low Level Degree
16 High Level Degree	16 Medium Level Degree
	17 High Level Degree

One final remark to the matrices: In some instances, stars are printed in the main diagonal cells. This means that either the corresponding value equals 1 (in the type 1 case), or that it has assumed a negative value (in the type 2 case). For actual computations, these values have been fixed by setting them equal to .999 and 0 respectively.

3. APPLICATIONS OF THE MODEL

The model is intended to serve three purposes: Description, prediction and simulation. I will now turn to description and then, in the following sections take up prediction and simulation.

3.1. Description of the Flow of Students through the Swiss University System

The matrices in APPENDIX A give a rather detailed description of the flow of the cohort 1965 through the Swiss University System. As a matter of fact, it is so detailed that it is difficult to capture the gist of it, although the model only takes into account a limited number of variables and is itself already the product of an elaborate condensation process. One way to make the information contained in these matrices more transparent is to graph it, another way is to represent it in yet another mathematical model.

3.1.1. The Flow between Universities

Roughly 20% of the cohort 1965 have ever changed the university; 11.8% once, 6.7% twice and 1.6% three or more times (see RIES/KRIESI 1974c). The question is: Do these changes follow any pattern?

To answer the question, I have gone back to the raw figures used to compute the transition matrices of type 1, i.e. to the n_{ij}'s. For each pair of state, I have counted the total number of moves between them in all the semesters covered by the model. Table 8 gives an overview over the total number of moves. In Table 9 these raw figures are transformed into percentages. Percentage basis is the total number of students originally enroled at the university, where the move originated. The tables show that both, in raw and in percentage terms, the flows between any two Swiss universities are not very impressive. In one fifth of the 90 possible directions no movements occur at all.

Table 8

Synopsis of The Total Number of Moves of the Cohort 1965 between the Swiss Universities

From/To	Basel	Bern	ETHL	ETHZ	Frib.	HSG	Genèv	Laus.	Neuch.	Zürich	Foreig	Interr.	Original Enrolement
Basel	-	16	-	4	10	-	9	6	1	15	41	128	407
Bern	9	-	1	8	6	1	3	14	3	30	45	174	512
ETHL	-	-	-	6	1	1	3	11	-	-	-	23	97
ETHZ	9	9	3	-	7	4	6	1	2	51	16	447	837
Fribourg	12	23	1	2	-	-	11	10	1	34	19	74	323
HSG	2	4	-	-	-	-	1	1	1	7	9	72	139
Genève	14	4	2	-	6	4	-	6	1	21	24	98	409
Lausanne	3	17	1	2	5	-	17	-	4	31	17	91	317
Neuchâtel	-	8	-	-	-	1	10	26	-	6	2	32	174
Zürich	14	27	-	9	19	13	22	27	4	-	105	297	880
Foreign U.	16	42	-	5	11	5	10	8	1	82	-	18	
Interruption	121	174	26	455	67	66	93	90	28	276	26	-	

Table 9

Total Number of Moves of the Cohort 1965 between the Swiss Universities: Percentages of the Original Enrolement at the Universities, where the Moves Originated

From/To	Basel	Bern	ETHL	ETHZ	Frib.	HSG	Genève	Lausan.	Neuch.	Zürich	Foreign	Interr.
Basel	-	3.9	-	1.0	2.5	-	2.2	1.5	.2	3.7	10.1	31.4
Bern	1.8	-	.2	1.6	1.2	.2	.6	2.7	.6	5.9	8.8	34.0
ETHL	-	-	-	6.2	1.0	1.0	3.1	11.3	-	-	-	23.7
ETHZ	1.1	1.1	.4	-	.8	.5	.7	.1	.2	6.1	1.9	53.4
Fribourg	3.7	7.1	.3	.6	-	-	3.4	3.1	.3	10.5	5.9	22.9
HSG	1.4	2.9	-	-	-	-	.7	.7	.7	5.0	6.5	51.8
Genève	3.4	1.0	.5	-	1.5	1.0	-	1.5	.2	5.1	5.9	24.0
Lausanne	.9	5.4	.3	.6	1.7	-	5.4	-	1.3	9.8	5.4	30.7
Neuchâtel	-	4.6	-	-	-	.6	5.7	14.9	-	3.4	1.1	18.4
Zürich	1.6	3.1	-	1.0	2.2	1.5	2.5	3.1	.5	-	11.9	33.8

The sheer number of moves has, of course, something to do with the size of the universities involved. Thus, most of the pairs that are blank involve one of the three smallest universities (ETHL, HSG, NEUCHATEL) as origin states. On the other hand, the cells with the greatest number of moves involve the largest university, the one of Zürich, either as origin or as destination state.

By far the most substantial flows percentagewise are those from Neuchâtel and the ETHL to Lausanne (14.9% and 11.3%) and the ones from Fribourg and Lausanne to Zürich (10.5% and 9.8%).

Figure 2 is a summary picture of the flow between the Swiss Universities. Only flows of more than 10 students are plotted. The width of the arrows is proportional to the number of moves in the corresponding direction. The color of the arrow, finally, represents the language spoken at the university, where the move originated: Arrows originating at French speaking universities are white, arrows starting from German speaking ones are black and arrows from Fribourg are hatched.

The figure allows a concise description of the flow of students between the universities. Comparing, first, the movements between the German speaking universities with those between the French speaking ones, it can be seen that the network of moves is much more intricate for the former: Neuchâtel sends students to Lausanne and Genève, Lausanne to Genève only, and there is no return movement from Genève to the other two universities. Between German speaking universities there are, on the other hand, reciprocal flows in all but one direction: There is no arrow from Bern to Basel.

Secondly, there are comparatively frequent moves between universities of the two language groups: Genève and Lausanne, on the one hand, and Zürich, on the other, are connected by reciprocal flows. There also exists a reciprocal relationship between Genève and Basel and an analogous one between Lausanne and Bern. Probably most of the students responsible for these flows between universities of different language groups are of German speaking origin. This

Figure 2

The Flow of Students between the Swiss Universities

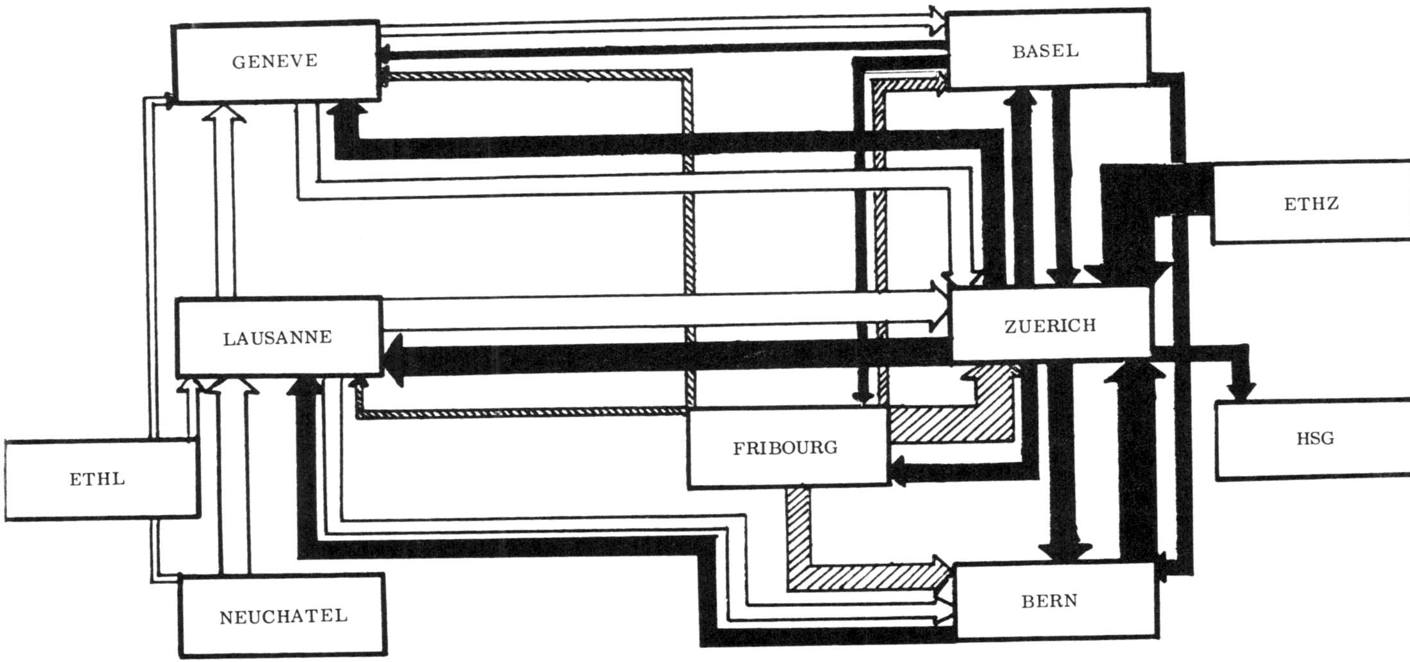

conjecture is supported by the fact that only German speaking students choose a certain university, because it is located in a language region other than their own (see RIES/KRIESI 1974b).

The figure also shows that Fribourg is more closely associated with German speaking universities, if the flow of students is taken as a criterion.

Finally, it points to the fact that the technical schools (ETHZ and ETHL) primarily lose students. It is, moreover, interesting to note that they lose students to the other university in the same city where they are located, the ETHL to the university of Lausanne and the ETHZ to the university of Zürich.

Figure 2 does not account for the movements to foreign universities taken as a whole and for interruptions. Whereas the flows to the foreign universities and the return from there are of the same order of magnitude as the flows between Swiss universities, the number of interruptions is much higher. This number reaches more than half of the original enrolment at the HSG (51.8%) and the ETHZ (53.4%). At the other universities it ranges from 18.4% of the original enrolment (Neuchâtel) to 34% (Bern). The very high percentages of the ETHZ and the HSG stem from the fact that the students at these universities have to do practical work as part of their studies for the completion of which they often take a semester off.

The most frequently cited reason for interrupting at Swiss universities, however, is military service: 35.9% of the interruptions are due to this fact. Practical work follows with 27.3% (see RIES/KRIESI 1974c).

3.1.2. The Flow between Subjects

Moves between subjects occur somewhat less frequently than moves between universities: 17.4% of the cohort 1965 have ever changed the subject; 14.9% once, 1.9% twice and .6% three or more times (RIES/KRIESI 1974c). Since for this paper the 60 odd subjects have been grouped into 12 fields of study (see Table 1), some of these changes are not noticable anymore, because they involve subjects that are grouped together now. Approximately one third of the moves thus goes unnoticed. The moves between groups of subjects recorded here are those that imply a more serious change of orientation than the ones within groups. Such changes involve considerable disadvantages for the students undertaking them, above all, they lead to a prolongation of the studies. These disadvantages are somewhat smaller for changes between related groups. It is, therefore, not astonishing that the most frequent changes occur between related groups, as can be seen from Table 10 and Table 11.

These tables are constructed in a way analogous to the construction of Tables 8 and 9. Figure 3 similarly corresponds to Figure 2. It shows the overall picture of the flow of students between groups of subjects. In this figure, as in the previous one, the width of the arrows is proportional to the number of moves in the corresponding direction. In comparing the two figures, it is immediately apparent that moves between groups of subjects are a lot less frequent than moves between universities. Two of the groups (theology and architecture) are not at all connected with the remaining ones by a noticeable flow of students (at least 10 students changing in any direction).

There are again three types of arrows: white ones for the flows originating from the natural sciences, black ones for those originating from 'professional' fields and hatched ones for flows starting from social and historical sciences and humanities, the fields that are traditionally subsumed under the so called 'Geisteswissenschaften'.

The two groups of subjects most closely interconnected with the rest are the social sciences and economics/business. Both are overall gaining students from all types of groups. Three other groups also have more incoming students than leaving ones: the historical sciences, law and the natural sciences II. They get their surplus from related fields: the historical sciences from the humanities, law from ecnomics/business and from medicine, the natural

Table 10

Synopsis fo the Total Number of Moves of the Cohort 1965 between the Groups of Subjects at Swiss Universities

From/to	Theol.	Med.	Law	Econ/ Bus	Soc Sci	Hum.	Hist Sci	Nat Sci I	Nat SciII	Engin.	Arch.	Others	Inter rupt	Original Enrolement
Theology	-	-	1	-	3	1	1	-	-	-	-	-	22	69
Medicine	4	-	11	5	9	9	4	3	18	1	2	-	165	682
Law	1	4	-	10	5	4	2	3	2	-	1	-	157	357
Econ/Business	-	1	11	-	10	2	1	3	2	2	2	2	172	452
Social Sci	3	10	7	9	-	11	4	1	2	3	1	2	67	276
Humanities	1	10	4	8	16	-	26	5	5	-	2	1	162	498
Historical Sci	1	2	2	4	5	8	-	-	8	-	-	1	56	161
Natural Sci I	1	3	5	12	11	3	5	-	26	13	1	2	119	396
Natural Sci II	1	17	5	8	7	3	2	11	-	5	1	2	163	521
Engineering	-	7	1	17	5	3	-	9	7	-	3	1	223	491
Architecture	-	3	2	2	3	2	3	3	-	5	-	-	129	127
Others	-	2	-	2	3	4	1	-	1	-	-	-	19	57
Interruptions	23	165	146	168	62	146	58	112	164	229	116	20	-	93

Table 11

Synopsis of the Total Number of Moves of the Cohort 1965 between the Groups of Subjects at Swiss Universities: Percentages of the Original Enrolement in the Group of Subjects, where the Move Originated

From/To	Theol	Med.	Law	Econ/ Bus	Soc Sci	Hum	Hist Sci	Nat Sci I	Nat Sci II	Engin	Arch	Other	Inter rupt
Theology	-	-	1.4	-	4.3	1.3	1.3	-	-	-	-	-	31.9
Medicine	.6	-	1.6	.7	1.3	1.3	.6	.4	2.6	.1	.3	-	24.2
Law	.3	1.1	-	2.8	1.4	1.1	.6	.8	.6	-	.3	-	44.0
Econ/Business	-	.2	2.4	-	2.2	.4	.2	.7	.4	.4	.4	.4	38.1
Social Sci	1.1	3.6	2.5	3.3	-	4.0	1.4	.4	.7	1.1	.4	.7	24.3
Humanities	.2	2.8	.3	1.6	3.2	-	5.2	1.0	1.0	-	.4	.2	32.5
Historical Sci	.6	1.2	1.2	2.5	3.1	5.0	-	-	5.0	-	-	.6	34.8
Natural Sci I	.3	.8	1.3	3.0	2.8	.8	1.3	-	6.6	3.3	.3	.5	30.1
Natural Sci II	.2	3.3	1.0	1.5	1.3	.6	.4	2.1	-	1.0	.2	.4	31.3
Engineering	-	1.4	.2	3.5	1.0	.6	-	1.8	1.4	-	.6	.2	45.4
Architecture	-	2.4	1.6	1.6	2.4	1.6	2.4	2.4	-	3.9	-	-	101.6
Others	-	3.5	-	3.5	5.3	7.0	1.8	-	1.8	-	-	-	33.3

Figure 3

The Flow of Students between the Groups of Subjects at Swiss Universities

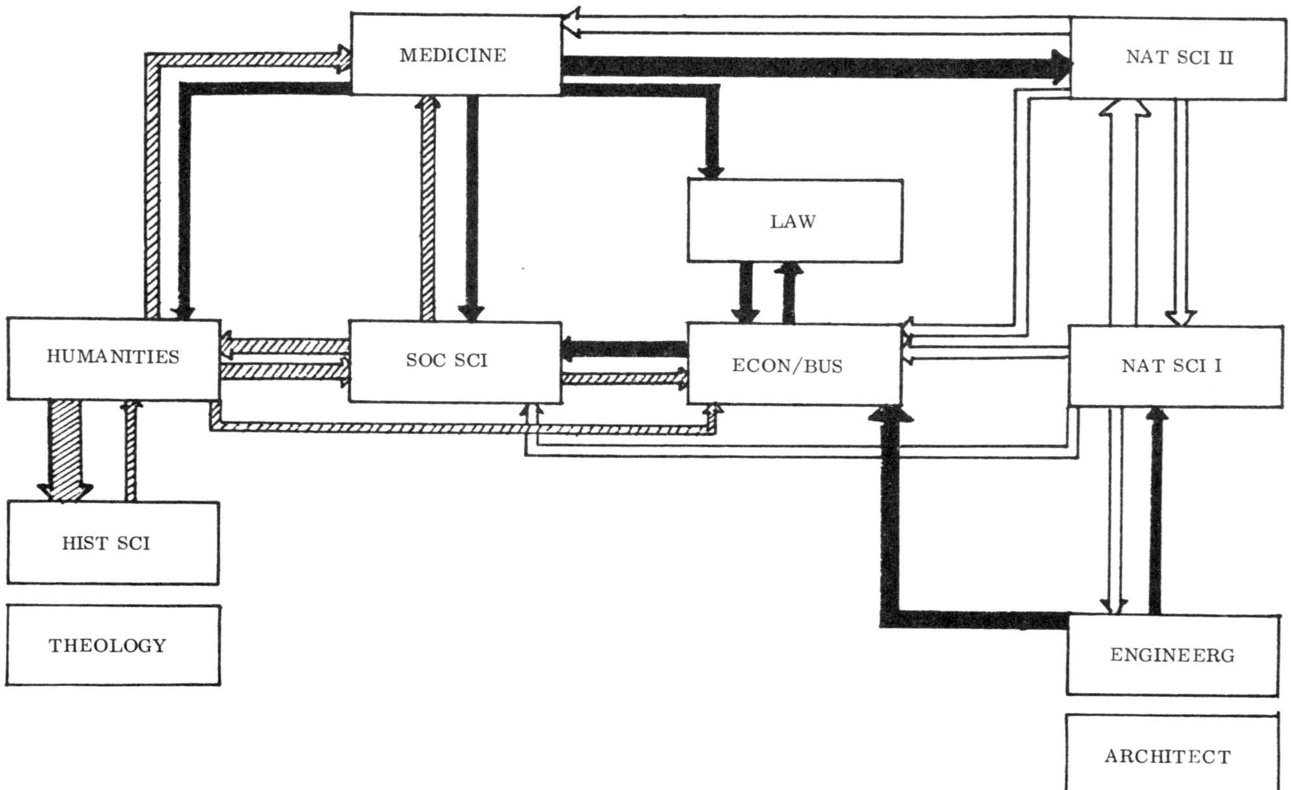

sciences II from medicine and the other natural sciences. The latter are the group with the greatest overall loss of students. Also losing are the humanities and engineering.

The interruptions are again not included in Figure 3. As Table 11 shows, there is one subject with an astonishing percentage of interruptions: In architecture practically every student interrupts his studies at least once. The percentages for engineering, law and for economics/business lie considerably above average, too. Students in these groups of subjects have to do a considerable amount of practical work. The high percentage in law is probably also attributable to the fact that law students more than any one else make a career in the army: 43.8% of the male law students become officers as compared with an average of 30.7% for students of all subjects (see KRIESI 1975).

3.1.3. Leaving the System: The University Model

3.1.3.1. Overall Rates

The first results that can be obtained from the model with regard to leaving the system are overall rates for leaving at different levels for different universities. Taking 100 students at a specific university, these rates will indicate how many of these will eventually graduate with what kind of degree. They are some kind of 'productivity index' for the universities in question.

Now, because of the flows between universities, it is necessary to distinguish between the university of entrance and the university of exit in computing these overall rates. The distribution of the students over the four levels of leaving is not the same for the two. Table 12 provides the distribution for the university of entrance, Table 13 the one for the university of exit.

Table 12 is derived from transition matrix 2.21 in the APPENDIX A. This matrix is the final adjusted matrix of type 2 covering the entire period of 22 semesters for the cohort 1965. The rows of this table do not sum to one, be-

Table 12

Level of Leaving the System by University of Entrance: Probability Distribution of the Cohort 1965 (1)

University of Entrance	Level of Leaving				
	Drop out	Low Level	Medium Level	High Level	Total
Basel	.144	.125	.374	.317	.96
Bern	.106	.179	.511	.207	1.003
ETHL	.183	.032	.856	.041	1.112
ETHZ	.127	.058	.652	.164	1.001
Fribourg	.133	.208	.434	.229	1.004
HSG	.105	.029	.611	.230	.975
Genève	.229	.139	.510	.122	1.000
Lausanne	.194	.041	.658	.120	1.013
Neuchâtel	.117	.236	.529	.086	.968
Zürich	.174	.145	.417	.278	1.014

(1) Source: Transition matrix 2.21 in the Appendix A,p. 303

Table 13

Level of Leaving the System by University of Exit: Probability Distribution of the Cohort 1965 (2)

University of Exit	Level of Leaving				
	Drop out	Low Level	Medium Level	High Level	Total
Basel	.156	.103	.407	.333	1.00
Bern	.113	.164	.503	.219	1.00
ETHL	.086	.025	.840	.049	1.00
ETHZ	.129	.035	.680	.156	1.00
Fribourg	.153	.239	.384	.224	1.00
HSG	.125	.007	.662	.206	1.00
Genève	.219	.120	.542	.120	1.00
Lausanne	.205	.041	.654	.098	1.00
Neuchâtel	.146	.308	.492	.054	1.00
Zürich	.164	.122	.437	.278	1.00

(2) Source: See Description in text, p. 36

cause of the adjustment process (see p. 36/37 above). There is one serious exception, the case of the ETHL. The simplicity of the adjusting scheme has led to this deviation at the ETHL, because of the unusually large correction factor (see Table 7). An arbitrary, but not unrealistic way to fix the situation is to reduce the probability of medium level graduation by the deviation of the total from unity. The probability of .856 for medium level degree at the ETHL has, thus, to be replaced by .744.

Table 13, on the other hand, is constructed on the basis of the raw figures for the adjusted matrices of type 1. For each cell in Table 13, the raw numbers in all the corresponding cells of the 22 transition matrices of type 1 have been summed over and divided by the row total for all four levels.

In comparing equivalent probabilities in the two tables, we notice that they are on the average rather similar. Only in a few number of cases they differ by more than .030: Dropping out at the ETHL is such a case; 18.3% of the students who enter the system at the ETHL are dropouts, but only 8.6% of the students leaving the system at the ETHL drop out. This means that more than half of those students who start their studies at the ETHL and drop out later on actually move to another university, most likely to Lausanne, before dropping out. Taking into account the correction made for medium level degrees above, there is also a considerable difference between the two medium level cells at the ETHL, a difference reciprocal to the one noted for dropping out.

At Fribourg and Neuchâtel, a significant group of students who start at these universities do not obtain their medium level degrees there. This leads to lower probabilities for medium level degrees in Table 13 and, correspondingly, to higher probabilities for lower level degrees. At Neuchâtel the proportion getting a doctorate is also lower, if Neuchâtel is taken as university of exit.

Responsible for these differences is the fact that students in medicine cannot complete their studies at these two universities, but have to move to another one for their clinical semesters. According to Figure 3, students from Fribourg primarily move to the German speaking universities, those from Neuchâtel to the French speaking ones.

The last pointed differences between Table 12 and Table 13 concern the medium level at Basel, Genève and Lausanne. At all three of these universities, the probability for graduation at the medium level is higher, if they are taken as university of exit.

The overall similarity of the probabilities in the two tables is another indicator for the rather limited scope of the flow between Swiss universities.

Figure 4 provides a graph for Table 13. It allows an easy comparison of the relative importance of any of the four levels at any of the 10 universities. The first fact that emerges from this figure is the salience of the medium level degree. At every university it is the most important level in absolute terms. There are, nevertheless, marked differences with respect to this level, too. It is most dominant at the technical schools and at the school for economics and business at St. Gallen (HSG) as well as at Lausanne. It is least dominant at Fribourg and the two German speaking universities of Basel and Zürich.

At most universities, the proportion of students graduating at a low level is rather insignificant. Only at Fribourg and at Neuchâtel there are larger shares of students graduating at this level.

With respect to the high level, it is interesting to note that it is more common among German speaking universities than among French speaking ones. Fribourg is again similar to the German speaking universities.

Finally, there are also marked differences between the universities in regard

Figure 4

Level of Leaving the System by University of Exit: Percentage Distribution of the Cohort 1965

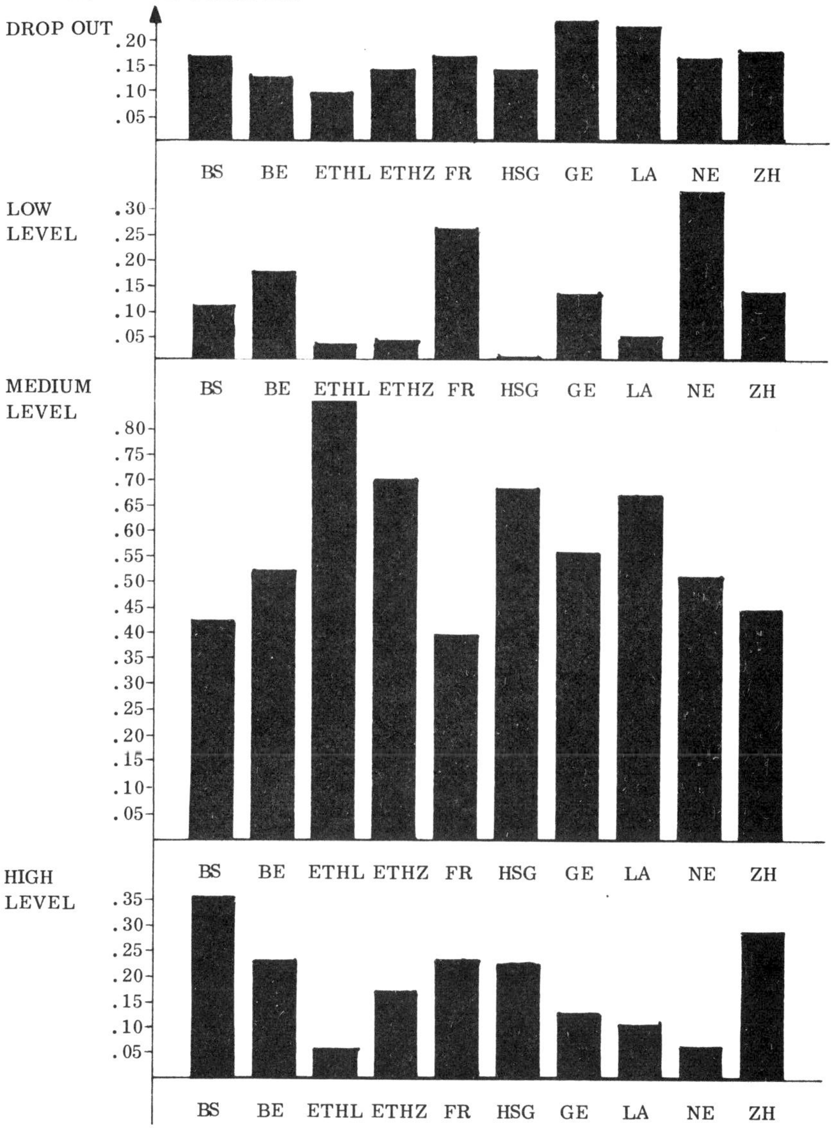

to dropping out, although these differences are not as great, on the average, as those for the other levels.

3.1.3.2. Duration of Studies

Overall rates are, of course, only very insufficient indicators for the productivity of a university. It is also interesting to know, how long it takes to get a certain degree at a specific institution. Speaking of duration, the number of semesters passed in the system (the duration status) has again to be distinguished from the number of semesters passed since the entrance of the cohort into the system (time measured in calendar time). It is the duration status at graduation that is probably more adequate in this connection.

I have computed the average number of semesters of actual enrolment in the system that are necessary to obtain a specific degree at a particular university. To do this, I have started out with the raw numbers of the transition matrices of type 1. I multiplied all the cells of the Z part of a matrix for a particular semester with the corresponding average duration status for that semester given in Table 3 (1). The resulting products were, then, divided by the total number of students ever leaving a certain university at the particular level. The program in APPENDIX B4 was used for this job.

In taking the same average duration status for all the 10 universities, this procedure is, of course, subject to some inaccuracies, because the percentage of students interrupting their studies varies considerably from one university to another (see Table 9). The interpretation of the results has to

---

1) For semesters subsequent to the SS73 I assumed that there were no interruptions anymore and, thus, just added one to the mean of the SS73 for the WS73, two for the SS74 and so on.

take into account these differences. Table 14 gives the results (2), Figure 5 shows the range of the results for each level.

Table 14

Average Number of Semesters Necessary to Obtain a Degree of a Specific Level at Swiss Universities

University	Level of Leaving			
	Drop out	Low Level	Medium Lev.	High Level
Basel	5.6	8.6	12.3	15.9
Bern	7.3	7.0	13.8	16.2
ETHL	3.0	—	9.9	16.5
ETHZ	6.4	—	10.3	17.0
Fribourg	5.5	6.6	10.0	15.6
HSG	4.0	—	9.1	16.5
Genève	5.6	8.0	11.9	17.4
Lausanne	5.3	6.6	11.0	16.2
Neuchâtel	5.9	7.0	10.7	15.9
Zürich	6.0	7.0	13.1	15.9

Figure 5

Range of the Average Number of Semesters Necessary to Obtain a Particular Degree

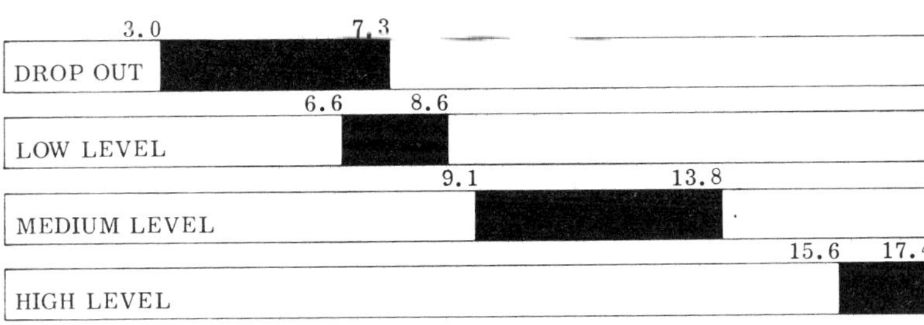

2) Three averages have not been computed, because the corresponding probabilities are too small.

All in all, it takes longer to get a high level degree than to get a medium level one, and it takes also longer to get a medium level degree than to get a low level one. With the exception of Bern, the average stay of a drop out is also shorter than the average time it takes to receive a low level degree. In Bern, however, dropouts stay in the system for a longer period of time than the graduates at a low level.

The ranges of the averages both, for low level and for high level degrees, are quite narrow, compared with those for dropping out and for medium level degrees. It takes about one and a half times as long, on the average, to get a medium level degree at Bern than it takes to get a similar degree at the School for Economics/Business at St. Gallen. Considering the fact that St. Gallen has a very high percentage of interruptions compared with an average one at Bern, this difference even underestimates the true discrepancy between the averages for medium level degrees at these two universities. One reason for this discrepancy lies in the different variety of subjects studied at the two universities: Whereas St. Gallen only offers degrees in economics and business and, to a limited extend, in law, all the subjects traditionally studied at a Swiss university are open to the students at Bern. But, as Table 17 will show, economics/business is the one group of subjects with the least average number of semesters necessary to obtain a medium level degree. In spite of this qualification not all the differences between universities can be attributed to differences in the variety of subjects studied at a particular one of them. Thus, all the French speaking universities with a distribution of students over the different subject areas that closely resembles the one of Bern (see RIES/KRIESI 1974b) have averages considerably shorter than Bern for medium level degrees.

From this discussion we can conclude that the Swiss universities are, on the average, about equally efficient in producing graduates at the low level and also in producing graduates at the high level, but that there are quite noticeable differences with respect to the medium level, the German speaking universities being least efficient on that level. This discussion was based on the estimates of the average duration status at graduation. It did not take into account possible differences in the variance of these figures at the different universities.

3.1.3.3. Times Series

The information contained in the transition matrices allows an even more detailed description of the leaving of the system. It is possible to describe the variation of the leaving at different levels over time. In Figures 6 to 9 the information contained in the transition matrices of the set 2 regarding the leaving of the system are graphed. Every one of these figures consists of 10 different cumulative percentage curves, which show the growth of the share of students in a particular absorbing state over time for the different universities. These curves pertain to the university of entrance, because they are constructed from matrices of the type 2.

Every curve contains at least four kinds of information:

- the starting point: It tells at what point in time the first students leave the system at what level.
- the slope: It tells about the rate of leaving. The steeper the slope, the higher the rate of leaving per unit time.
- the relative maximum: This is the point at which the curve starts to level off. It tells at which point in time most of the students who will ever graduate at a certain level have obtained their degrees.
- the end point: It tells about the relative importance of the degree in question at the specific universities. This information was already contained in Table 12.

I do not intend to review all of this information contained in the figures, but want to point out only a few obvious results:

As to the starting point, it can be noted that the higher the level of leaving,

Figure 6

Dropping out by Universities: Cumulative Percentage Curves

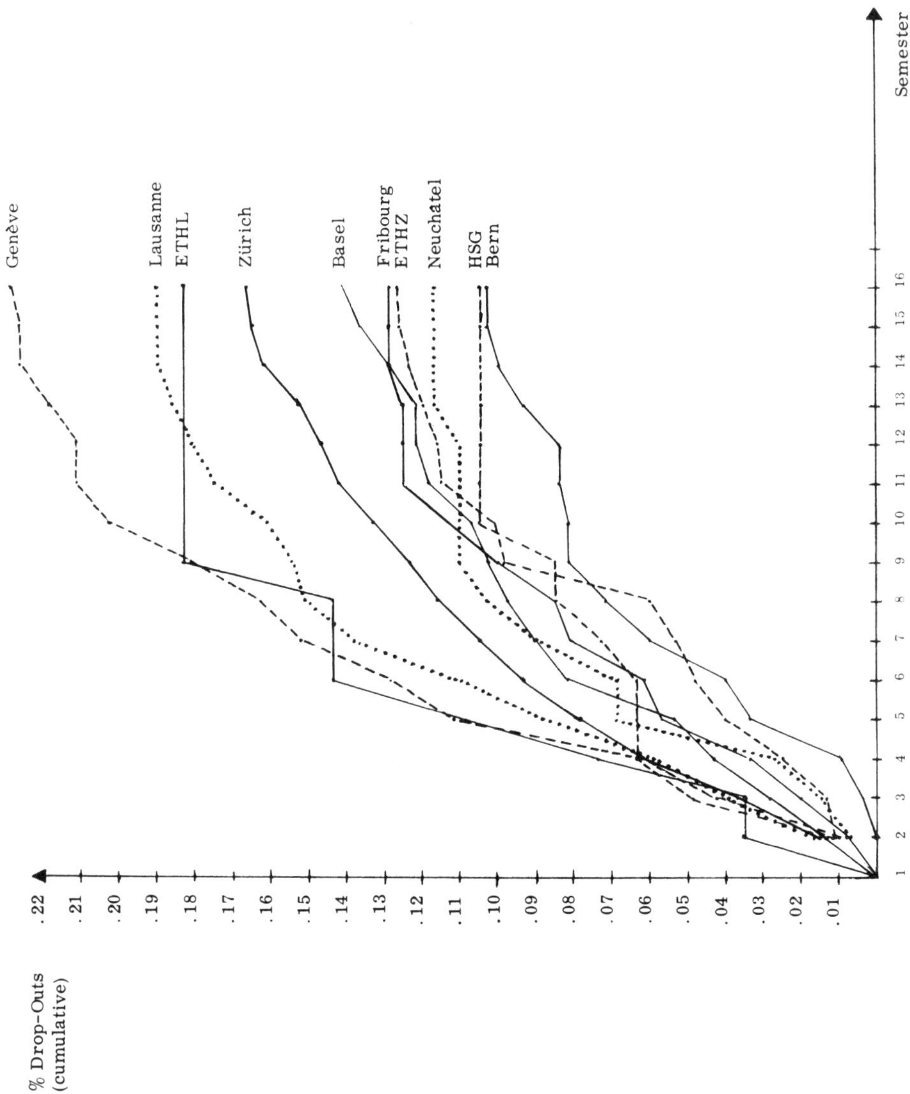

Figure 7

Graduating at a Low Level by Universities: Cumulative Percentage Curves

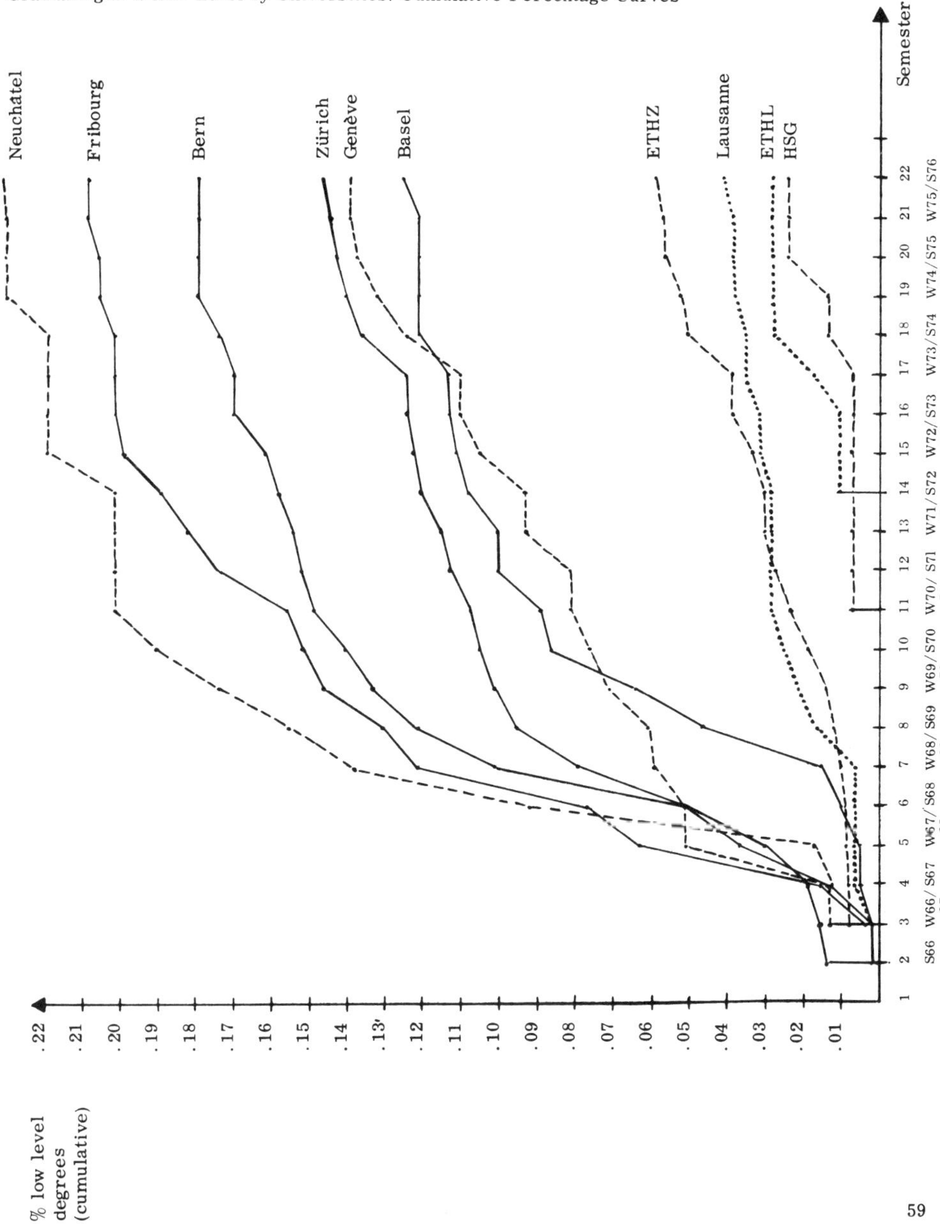

Figure 8

Graduating at a Medium Level: Cumulative Percentage Curves

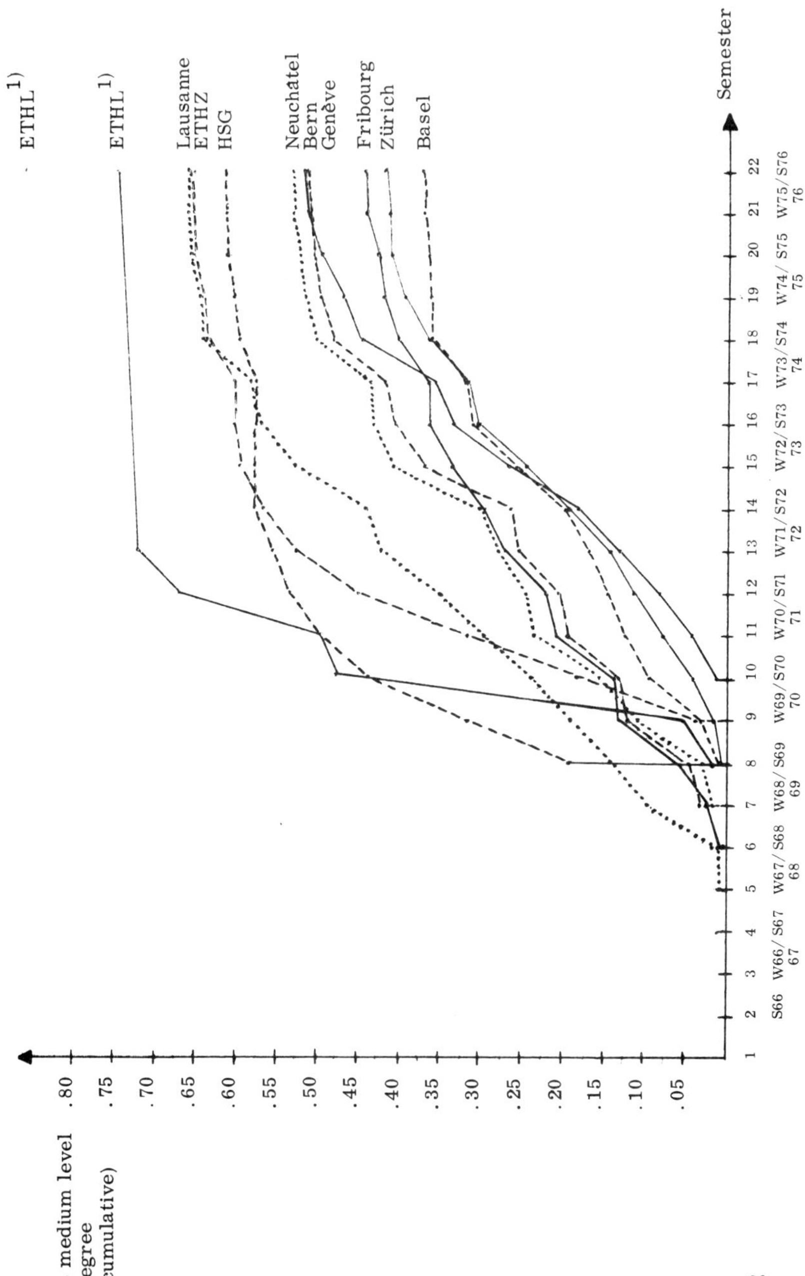

Figure 9

Graduating with a Ph. D. at the Swiss Universities: Cumulative Percentage Curves

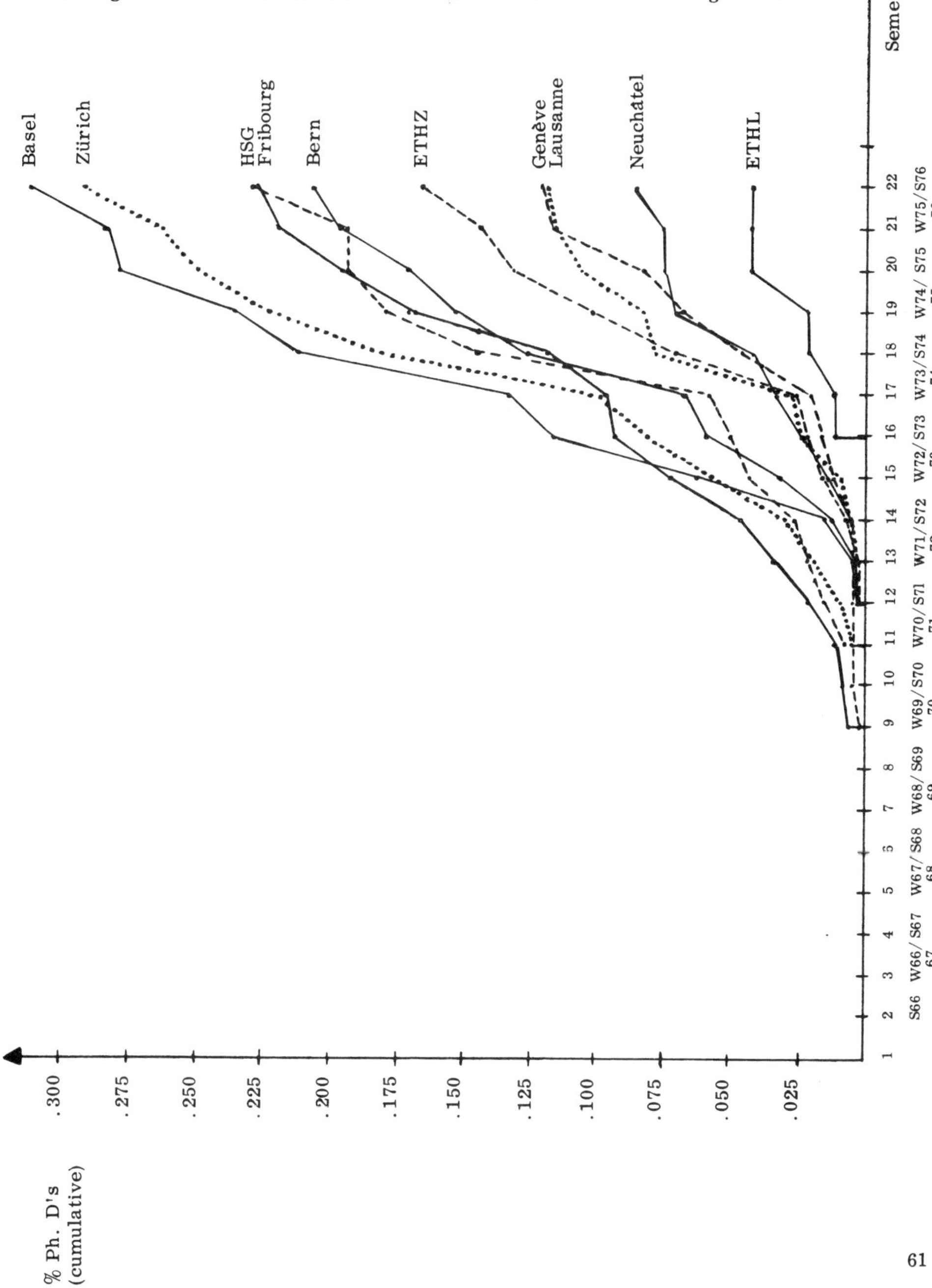

the later the starting point of the curve (3). Extremely early starting points relative to the other starting points for the same level should be taken with a grain of salt. The students in question might have been mistaken in giving the date of their graduation retrospectively. Thus, the student indicating that he graduated with a medium level degree at Lausanne after only five (possibly even four semesters) can hardly be trusted.

All the curves start at a very low percentage point, with one exception, the curve for medium level degrees at the HSG. By the end of the 7th semester (only fall-starters at the HSG), almost one fifth of the students entering the system at the HSG have graduated at the medium level.

There are, in general, two types of curves with respect to the slope and its variation over time: logarithmic type curves with a steadily decreasing slope, on the one hand, and S-shaped curves, with a slope that is first increasing and then decreasing, on the other. Dropping out and graduation at a low and medium level generally take the shape of the former, graduation at a high level the one of the latter. There are, of course, exceptions to this general scheme: Thus, graduation at a low level in Basel and Zürich pick up only slowly and turn out to be S-shaped. A similar remark applies to medium level graduation at Bern and Zürich.

With respect to the relative maximum we note that the lower the level of leaving, the earlier the point of leveling off is reached. The curves for the Ph. D. at German speaking universities and at Fribourg do not level off at all.

Some of the curves are fairly smooth and it is tempting to fit them by a math-

---

3) The numbers along the horizontal axis refer to the number of semesters, measured in calendar time. They are somewhat ambiguous, because of the difference between spring- and fall-starters; a '2', for example, marks the second semester for spring-starters and the beginning of the same semester for fall-starters.

ematical model. A model that seems suitable for the logarithmic type curves, the majority of the curves in question, is the one of immanent change.

This rather simple model assumes that change does not only depend on time alone, but also on the value of the variable itself undergoing change. Call the variable undergoing change, i.e. the percentage of students graduating at a certain level, x_1. If dx_1/dt, the rate of change, depends on the value of x_1 in a linear fashion, then the equation for such a dependence is:

$$\frac{dx_1}{dt} = a + bx_1$$

According to COLEMAN (1969), the solution of this equation gives x_{1t} as a function of x_{10} and the two coefficients a and b. Letting $\triangle t = t - 0$:

$$x_{1t} = \frac{a}{b}(e^{b\triangle t} - 1) + x_{10}e^{b\triangle t}$$

Following the procedure outlined by COLEMAN, it is possible to estimate the values of a and b, which constitute the fundamental parameters of change, from the data contained in the curves of Figures 6 to 9. I have done so for the curve for dropping out from the university of Genève. The computation has yielded the following values for a and b:

$$a = .0429$$
$$b = -.1690$$

The value for $-\frac{a}{b}$ (.254) is the one which the curve of the model approaches asymptotically, indicating the maximum possible drop-out rate for Genève according to the model. The percentages of the model develop as follows (in brackets the actual figures for Genèves as plotted in Figure 6:

t_0	.011	(.011)	t_8	.1911	(.203)
t_1	.0488	(.049)	t_9	.2009	(.211)
t_2	.0806	(.062)	t_{10}	.2091	(.211)
t_3	.1076	(.111)	t_{11}	.2161	(.218)
t_4	.1304	(.129)	t_{12}	.2220	(.226)
t_5	.1496	(.152)	t_{13}	.2270	(.226)
t_6	.1659	(.162)	t_{14}	2312	(.229)
t_7	.1794	(.182)			

Figure 10

Actual and Fitted Curve for Dropping out at the University of Genève

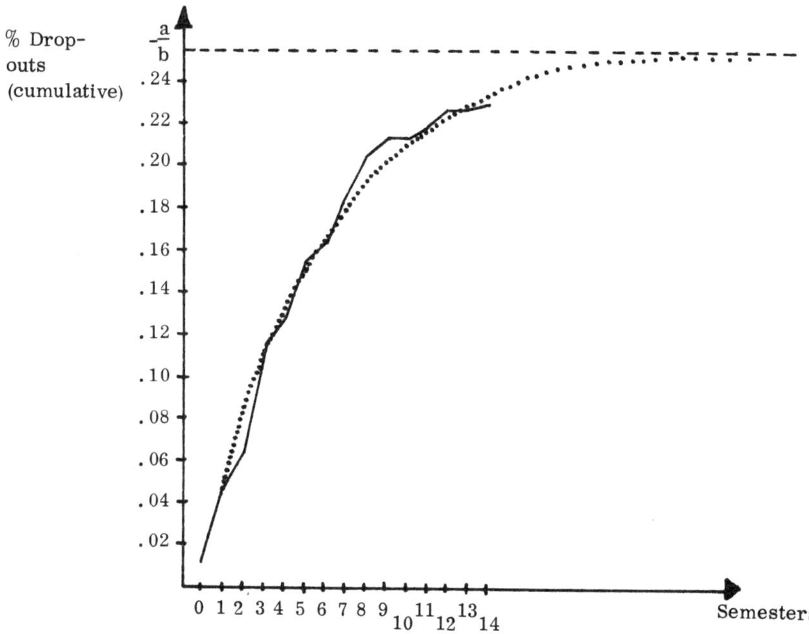

Figure 10 provides a plot for the actual curve and the fitted curve of the model of immanent change (dotted line). The correspondance of the two curves is satisfyingly close. We can, therefore, conclude that the three assumptions of the model are met by the data:

- a and b are constant over time
- dx_1/dt is linearly dependent on x
- dx_1/dt is dependent upon the present value of x_1

This means that there are no lagged effects and, above all, no exogenous variables influencing the rate of change in the percentage of students dropping out at the university of Genève.

The curve chosen, dropping out at Genève, was a rather smooth one to begin with. In principle, however, the same procedure can be applied to all curves of a logarithmic type shape to yield a model of immanent change represented by two parameters a and b, whereby a is always positive and b is always negative.

3.1.4. Leaving the System: The Subject Model

The description for this model exactly follows the procedure outlined for the university model. In order to avoid repetitions, I try to be as brief as possible.

3.1.4.1. Overall Rates

A distinction analogous to the one in 3.1.3.1. can be made between the group of subjects of entrance and the group of subjects of exit. Table 15 pertains to the former, Table 16 to the latter. Table 15 is the adjusted second part of transition matrix 4.21 in APPENDIX A. This is the final matrix of type 2 covering the entire period of 22 semesters for the cohort 1965. The rows do not sum to unity for the same reason as in Table 12. Here, too, there is one serious deviation from unity, but it concerns the residual category, which is of no intrinsic interest anyway. Table 16 is constructed like Table 13.

The two tables for the groups of subjects are again very similar. Only in a few number of cases do equivalent cells differ by more than .03. Thus, a higher percentage of students who start in medicine graduate on a high level than students who finish in medicine. In the historical sciences it is just the other way around. In theology, the probability of a low level graduation is higher for those who start in the field than for those who graduate in it. On the medium level, the percentage increases for architecture and it decreases for the humanities, if these fields are taken as subject of exit. - The limited differences between the two tables can again be taken as another indication of the narrow scope of the flow between the groups of subjects at Swiss universities.

Table 15

Level of Leaving by Group of Subject of Entrance: Probability Distribution of the Cohort 1965 (1)

Group of Subject of Entrance	Level of Leaving				
	Drop out	Low Level	Medium Level	High Level	Total
Theology	.133	.116	.739	.043	.961
Medicine	.097	.025	.698	.243	1.063
Law	.069	.056	.504	.364	.993
Econ/Business	.170	.027	.606	.206	1.009
Social Sci	.189	.300	.391	.130	1.010
Humanities	.216	.344	.308	.139	1.007
Historical Sci	.231	.112	.410	.230	.983
Natural Sci I	.180	.147	.481	.182	.990
Natural Sci II	.125	.119	.294	.434	.972
Engineering	.147	.043	.771	.071	1.032
Architecture	.174	.055	.779	.024	1.032
Others	.258	.474	.316	.053	1.101

(1) Source: Transition matrix 4.21 in the Appendix A

Table 16

Level of Leaving the System by Group of Subject of Exit: Probability Distribution of the Cohort 1965 (2)

Group of Subject of Exit	Level of Leaving				
	Drop out	Low Level	Medium Level	High Level	Total
Theology	.145	.092	.711	.053	1.00
Medicine	.085	.017	.718	.180	1.00
Law	.062	.048	.514	.376	1.00
Econ/Business	.182	.021	.607	.190	1.00
Social Sci	.195	.316	.379	.110	1.00
Humanities	.229	.370	.274	.127	1.00
Historical Sci	.228	.096	.407	.269	1.00
Natural Sci I	.207	.152	.468	.173	1.00
Natural Sci II	.130	.138	.291	.441	1.00
Engineering	.147	.028	.769	.056	1.00
Architecture	.147	.026	.810	.017	1.00
Others	.271	.525	.186	.017	1.00

(2) Source: See Description in text, p. 50

Figure 11 pictures the numbers of Table 16. In contrast to the universities, where the medium level was uniformly dominant, this is not the case for the groups of subjects. For two of the groups another level is the most prevalent one; for the humanities it is the low level and for the natural sciences II it is the high level. In all the professional fields (with the possible exception of law), the medium level is, however, clearly the most important one. The field of law is an exception insofar as it has a high percentage of high level graduates, too.

The low level is insignificant for most of the groups of subjects as it was for most of the universities. Only in the humanities, as just mentioned, and in the social sciences it is of considerable importance. The high percentage in the latter is probably due to the School of Psychology and Education (Ecole de Psychologie et de Science de L'Education) at Genève, which educates teachers of primary and lower secondary schools.

At the high level, it is, apart of the importance of this level for law and the natural sciences II, remarkable to note how insignificant it is for theology, architecture and engineering. The reader acquainted with the Swiss University System might also have expected the percentage of high level degrees to be much higher in medicine than it actually turns out to be. Most of the physicians in Switzerland have received not only a professional degree, but also a doctorate. Since the doctorate in medicine is a rather low key affair, it is, however, possible that students turn in the thesis at some point after their official graduation and do not consider it as part of their studies anymore. Thus, the differentiation between medium and high level is probably of a superficial kind in medicine.

There are, finally, also marked differences with regard to dropping out between the groups of subjects. It is astonishing to note that the proverbial unsuccessful law-student is not as common as the stereotype has it. Law students are, on the contrary, the most successful ones, if dropping out is taken as the criterion.

Figure 11

Level of Leaving the System by Group of Subject of Exit: Percentage Distribution of the Cohort 1965

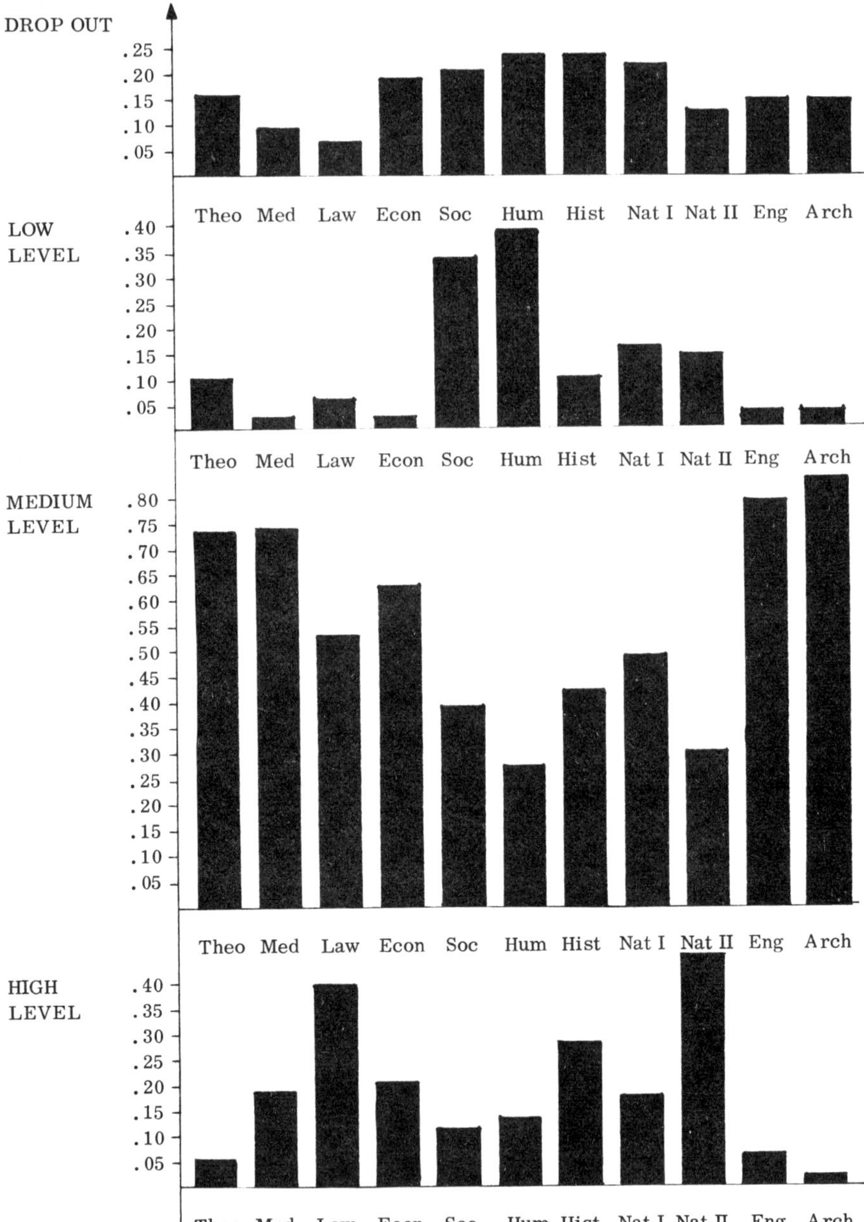

3.1.4.2. Duration of Studies

Table 17 gives the average number of semesters necessary to obtain a degree of a specific level in the different groups of subjects (4). The same limitations apply to these estimates as to the ones in Table 14, because they were constructed analogously. Figure 12 shows the range of the averages for each level.

Table 17

Average Number of Semesters Necessary to Obtain a Degree of a Specific Level in Different Groups of Subjects

Group of Subject	Level of Leaving			
	Drop out	Low Level	Medium Lev.	High Level
Theology	2.7	11.3	11.5	15.5
Medicine	5.3	—	13.9	14.9
Law	6.0	13.8	10.8	16.1
Econ/Bus	6.3	—	10.0	16.1
Soc Sci	5.3	7.3	11.2	15.9
Humanities	5.7	7.2	12.9	15.8
Hist Sci	6.7	10.2	13.3	16.6
Nat Sci I	5.9	7.3	11.3	17.0
Nat Sci II	6.2	8.6	11.9	16.6
Engineering	6.9	—	10.0	17.4
Architecture	4.0	—	11.5	—
Others	5.3	8.2	9.5	14.3

4) Five averages have not been computed, because the corresponding probabilities are too small.

Figure 12

Range of the Average Number of Semesters to Obtain a Particular Degree

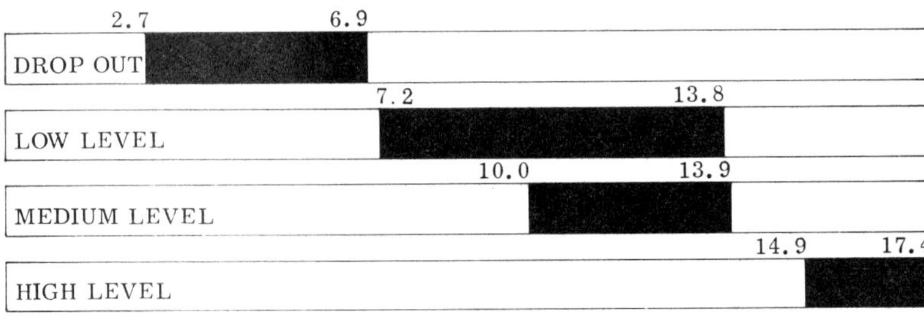

The distinction between the ranges for the three degree levels is not as clear cut as it was in the case of the university model, because there is some overlap between the low and the medium level. In theology, law and in the historical sciences it takes excessively long, on the average, to obtain a low level degree. Now, at least in theology and law it is very unusual to obtain a low level degree in the first place. It is quite probable that there are some miss-classified students in these categories, distorting the range for the low level. It is conceivable that these are students with very unusual university 'careers', possibly with a degree from a foreign university. - Apart from these three fields, there is not much spread for the averages on the low level.

The range for the medium level again is broader than the one for the high level. On the average, one is fastest to graduate on a medium level in economics/business and in engineering. Taking into account that the students in engineering interrupt their studies much more often than the average student in other subjects, the average for engineering probably lies even below the 10.0 semesters indicated. This remark applies also and above all to architecture. In medicine, a medium level degree is hardest to get: On the average it takes 13.9 semesters for a professional degree in medicine. If, on the one hand, it takes comparatively long to graduate on a medium level in medicine, the doctorate in this subject is, on the other hand, the one to be obtained after the shortest average number of semesters. Now, this does not necessarily mean that the doctorate in medicine is the one that affords the least amount of effort to get it (measured in input of time), because the intensity of the studies in medicine probably lies far above the average for all groups of subjects in Switzerland.

The small difference between the averages for the number of semesters necessary to obtain a medium and a high level degree in medicine underlines the superficiality of that distinction for this group of subjects. In general, the difference in question could be taken as an indicator of the 'value' of a doctorate in the respective groups of subjects: The higher the difference between the two averages, the higher the 'value' of the doctorate. Ranked according to this difference, the following order emerges from among the groups of subjects:

1. Engineering	7.4
2. Economics/Business	6.1
3. Natural Sciences I	5.7
4. Law	5.3
5. Social Sciences	
Natural Sci II	4.7
6. Theology	4.0
7. Historical Sciences	3.3
8. Humanities	2.9
9. Medicine	1.0

Whereas it takes, on the average, 7.4 semesters more receive a doctorate in engineering than to get a medium level degree in that subject, the corresponding difference in medicine is only 1.0 semester. Judging from this rank order, a doctorate in engineering would be comparatively most valuable, one in medicine least valuable.

The duration for dropping out varies on the average between 2.7 semesters (theology) and 6.9 semesters (engineering). The rather high average in engineering is due to a peculiarity at the ETHZ: The studies at the ETHZ in engineering are organized in a way that the student has to pass two preliminary exams, the first one usually after two semesters and the second one after four, before he can go on working for his degree in engineering. Now, those students who have failed twice in the second prelim cannot get a diploma in engineering, but they can complete the required course work and can call themselves 'engineers' afterwards. There are quite a few students who make use of this possibility. Since they have been classified as dropouts, they raise the pertinent average.

3.1.4.3. Time Series

In Figure 13 to 16 cumulative percentage curves are provided for the leaving of the system at different levels in each of 11 groups of subjects. Parallel to the Figures 6 to 9 they are intended to give some insight into the development of the leaving of the system over time. They are constructed like the earlier figures, based on the adjusted transition matrices of type 2 for groups of subjects (4.1.-4.21 in APPENDIX A).

The findings for the university model also apply here, with a few qualifications. There are some special cases with respect to the shape of the curves. First of all, the curve for dropping out in engineering is not logarithmic, but S-shaped, cutting across some of the other curves in Figure 13. The reason for this deviant shape lies in the peculiarity of the ETHZ just described above. - The curve for dropping out in law is also S-shaped, but to a lesser extent.

There are, secondly, two more S-shaped curves on the medium level, cutting across all the rest of the curves: the curves for architecture and medicine. In line with the discussion in the previous section with respect to medicine, it is, finally, not surprising that the medium level and the high level curves for that group of subjects level off at about the same point in time, i.e. around the 17th semester.

I have again tried to fit the model of immanent change to one of the curves. To

Figure 13

Dropping out by Groups of Subjects: Cumulative Percentage Curves

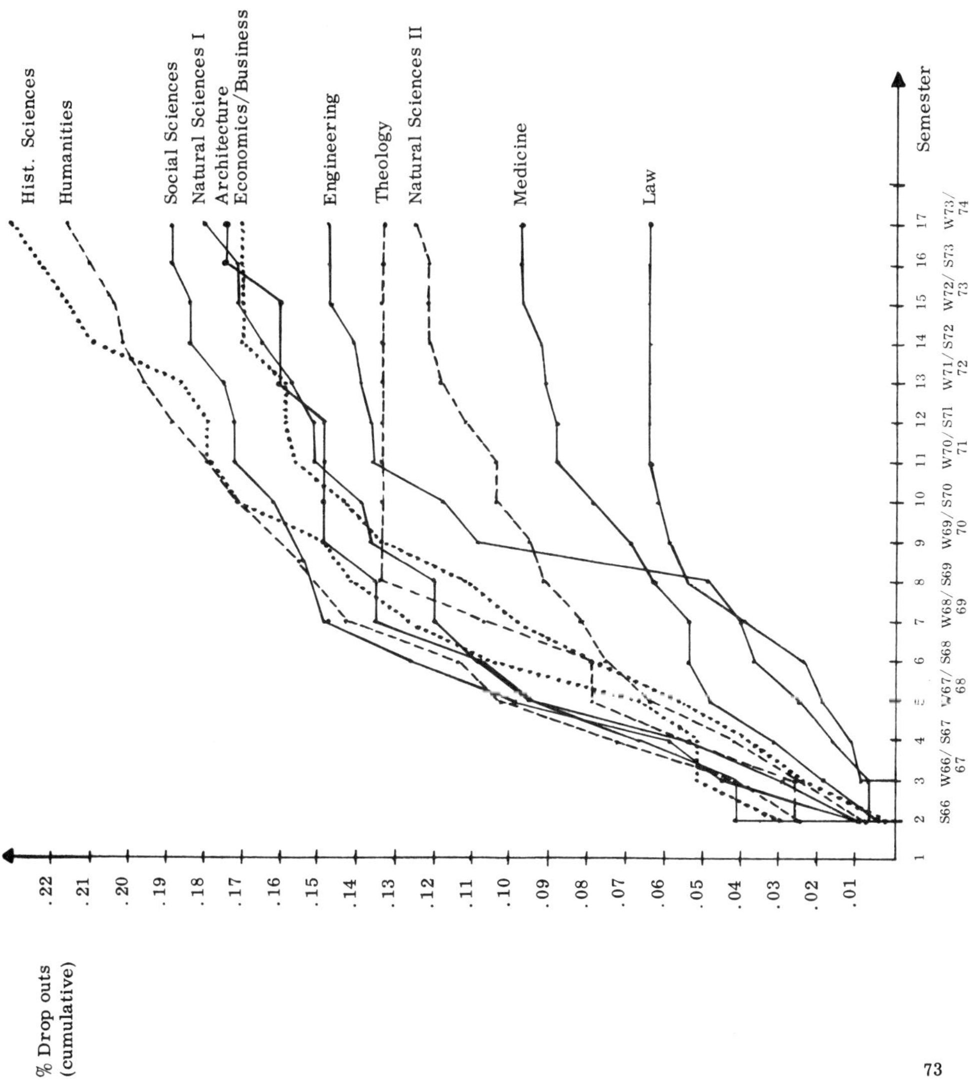

Figure 14

Graduating with a Low Level Degree by Groups of Subjects: Cumulative Percentage Curve

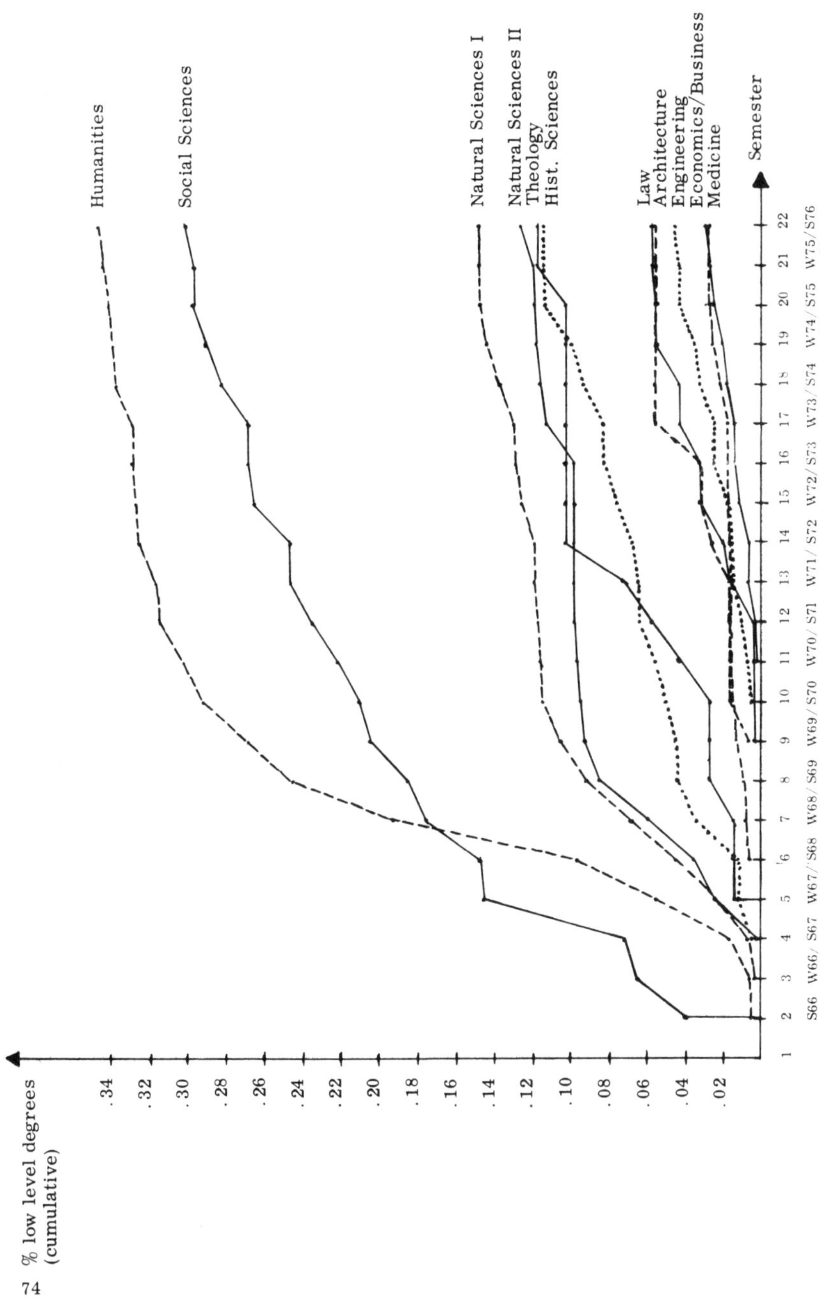

Figure 15

Graduating with a Medium Level Degree by Groups of Subjects: Cumulative Percentage Curves

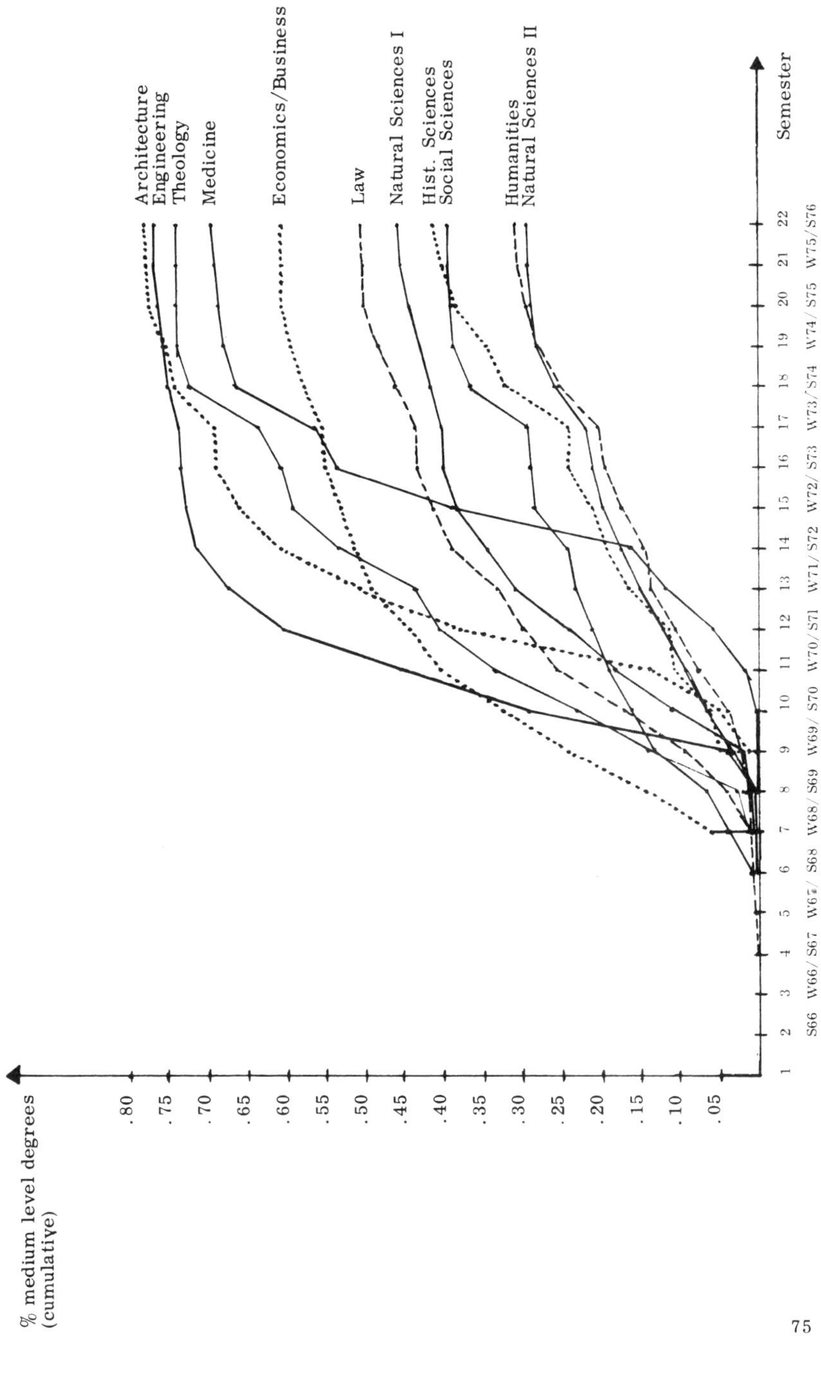

Figure 16

Graduating with a Ph. D. in the Different Groups of Subjects: Cumulative Percentage Curves

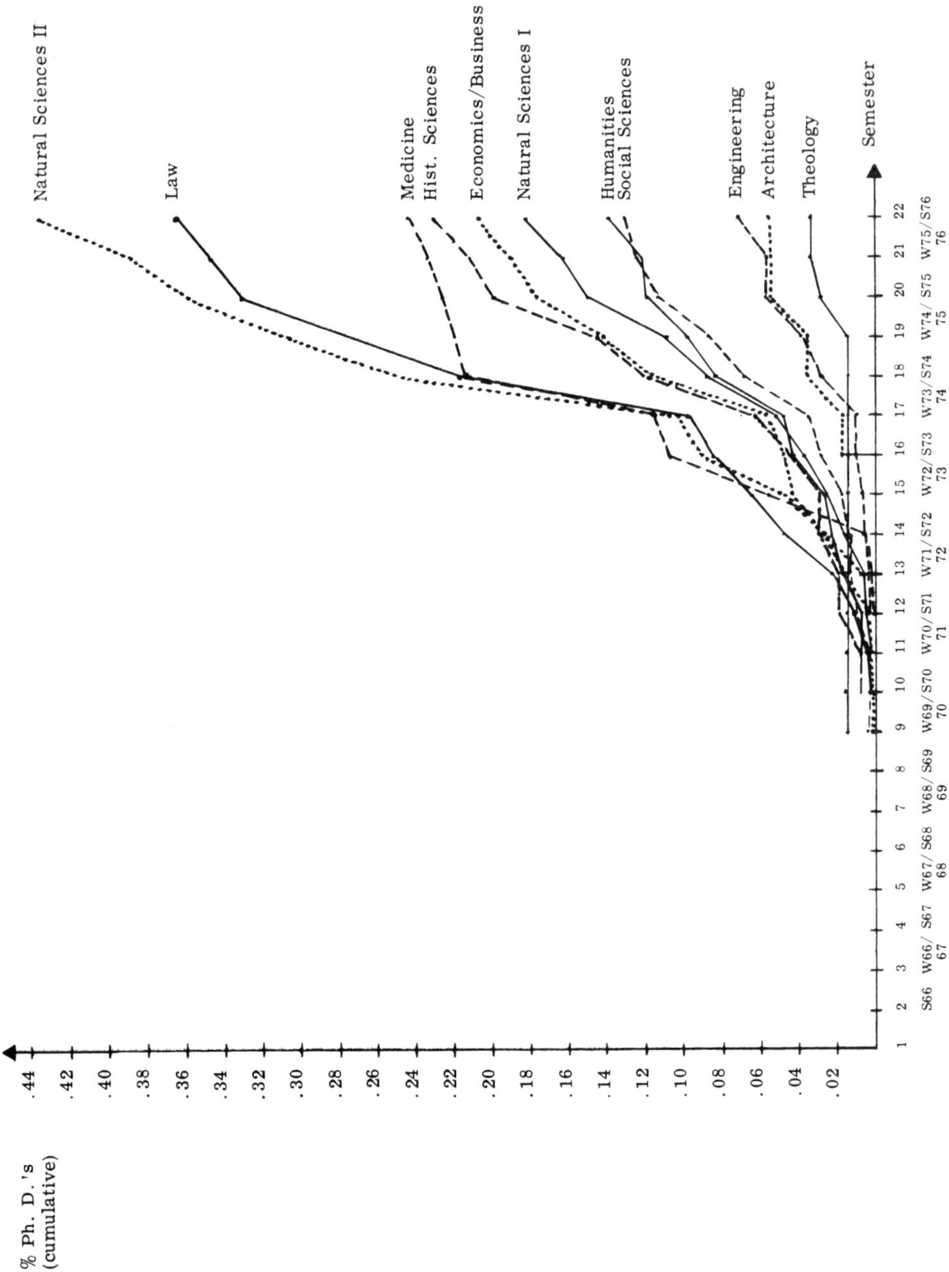

do so I have chosen the medium level curve for <u>economics/business</u>. The computational procedure has yielded the following values for the fundamental parameters of change:

$a = .1304$
$b = -.2036$

The curve for the model asymptotically approaches the value for $-\frac{a}{b}$, i.e. .6406. According to the model, the percentages develop as follows (in brackets the actual figures):

t_0	.06	(.06)
t_1	.1669	(.146)
t_2	.2542	(.245)
t_3	.3254	(.325)
t_4	.3835	(.402)
t_5	.4308	(.446)
t_6	.4695	(.486)
t_7	.5010	(.517)
t_8	.5267	(.530)
t_9	.5476	(.550)
t_{10}	.5648	(.553)
t_{11}	.5788	(.579)
t_{12}	.5901	(.597)
t_{13}	.5995	(.606)
t_{14}	.6071	(.606)
t_{15}	.6132	(.606)

In Figure 17 the actual curve and the fitted curve (dotted line) can easily be compared. The correspondance between the two is even closer than in the example used in the case of the university model. This is further evidence that the model of immanent change fits the development of the leaving of the system in a majority of cases (the cases with the logarithmic type curve). The S-chaped curves cannot be reproduced by similarly simple models.

3.1.5. Conclusion

The possibilities for the description of the development of the studies of the cohort 1965 are by no means exhausted by now. The examples given in this chapter should, however, suffice to prove the usefulness of the model as a descriptive device.

Figure 17

Actual and Fitted Curve for Graduating at a Medium Level in Economics/Business

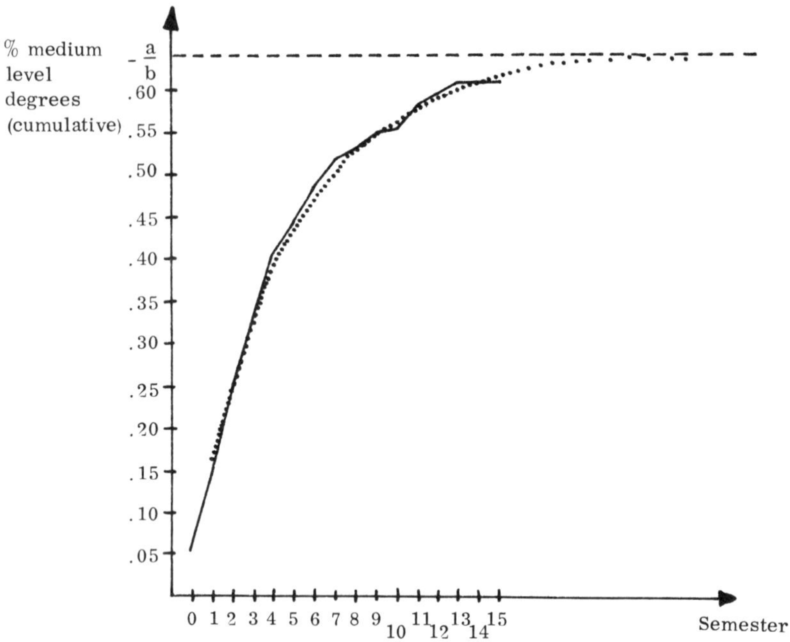

3.2. Prediction

In order to use the model for predictive purposes an additional important assumption has to be made: It has to be assumed that the transition rates are stable over time. That is, one has to assume that the flow of future cohorts through the Swiss University System will be governed by the same rates as the flow of the cohort 1965. As COLEMAN (197?) points out, this assumption is not far fetched, but a good approximation to the empirical world, because flow rates are, in general, much more stable than stocks.

Under this assumption, the model allows us to predict future total numbers of students and graduates on the basis of input data for future stocks of first enrolments.

3.2.1. Prediction of the Total Number of Students

Table 18 provides a simple scheme for the projection of given stocks of first enrolments to future winter-semesters. A similar scheme could be provided for summer-semesters. According to this scheme, one proceeds as follows to, say, transform the stocks for the cohort 1965 into stocks in the winter-semester 1975:

First, one computes the stocks in winter 1965 (S(65)) by transforming the summer stocks 1965 (S(SS65)) into winter stocks and adding them to the winter-starters 1965 (S(WS65)). This transformation is done by use of T_{11}, the T part of the first transition matrix of type 1.

The resulting stocks are, then, multiplied by the T part of the twentieth transition matrix of type 2 to yield the stocks we are looking for.

The decision taken in 2.4.1. implies that it takes a cohort 23 semesters to completely pass through the system. This means that in order to predict the total number of students in one particular semester, all the cohorts that have

Table 18

Scheme for the Projection of Stocks of Students to Future Winter Semesters

Stock Data	Transition Data 65	66	67	68	69	70	71	72	73	74	75	76	77	78	79
$S(65) = S(SS65) \cdot T_{11} + S(WS65)$	I	T_{22}	T_{24}	T_{26}	T_{28}	T_{210}	T_{212}	T_{214}	T_{216}	T_{218}	T_{220}				
$S(66) = S(SS66) \cdot T_{11} + S(WS66)$		I	T_{22}	T_{24}	T_{26}	T_{28}	T_{210}	T_{212}	T_{214}	T_{216}	T_{218}	T_{220}			
$S(67) = S(SS67) \cdot T_{11} + S(WS67)$			I	T_{22}	T_{24}	T_{26}	T_{28}	. . .	. . .	. . .	T_{216}	T_{218}	T_{220}		
" "				I							T_{214}	T_{216}	T_{218}	T_{220}	
" "					I						T_{212}	T_{214}	. . .	. . .	. . .
" "						I					T_{210}				
" "							I				T_{28}				
" "								I			T_{26}				
" "									I		T_{24}				
" "										I	T_{22}				
$S(75) = S(SS75) \cdot T_{11} + S(WS75)$											I				
" "															
" "															
" "															

T_{11} = first transition matrix of type 1
T_{24} = fourth transition matrix of type 2 etc.
I = Identity matrix

started their studies within the period of the 23 semesters which terminates with the semester we intend to make the prediction for will have to be transformed accordingly. This includes 11 cohorts plus part of a 12th cohort. I will only take into account the 11 cohorts and neglect the part of the earliest one, because the remaining number of students of this earliest cohort are negligible for all practical purposes.

Now, to obtain the total number of students in a future winter-semester, say in WS75, all one has to do is to sum over all the transformed stocks:

$$N(WS75) = S(75) \cdot I + S(74) \cdot T_{22} + S(73) \cdot T_{24} + \ldots\ldots + S(65) \cdot T_{220}$$

(N = row vector for the total number of students)

The procedure is identical for both, the university model and the subject model.

3.2.2. Prediction of Dropouts and Graduates

There are two interesting aspects to the prediction of the number of students in the different absorbing states:

- prediction of the dropouts and graduates of a single cohort
- prediction of dropouts and graduates in a single semester

For both of these aspects there are two solutions, one for the university (group of subject) of entrance and one for the university (group of subject) of exit.

The solution for the prediction for a single cohort is simple in the case of the university (group of subject) of entrance. In a first step, one obtains the stocks in the winter-semester of the cohort in question in a way analogous to the one described in the previous section. Then, the stocks of the winter-semester are multiplied with the Z part of the final transition matrix of type 2 to yield the desired figures. In the case of the cohort 1965:

Step 1: $S(65)$	=	$S(SS65) \cdot T_{11} + S(WS65)$
Step 2: $G(65)$	=	$S(65) \cdot Z_{221}$

(G = row vector for dropouts and three levels of graduates)

In the case of the university of exit, the steps taken for the prediction of a single cohort are closely related to the procedure employed in constructing the tables 13 and 16: The first step is identical with the first step above. In a second step the initial stocks are projected to all the subsequent semesters up to the final one using type 1 matrices to yield transformed stocks (5). In a third step these transformed stocks are converted into diagonal matrices which are to be multiplied with the appropriate Z part of the type 1 transition matrices to yield 4x12 matrices of type G. Finally, entries in corresponding cells of the G matrices are added to give the desired figures. In the case of the cohort 1965:

Step 1:	$S(65)$	=	$S(SS65)\cdot T_{11} + S(WS65)$
Step 2:	$S_{SS66}(65)$	=	$S(65)\cdot T_{12}$
	$S_{WS66}(65)$	=	$S_{SS66}(65)\cdot T_{13}$
	$S_{SS67}(65)$	=	$S_{WS66}(65)\cdot T_{14}$
	$\vdots$		$\vdots$
	$S_{WS75}(65)$	=	$S_{SS75}(65)\cdot T_{121}$
	$S_{SS76}(65)$	=	$S_{WS75}(65)\cdot T_{122}$
Step 3:	$G_1(65)$	=	$S(SS65)\cdot Z_{11}$
	$G_2(65)$	=	$S(65)\cdot Z_{12}$
	$G_3(65)$	=	$S_{SS66}(65)\cdot Z_{13}$
	$\vdots$		$\vdots$
	$G_{22}(65)$	=	$S_{WS75}(65)\cdot Z_{122}$
Step 4:	$G_{cohort}(65)$	=	$\sum_{i=1}^{22} G_i(65)$

The prediction of dropouts and graduates in a single semester in the case of the university of entrance is rather intricate, because it involves taking dif-

5) These transformed stocks can be called $S_{SSn}(m)$; to be read as: Stocks of the cohort m, transformed to stocks in the summer-semester n.

ferences between corresponding cells of Z parts of adjacent type 2 matrices. Although the task is quite possible to solve in principle, I will not discuss it any further at this point.

The same task for the university of exit is quite complicated, too. It follows, however, the procedure outlined for the first kind of prediction in the case of the university of exit. The only difference is, that it involves more than one cohort. Thus to predict the total number of dropouts and graduates by university of exit in the WS75, one would have to take the following steps:

Step 1:

$$S(65) = S(SS65)\cdot T_{11} + S(WS65)$$
$$S(66) = S(SS66)\cdot T_{11} + S(WS66)$$
$$\vdots$$
$$S(75) = S(SS75)\cdot T_{11} + S(WS75)$$

Step 2:

$$S_{SS66}(65) = S(65)\cdot T_{12}$$
$$S_{WS66}(65) = S_{SS66}(65)\cdot T_{13}$$
$$\vdots$$
$$S_{WS75}(65) = S_{SS75}(65)\cdot T_{121}$$

analogously for all the other cohorts involved:

$$S_{WS75}(66) = S_{SS75}(66)\cdot T_{119}$$
$$S_{WS75}(67) = S_{SS75}(67)\cdot T_{117}$$
$$\vdots$$
$$S_{WS75}(74) = S_{SS75}(74)\cdot T_{13}$$

Step 3:

$$G_1(75) = S(75)\cdot Z_{12}$$
$$G_2(75) = S_{WS75}(74)\cdot Z_{14}$$
$$G_3(75) = S_{WS75}(73)\cdot Z_{16}$$
$$\vdots$$
$$G_{11}(75) = S_{WS75}(65)\cdot Z_{122}$$

Step 4:

$$G_{semester}(75) = \sum_{i=1}^{11} G_i(75)$$

3.2.3. Input Data

The prediction procedures just described not only assume that past input data are known, but also that future numbers of first enrolments can be extrapolated. Both assumptions are quite questionable:

The first time a first enrolment statistics was compiled in Switzerland by the ESTA (6) was in winter 1967/68. Since then, first enrolment statistics have been made available for the following periods:

- SS68
- WS68/69 + SS69
- WS69/70 + SS70
- WS70/71 + SS71
- WS71/72 + SS72
- WS72/73
- WS73/74
- SS74

In the four cases, where the period covered includes a winter-semester and the following summer-semester, there are no separate figures for the two semesters included. In addition to these figures, the research project at AARAU provided the numbers for the two semesters of the year 1965.

This is quite a meagre basis, but it is, nevertheless, possible to estimate the missing numbers on the basis of inter- and extrapolation. Luckily, in all but one case it could be assumed that the proportions of spring- and fall-starters remained fairly constant over time. This assumption was possible on the basis of the three years for which separate figures are known. The exception is the University of Basel, where the proportion of spring-starters declined by one third between 1965 and 1974, from roughly two thirds of the cohort to about one third. In the process of interpolation I assumed that this decline followed a linear pattern.

Tables 19 and 20 provide the numbers of students enrolling for the first time by

---

6) ESTA = Eidgenössisches Statistisches Amt (Federal Bureau of Statistics)

Table 19

Input Data: Number of Students Enroling for the First Time by University and Semester of Entrance

University	S63 1)	W63/64 1)	S64 1)	W64/65 1)	S65 2)	W65/66 2)	S66 3)	W66/67 3)	S67 3)	W67/68 4)	S68 4)	W68/69 5)
Basel	305	153	327	164	327	164	322	181	318	197	313	268
Bern	88	478	94	512	94	512	98	590	100	667	101	723
ETHL	1	118	1	126	1	126	1	150	1	173	0	144
ETHZ	1	954	1	1022	1	1022	2	1029	3	1035	4	984
Fribourg	36	354	37	379	37	379	43	397	47	414	51	498
HSG	8	147	9	157	9	157	5	186	3	215	0	280
Genève	20	507	21	543	21	543	52	664	73	785	94	914
Lausanne	15	384	16	411	16	411	19	421	21	430	23	428
Neuchâtel	7	203	7	218	7	218	14	226	18	233	21	254
Zürich	243	753	260	307	260	807	287	876	302	944	316	967
Total	724	4051	773	4339	773	4339	843	4720	886	5093	923	5460

1) Arrived at by extrapolation on the basis of figures supplied by RIES/Kriesi (1974a)
2) Data supplied by the project of the University of AARGAU
3) Arrived at by interpolation of the data supplied by the project of the University of AARGAU, on the one hand, and of ESTA figures on the other
4) ESTA figures
5) Arrived at by splitting up yearly figures supplied by ESTA into spring- and fall-starters
For further comments on the input data see the text on p.68 ff.

Table 19 (Continued)

Input Data: Number of Students Enroling for the First Time by University and Semester of Entrance

University	S69 5)	W69/70 5)	S70 5)	W70/71 5)	S71 5)	W71/72 5)	S72 5)	W72/73 4)	S73 6)	W73/74 4)	S74 4)
Basel	350	298	322	378	338	331	245	427	266	579	268
Bern	100	759	105	755	105	846	117	1323	131	1155	145
ETHL	5	190	6	189	6	224	7	223	13	231	16
ETHZ	10	994	10	1063	11	1122	11	1156	18	1353	24
Fribourg	50	502	51	520	53	536	54	579	53	603	52
HSG	0	282	0	276	0	272	0	221	1	287	2
Genève	103	512	58	695	78	752	84	875	100	1093	115
Lausanne	31	455	33	504	36	491	35	808	56	845	76
Neuchâtel	32	263	33	172	22	213	27	268	34	241	40
Zürich	324	974	327	1001	336	1037	347	1279	470	1690	566
Total	1005	5229	945	5553	985	5824	927	7159	1142	8077	1304

4) ESTA figures
5) Arrived at by splitting up yearly figures supplied by ESTA into spring- and fall-starters
6) Arrived at by interpolation of ESTA figures
For further comments on the data see the text

Table 20

Input Data: Number of Students Enroling for the First Time by Group of Subjects and Semester

Group of Subject	S63[1)]	W63/64[1)]	S64[1)]	W64/65[1)]	S65[2)]	W65/66[2)]	S66[3)]	W66/67[3)]	S67[3)]	W67/68[4)]	S68[4)]	W68/69[5)]
Theology	18	63	19	67	19	67	21	78	22	89	22	113
Medicine	129	675	138	723	138	723	148	814	154	905	157	1135
Law	77	341	83	366	83	366	92	404	97	442	101	433
Econ/Business	67	465	72	498	72	498	87	648	93	798	98	860
Social Sci	66	272	71	291	71	291	70	290	69	288	68	381
Humanities	168	417	180	447	180	447	208	511	222	574	235	633
Hist. Sci	41	152	44	168	44	168	48	160	50	151	51	154
Natural Sci I	50	374	54	401	54	401	45	432	40	463	34	450
Natural Sci II	112	482	120	517	120	517	114	527	110	537	107	579
Engineering	1	581	1	623	1	623	1	626	1	629	1	602
Architecture	0	161	0	176	0	176	0	157	0	138	0	121
Others	15	52	16	56	16	56	24	68	30	79	35	81
Total	744	4035	798	4333	798	4333	858	4715	888	5093	909	5542

1) Arrived at by extrapolation on the basis of figures supplied by RIES/KRIESI (1974a)
2) Data supplied by the project of the University of AARAU
3) Arrived at by interpolation of the data supplied by the project of the University of AARAU, on the one hand, and of ESTA figures, on the other
4) ESTA figures
5) Arrived at by splitting up yearly figures supplied by ESTA into spring- and fall-starters
For further comments on the input data see the text on p.68 ff.

Table 20 (Continued)

Input Data: Number of Students Enroling for the First Time by Group of Subjects and Semester

Group of Subject	S69 5)	W69/70 5)	S70 5)	W70/71 5)	S71 5)	W71/72 5)	S72 5)	W72/73 5)	S73 6)	W73/74 4)	S74 4)
Theology	32	77	21	77	11	69	19	100	26	102	32
Medicine	150	1010	134	1066	141	859	114	1332	134	1458	154
Law	94	450	98	504	110	544	118	743	150	847	181
Econ/Bus	112	825	107	836	109	778	101	883	105	827	109
Social Sci	82	328	71	422	89	469	102	519	122	711	141
Humanities	205	654	212	644	209	800	260	1061	289	1184	317
Hist. Sci	46	153	46	142	42	143	42	243	59	286	76
Natural S I	63	424	60	477	67	507	71	583	89	537	107
Natural SII	110	492	94	530	101	580	110	666	125	789	139
Engineering	2	633	2	621	2	697	2	664	10	764	18
Architectur	1	169	1	195	1	182	1	221	2	257	3
Others	27	88	29	80	26	109	36	134	42	167	48
Total	924	5303	875	5594	908	5737	976	7149	1153	7929	1325

4) ESTA figures
5) Arrived at by splitting up yearly figures supplied by ESTA into spring- and fall-starters
6) Arrived at by interpolation of ESTA figures
For further comments on the input data see text

university and group of subjects of entrance, on the one hand, and by semesters, on the other. These numbers pertain to Swiss students only. The semesters covered range from SS63 to SS74. The four semesters prior to 1965 are included for reasons to be explained shortly. Their numbers are extrapolations based on the total number of students graduating from higher secondary schools in the respective years (see RIES/KRIESI 1974c, p. 150). It is assumed that the distributions of the students over the universities and groups of subjects were the same in 1963 and 1964 and that they are equal to the ones in 1965.

In looking at the figures in the two tables, the reader may note that the totals in corresponding years are not always exactly the same. These discrepancies are, on the one hand, due to the interpolations, on the other hand, they are the result of the fact that the ESTA does not group the subjects as does this paper. In the process of regrouping the ESTA figures for this paper, it was in some cases impossible to classify them according to the categories presently employed.

Another look at the totals reveals a steady increase of first enrolments in the period covered. This does, however, not mean that first enrolments have universally increased in that period. The University of Genève, for example, experienced a sharp decline in first enrolments between W68/69 and W69/70: the number fell from 914 to 512. Medicine provides another example of an erratic pattern: First enrolments in this group of subjects decreased at the same time as Genève experienced its reduction, they picked up again in W70/71, only to fall back anew in W71/72. Afterwards they grew dramatically.

Such unsystematic patterns make it difficult to extrapolate into the future. But, before turning to such delicate matters, one additional drawback of existing figures has to be mentioned: The project at AARAU has shown that the official enrolment statistics do not always make a distinction between first enrolments and re-matriculations (see RIES 1974). Thus, only a fraction of the first enrolments according to the university archive data actually turn out to be first enrolments. Table 21 gives the estimates of these fractions for universities and groups of subjects. The estimates for the universities are based on questionnaire responses, those for the groups of subjects are weighted averages on the basis of the data for the universities. At least for three of the universities, Fribourg, Neuchâtel and, above all, Bern, the reductions are quite substantial.

Table 21

Estimated Number of First Enrolments: Percentages of First Enrolments according to University Archive Data (for Universities and Groups of Subjects)

University	Estimated Percentage of First Enrolements	Groups of Subjects	Estimated Percentage of First Enrolements
Basel	98.6	Theology	90.9
Bern	78.5	Medicine	91.7
ETHL	98.5	Law	91.9
ETHZ	98.6	Econ/Bus	93.0
Fribourg	86.0	Social Sci	92.1
HSG	98.8	Humanities	92.8
Genève	96.7	Hist Sci	92.0
Lausanne	94.4	Nat Sci I	93.4
Neuchâtel	87.3	Nat Sci II	93.9
Zürich	96.7	Engineering	98.5
		Architecture	98.4
		Others	96.4

In WS72/73 the ESTA started to completely reorganize the students' statistics of Switzerland (see ESTA 1974). The new system is much more efficient than the old one, and it is probable that since WS72/73 these earlier mistakes have been eliminated.

Now, as to extrapolation of first enrolment figures, this is not only questionable because of non-linear patterns in the past. There are at least three additional reasons which may make extrapolations look too simplistic:

First of all, there is the debate on the 'numerus clausus', the restriction of the total number of students in different fields. A number of subjects have already been placing restrictions on the number of entrants. More subjects might turn to this measure in the future.

Secondly, with employment prospects for university graduates getting worse, a growing fraction of graduates of higher secondary schools might prefer not to enter the university system at all. This tendency might be endorsed by the setting up of more and more alternative educational institutions for graduates of a higher secondary school (schools for social work, teacher's universities, technical schools).

Finally, the value system in the society might turn to a more pronounced anti-intellectualism under the growing impact of the energy crisis and religious ideologies.

All these factors being very elusive, they are difficult to measure. They all point in the same direction, though, suggesting that the model's predictions based on an extrapolation of the input data will exaggerate the stocks of students in the future.

An alternative to the extrapolation of the first enrolments is an extrapolation of the graduates of higher secondary schools. They are easier to extrapolate, because future graduates are already enrolled in the higher secondary schools. Making additional assumptions about the fraction of these graduates getting enrolled at the universities, assumptions that would take into account informations of the kind discussed above, it would be possible to get a more realistic figure for the total of future first enrolments. It would, however, take still additional assumptions with respect to the distribution of these numbers over the different universities and groups of subjects to arrive at the desired figures.

This is the major shortcoming of the model presented here: It replaces the difficulty to estimate future stocks of students and graduates from universities by the difficulty to estimate future stocks of entrants into the university system and graduates from the higher secondary schools. In other words, it presupposes a similar kind of model for the lower levels in the hierarchy of the Swiss school system.

3.2.4. A Test of the Predictive Power of the Model: Prediction of the Total Number of Students in W73/74

In order to test the predictive power of the model, the total number of students in WS73/74 is predicted. This is the most recent semester for which the ESTA has already published official figures (see ESTA 1975). As outlined in 3.2.1., 11 cohorts have to be taken into account in predicting the total number of students in any particular semester. For the WS73/74 these 11 cohorts include all the Swiss students who first enrolled at Swiss universities between 1963 and 1973 (7).

Five kinds of predictions have been tried both, for the university model and the subject model. All the predictions follow the general procedure outlined in 3.2.1. They differ, however, with respect to the transition matrices employed and/or the input data that is transformed:

- the simple prediction: For this first kind of prediction the unadjusted matrices of type 2, 2.1.-2.21. (university model) and 4.1.-4.21. (subject model) are used. The input data is the one given in Tables 19 and 20.
- the prediction adjusting for stocks: In this prediction the input data of the Tables 19 and 20 are discounted by the estimated percentages of first enrolments given in Table 21. The transition matrices are the same as above.
- the prediction adjusting for dropouts: Here, the transition matrices are replaced by the adjusted series of type 2, derived from the above mentioned two sets 2.1.-2.21. and 4.1.-4.21. The input data are the original ones.
- the prediction adjusting for stocks and dropouts: In this prediction both, the transition matrices and the stocks are adjusted.

---

7) This is why the semesters for 1963 and 1964 are also included in the Tables 19 and 20.

- the final prediction: This prediction differs from the previous one in only one respect: It allows for the fact that the ESTA has reorganized and improved the students' statistics in WS72/73 (see p. 75 above). Thus, the input data for the WS72/73 and the following semesters are not discounted in this prediction, whereas all earlier data are treated as before when adjusting for stocks.

APPENDIX B5 provides the program for one of these procedures, the one adjusting for stocks only.

The results are given in Tables 22 and 23 for the university model and in Tables 24 and 25 for the subjects model (8). The first table in each of the two pairs gives the raw figures, the second one gives the predicted figures as percentages of the official total number of students in WS73/74.

In the case of the university model, the simple predictions are, in general, too big. This was to be expected, since this first kind of prediction overestimates the number of entrants and underestimates the number of dropouts. There are, nevertheless, three exceptions to this general pattern: Both, at Basel and at Neuchâtel, the simple predictions already turn out to be too small, and at the HSG, the simple prediction is deadly accurate.

Adjusting for stocks and dropouts, of course, reduces the predicted numbers. The extent of the reductions, however, varies considerably from one university to the next:

8) Since the subject model does not make any provisions for going abroad, the figures resulting from this model's predictions had to be adjusted to include students at the Swiss universities only. Assuming that the students in all the subjects are equally likely to go abroad, I have multiplied all the predictions by the fraction of students staying at Swiss universities as it emerges from the corresponding predictions in the university model. Although this assumption is somewhat incorrect, it does not greatly influence the results, because the fraction of the total number of students staying at Swiss universities approaches 1 in the first place (.982 in the case of the final prediction).

Table 22

Prediction of the Total Number of Students at Swiss Universities in the Winter-Semester 1973/74: Raw Figures

University	Total according to ESTA	Simple Prediction	Prediction adjusting for stocks	Prediction adjusting for drop o.	Prediction adjusting for stocks and drop o.	Final Prediction
Basel	4244	4183	4108	4062	3989	4006
Bern	5790	6835	5459	6656	5319	5868
ETHL	994	1081	1062	955	937	944
ETHZ	5797	6058	5963	5817	5726	5758
Fribourg	2354	2884	2501	2775	2406	2573
HSG	1413	1416	1396	1355	1336	1342
Genève	4480	5138	4954	4695	4525	4590
Lausanne	3057	3566	3353	3427	3223	3314
Neuchâtel	1172	1152	1008	1114	975	1041
Zürich	9057	9587	9235	9210	8870	8979
Total	38358	41900	39041	40066	37306	38415
Foreign U.	-	721	674	721	674	676
Interruptions	-	1939	1832	1939	1832	1854

Table 23

Prediction of the Total Number of Students at Swiss Universities in the Winter-Semester 1973/74: Percentages of the Official Figures

University	Total according to ESTA (=100%)	Simple Prediction	Prediction adjusting for stocks	Prediction adjusting for drop o.	Prediction adjusting for stocks and drop o.	Final Prediction
Basel	4244	98.6	96.8	95.7	94.0	94.4
Bern	5790	118.0	94.3	115.0	91.9	101.3
ETHL	994	108.8	106.8	96.1	94.3	95.0
ETHZ	5797	104.6	102.9	100.4	98.8	99.4
Fribourg	2354	122.5	106.2	117.9	102.2	109.3
HSG	1413	100.2	98.8	95.9	94.6	95.0
Genève	4480	114.6	110.6	104.8	101.0	102.5
Lausanne	3057	116.7	109.7	112.1	105.4	108.4
Neuchâtel	1172	98.3	86.0	95.1	83.2	88.8
Zürich	9057	105.9	102.0	101.8	98.0	99.2
Total	38358	109.2	101.8	104.5	97.3	100.1

Table 24

Prediction of the Total Number of Students in the Groups of Subjects in the Winter-Semester 1973/74: Raw Figures

Groups of Subjects	Total according to ESTA	Simple Prediction	Prediction adjusting for stocks	Prediction adjusting for drop o.	Prediction adjusting for stocks and drop o.	Final Prediction
Theology	532	674	614	630	573	594
Medicine	7843	8616	7904	8377	7685	7914
Law	4205	4528	4160	4455	4093	4225
Econ/Business	4550	5171	4809	4962	4615	4731
Social Sci[1)]	2535	2796	2576	2678	2468	2565
Humanities	5030	5339	4950	4994	4630	4797
Hist. Sci	1544	1759	1621	1704	1570	1616
Nat Sci I	2970	3028	2827	2896	2703	2775
Nat Sci II	4314	5014	4700	4822	4520	4610
Engineering	3353	3434	3370	3270	3210	3230
Architecture	1081	987	966	910	891	897
Others	401	543	522	472	453	463
Totals	38358	41889	39019	40170	37411	38417
Interruptions	-	2070	1944	2070	1944	1974

1) Contrary to the Convention of the ESTA that includes 'Sociology' among the group 'Econ/Bus' this subject is included among the 'Social Sciences' in the present study.

Table 25

Prediction of the Total Number of Students in the Groups of Subjects in the Winter-Semester 1973/74: Percentages of the Official Figures

Groups of Subjects	Total According to ESTA	Simple Prediction	Prediction Adjusting for Stocks	Prediction Adjusting for drop out	Prediction Adjusting for stocks and drop o.	Final Prediction
Theology	532	127.6	115.4	118.4	107.7	111.7
Medicine	7843	109.9	100.8	106.8	98.0	100.9
Law	4205	107.7	98.9	105.9	97.3	100.5
Econ/Business	4550	113.6	105.7	109.1	101.4	104.0
Social Sci[1)]	2535	110.3	101.6	105.6	97.4	101.2
Humanities	5030	106.1	98.4	99.3	92.0	95.4
Historical Sci	1544	113.9	105.0	110.4	101.7	104.7
Natural Sci I	2970	102.0	95.2	97.5	91.0	93.4
Natural Sci II	4314	116.2	108.9	111.8	104.8	106.9
Engineering	3353	102.4	100.5	97.5	95.7	96.3
Architecture	1081	91.3	89.4	84.2	82.4	83.0
Others	314	135.4	130.2	117.7	113.0	115.5
Total	38358	109.2	101.7	104.7	97.5	100.2

1) Contrary to the convention of the ESTA that includes 'Sociology' among the group 'Econ/Bus', this subject is included among the 'Social Sciences' in the present study.

	adjusting for stocks	adjusting for dropouts
Basel	- 1.8	- 2.9
Bern	- 23.7	- 3.0
ETHL	- 2.0	- 12.7
ETHZ	- 1.7	- 4.2
Fribourg	- 16.3	- 4.6
HSG	- 1.4	- 4.3
Genève	- 4.0	- 9.8
Lausanne	- 7.0	- 4.6
Neuchâtel	- 11.7	- 3.2
Zürich	- 3.9	- 4.1

First of all, the two adjustment procedures are not equally important at all the universities. In four cases (Bern, Fribourg, Neuchâtel, Lausanne, in that order), the adjustment for stocks is the decisive one, whereas in another four cases it is just the other way around: For the ETHL, ETHZ, HSG and Genève adjusting for dropouts makes a greater difference than adjusting for stocks. For Basel and Zürich, finally, both adjustment procedures are of about equally little importance.

By far the greatest reductions overall are the ones for the adjustment of stocks at Bern and Fribourg. While the reduction for Bern goes too far, the one for Fribourg is still not large enough. Also rather substantial are the reductions for adjusting for dropouts in the cases of the ETHL and Genève, and those for adjusting for stocks at Neuchâtel and Lausanne.

Combining both types of adjustments in the fourth prediction leads to too large reductions in all but three cases and to a too small total, too. One of the reasons for this unexpected result lies most likely in the reorganization of the students' statistics in WS73/74. Taking this reorganization into consideration leads to an overall improvement of the predictions as can be seen by comparison of the last two columns of Table 23.

The final prediction results in the following deviations from the official figures for the different universities:

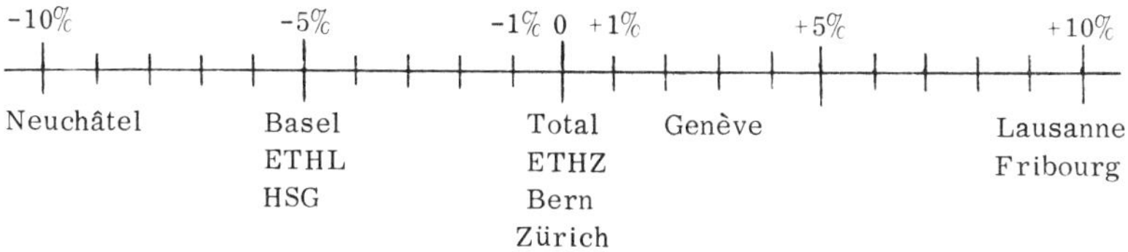

The prediction for the total number of students is surprisingly accurate. The next most accurate figures in percentage terms are those for the ETHZ, the university of Zürich and the university of Bern, the largest three universities, followed by the fourth largest one, the university of Genève. There is, clearly, a correlation between accuracy of prediction and size of the university for which the predictions is made. The university with the least accurate prediction (Neuchâtel), however, is not the smallest, but the second to the smallest one.

With the exceptions of Neuchâtel, Lausanne and Fribourg, the final predictions are quite satisfactory. In these three cases, moreover, it is not necessarily the model that is incorrect. At least equally plausible are inaccuracies in the input data.

The fact that the model underestimates the dropouts (see p. 20 above) has little influence on this type of prediction, because the additional dropouts among the long-term students (the 3.5% from p. 35) come from cohorts that have the smallest impact on the prediction of the total number of students.

For the subject model as for the university model, the simple predictions are in general too big. The one notable exception is the case of architecture, where even the simple prediction underestimates the official figure.

The extent of the reductions due to the two types of adjustment again varies considerably from one group of subjects to another:

	adjusting for stocks	adjusting for dropouts
Theology	- 12.2	- 9.2
Medicine	- 9.1	- 3.1
Law	- 8.8	- 1.8
Economics/Business	- 7.9	- 4.5
Social Sciences	- 8.7	- 4.7
Humanities	- 7.7	- 8.8
Historical Sciences	- 1.1	- 3.5
Natural Sciences I	- 6.8	- 4.5
Natural Sciences II	- 7.3	- 4.4
Engineering	- 1.9	- 4.9
Architecture	- 1.9	- 7.1
Others	- 5.2	- 17.7

In seven instances, the adjustment for stocks makes the greater difference (for theology, medicine, law, economics/business, the social sciences and both of the natural sciences), and four times it is the adjusting for dropouts which is the more important one. In the case of the humanities, both procedures lead to about the same amount of reduction.

Apart from the residual category, the largest reductions are those for theology in both procedures. Although these reductions are quite substantial, they do not reach the size of the ones for Bern and Fribourg in the university model, and they do not go far enough either.

The final prediction results in the following deviations from the official figures for the different group of subjects:

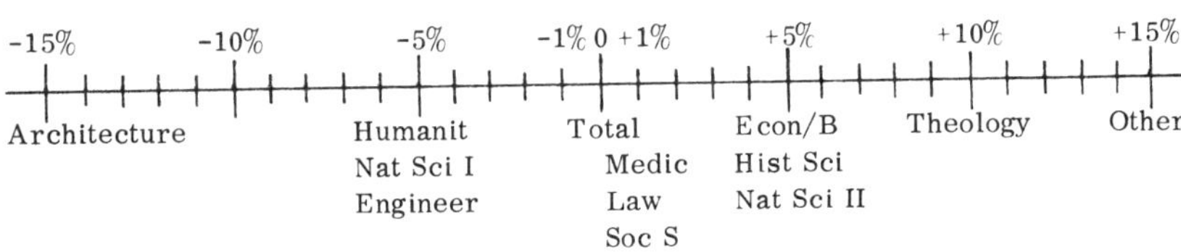

The prediction for the total number of students is again extremely accurate. The predicted figure for the largest group of subjects, medicine, is also very close to the official one. The prediction is worse, on the other side, for the three smalles groups of subjects - architecture, theology and the residual category. For the rest of the groups, the correspondance between accuracy and size is not as close as for the university model. Inaccuracies in the predictions are again not necessarily due to shortcomings of the model, but have also something to do with inaccuracies in the input data.

Except for the three smallest groups of subjects the fit between official and predicted figures is again quite satisfactory.

In addition to the figures for universities and groups of subjects, the models also provide estimates of the total number of Swiss students studying abroad in WS73/74 and of the total number of interruptions in this semester: A very small proportion of the Swiss students were enrolled at foreign universities in WS73/74, all in all approximately one percent and three quarters. The estimate for interruptions lies in the vicinity of 5% for the same semester.

In predicting the total number of students as above, the interruptions have not been included in the figures for universities and subjects. Thus, the predicted figures refer to the numbers of students actually studying at a certain university in a particular group of subjects. The official figures, on the other hand, refer to the number of matriculations. Now, the number of matriculations does not necessarily correspond to the number of students actually studying, because at Swiss universities one can be matriculated without actually attending any courses. In the questionnaire, the students of the cohort 1965 have been asked, if they had interrupted their studies and, if yes, at what point in time. They have, however, not been asked, if they had been matriculated during their interruptions or not. It is, therefore, possible that some of those students who indicate that they had interrupted their studies (because they did not attend courses) have actually been matriculated (in order to benefit from the status of being a student). I, therefore, added the estimated number of interruptions in WS73/74 to the total numbers of students given by the final prediction in Tables 22 and 24 to get the adjusted figures including interruptions. These figures reflect some kind of a maximum possible number of students in the different universities and groups of subjects. They are given in the Tables 26 (for the university model) and 27 (for the subject model).

In most cases, the fit is worse under the new circumstances. In some instances, however, it is improved, most dramatically in the case of the group of architecture. Remembering the tremendous percentage of interruptions in this field (see Table 11), it is quite possible that the final prediction (not including interruptions) is seriously underestimating the actual enrolment.

To give an overall assessment of the predictive power of the model, it seems that it has done comparatively well in estimating the total number of students in WS73/74 in the different groups of subjects and at the different universities. In the light of these results it promises to be a useful instrument in predicting future numbers of students.

3.3. Simulation

We can perform specific experiments on the observed transition matrices of the model. Asking questions of the sort 'what would happen, if .... ?', it is possible to introduce specific assumptions about the flows into the model and then to work out their implications.

This strategy has, for example, been employed by LIEBERSON/FUGUITT (1967), who asked in the context of U.S. race relations what would happen, if racial discrimination were to be eliminated at once. Using transition matrices similar to the ones developed here, they were able to show that only after seven generations all effects of prior discrimination would have been eliminated.

In the present context, I suggest to ask two questions to illustrate the fruitfulness of this technique:

Table 26

Prediction fo Total Number of Students at Swiss Universities in the Winter-Semester 1973/74: Adjusted Figures, Including Interruptions

University	Adjusted Prediction (see Table 22)	Interruptions	Adjusted Total including Interruptions	in % of ESTA Figures
Basel	4006	148	4154	97.9
Bern	5868	184	6052	104.5
ETHL	944	41	985	99.1
ETHZ	5758	533	6291	108.6
Fribourg	2573	101	2674	113.6
HSG	1342	117	1459	103.3
Genève	4590	190	4780	106.7
Lausanne	3314	136	3450	112.9
Neuchâtel	1041	33	1074	91.6
Zürich	8979	372	9351	103.3
Total	38415	1855	40270	105.0

Table 27

Prediction of the Total Number of Students in the Groups of Subjects in the Winter-Semester 1973/74: Adjusted Figures Including Interruptions

Groups of Subjects	Adjusted Prediction of Table 24	Interruptions	Adjusted Prediction including inter.	in % of ESTA
Theology	594	28	622	116.9
Medicine	7914	199	8113	103.4
Law	4225	214	4439	105.6
Economics/Business	4731	252	4983	109,5
Social Sciences[1)]	2565	151	2716	107.1
Humanities	4797	262	5059	100.6
Historical Sciences	1616	56	1672	108.3
Natural Sciences I	2775	139	2914	98.1
Natural Sciences II	4610	172	4782	110.8
Engineering	3230	280	3510	104.7
Architecture	897	156	1053	97.2
Others	463	34	497	123.9
Total	38417	1944	40361	105.2

1) Contrary to the convention of the ESTA that includes 'Sociology' among the group 'Econ/Bus', this subject is included among the 'Social Sciences'in the present study.

- What would happen, if all the flows between the universities (groups of subjects) were eliminated?
- What would happen, if all the universities (groups of subjects) had the same pattern of leaving as some specific university (group of subject)?

As a criterion to judge the outcome of these experiments, it is convenient to rely on the total number of students in the WS73/74 as predicted by the model in the previous section. Thus, the question to be asked more specifically reads as this:

How would the total number of students predicted by the model be affected, if there were no inter-university (inter-subject) flows, or, if there was just one pattern of leaving the system? Of course, it is possible to ask many more questions of this kind within the framework of the model presented and only time and space prevent us from doing so.

3.3.1. Elimination of Flows between Swiss Universities or between Groups of Subjects

If there were no flows between Swiss universities (groups of subjects), all the students would remain at their university of origin (original group of subject), except for movements abroad. This implies that all the pertaining off-diagonal elements in the respective sub-matrices $T(t, t+n)$ would have to be set to zero and all the diagonal elements would have to be incremented by the sum of these off-diagonal elements in their row. These changes can easily be introduced into the observed set of transition matrices.

Table 28 shows the outcome of this experiment in terms of the number of students predicted for the WS73/74 for the case of the universities. The numbers are presented as raw figures and as percentages of the figures of the final prediction in the previous section (see Table 22). The universities are ranked according to whether they would gain or lose students under the present assumption.

Table 28

Predicted Number of Students at Swiss Universities in the WS73/74 under the Assumption of no Inter-University Flows: Raw Figures and Percentages

University	Number of Students in WS73/74	
	abs	in % of final prediction (see Table 22)
Bern	5488	94
Zürich	8470	94
Lausanne	3190	96
Basel	4014	100
HSG	1344	100
Genève	4614	101
ETHL	1021	108
ETHZ	6195	108
Fribourg	2786	108
Neuchâtel	1293	124
Total	38415	100

As the table shows, three universities (BERN, ZUERICH and LAUSANNE) would wind up having less students in the absence of inter-university flows, BASEL and the HSG would have exactly the same amount of students, whereas the remaining five universities, above all NEUCHATEL, would experience an increase in the numbers enrolled. Stated differently, this result provides a short-hand summary of Fig. 2 given above. For each university it gives the net gain or loss of students in the actual process of inter-university flows. The losers in the experiment are the actual gainers and vice versa. In the second part of this study, I will try to explain this outcome.

Analogously, Table 29 reports the results for the subject model. Under the assumption of no inter-subject flows, we have six groups of subjects losing in numbers, above all theology and the historical sciences, and five groups gaining, especially the residual category and the natural sciences I. The number of students studying medicine would remain unchanged.

Table 29

Predicted Number of Students in the Groups of Subjects in the WS73/74 under the Assumption of no Inter-Subject Flows: Raw Figures and Percentages

Group of Subject	Number of Students in WS73/74	
	abs	in % of final prediction (see Table 24)
Theology	518	87
Historical Sci	1401	87
Law	4080	97
Economics/Business	4569	97
Social Sci	2496	97
Natural Sci II	4504	98
Medicine	7916	100
Humanities	5034	105
Engineering	3397	105
Architecture	897	106
Natural Sci I	3027	109
Others	537	116
Total	38432	100

The reciprocals of the percentages in Table 29 again provide an indicator of the net gains or losses of the different groups of subjects from the actual flows among them as reported in Fig. 3 above. Theology and the historical sciences are the actual big gainers, the natural sciences I and the residual category the actual losers.

3.3.2. Assumptions about the Pattern of Leaving

The pattern of leaving may differ from one university (group of subjects) to another in two respects:

- in the distribution of the degrees obtained (see Fig. 4 and Fig. 11)
- in the average duration taken to obtain a specific degree (see Table 14 and Table 17)

The differential impact of each one of these aspects on the outcome of experiments involving assumptions about the pattern of leaving is difficult to sort out.

What would happen, if all the universities had the same leaving pattern as the School for Economics and Business at St.Gallen (the HSG)? This specific form of the second question raised above is primarily concerned with the duration aspect. The average duration to obtain a medium level degree at the HSG, which is the most prominent degree there, is the lowest in the country. This can at least partly be attributed to the strictly structured schedule enforced at this university. To ask the above question, then, is to ask about the impact of structural change in the organization of studies on the average duration necessary to obtain a degree at the different Swiss universities.

The distributive aspect, of course, obscures the relationship between the organization of studies and their average duration. Thus, the HSG shows a conspicuous absence of low level degrees and an above average share of medium level ones. The average durations are also affected by the different subject compositions of the universities (see p. 56). The HSG, which only offers economics and business courses, is, of course, not representative of any other university with respect to the subject composition. Economics and Business are, however, characterized by a particularly low average duration required to obtain a medium level degree.

Luckily, the distributive aspect of leaving and the subject composition work in opposite directions in obscuring the relationship in question. As a rough approximation we may assume that their opposite impacts cancel and that, consequently, the result of the experiment will justify some conclusions about this relationship.

If all the universities have the same pattern of leaving as the HSG, all the rows in the Z(t,t+n) sub-matrices will have to be identical. In order to make the resulting rows of the entire M(t,t+n) matrices sum to 1, the main diagonal elements will have to be adjusted accordingly. These transformations are again easily undertaken by a small program.

Table 30 shows the outcome of the experiment, again in terms of the number

Table 30

Predicted Number of Students at the Swiss Universities in the WS73/74 under the Assumption that all Universities have the same Pattern of Leaving as the HSG: Raw Figures and Percentages

University	Number of Students in WS73/74	
	abs	in % of final prediction (see Table 22)
Basel	3491	87
Bern	5238	89
Zürich	8204	91
ETHZ	5472	95
HSG	1311	98
Genève	4582	100
Neuchâtel	1045	100
Fribourg	2608	101
Lausanne	3435	104
ETHL	1051	111
Total	36437	95

of students predicted in the WS73/74 for each university. The universities are once more ranked according to whether their student body would be reduced or enlarged under the present assumption. The results of section 3.1.3.2. are clearly replicated. It is the Swiss German speaking universities, which could reduce the number of students enrolled by processing them at the rate of the HSG. More specifically, this table shows that their student bodies could be reduced by roughly 10%. Interestingly enough, the French speaking universities are in much better shape. Apparently, their courses of study are much better organized than the ones of their Swiss German counterparts (8).

8) It is interesting to note the lack of awareness of the said difference between the Swiss universities even among educational planners in the Swiss National Science Council ('Wissenschaftsrat'). In personal discussions I was struck by the extent to which French speaking Swiss planners are unable to realize how unstructured the Swiss German universities actually are with respect to the courses of studies offered in a large number of subjects.

The difference between the universities of the two language regions is replicated on a different level by the difference between the two technical schools. Both process their students faster than the universities in their region, but again, the French speaking school is considerably more efficient than the German speaking one with respect to the average duration. Overall, there would be a decrease in the total number of students enrolled.

The analogous question for the subject model shall be phrased as follows: What would happen, if all the subjects had the same pattern of leaving as the subject of medicine? A popular notion in Switzerland is that students in medicine study longer than anyone else, and, therefore, have to be remunerated accordingly. As was shown in 3.1.4.2., the average amount of time required to get a medium level degree in medicine is comparatively long, whereas the average amount of time required to receive a high degree in this subject is comparatively short. But what is the overall picture? The present experiment allows to clarify this question.

As Table 31 shows, the outcome of the experiment substantiates the common belief to some extent. All groups of subjects would experience an increase in enrolment, if their students were to adopt the same leaving pattern as the students in medicine, although this increase would turn out to be negligible for natural sciences II, for law and for the historical sciences. These are, however, precisely those three fields with the largest shares of doctorates among their graduates (see Fig. 11). Those groups of subjects with large shares of low level leavers (the social sciences and the humanities) or the ones with large shares of medium level graduates (engineering) would undergo the largest growth in numbers of students enrolled, which is to say that by drawing on a rank order like the one represented by Table 31, the popular notion of the leaving pattern neglects the distributive aspect of leaving. If it were to take the more differentiated point of view of Table 17, it would probably be less impressed by medical doctors.

Table 31

Predicted Number of Students in the Groups of Subjects in the WS73/74 under the Assumption that all Groups of Subjects have the same Pattern of Leaving as the Subject of Medicine: Raw Figures and Percentages

Group of Subjects	Number of Students in WS73/74	
	abs	in % of final prediction (see Table 24)
Medicine	7878	100
Natural Sci II	4634	101
Law	4332	103
Historical Sci	1687	104
Natural Sci I	3288	118
Architecture	1078	120
Theology	742	125
Economics/Business	5910	125
Humanities	6170	129
Engineering	4155	129
Social Sci	3380	132
Others	802	173
Total	44056	115

4. SUMMARY AND DISCUSSION

In this first part, a model for the flow of students through the Swiss University System has been proposed. On the basis of detailed data on the total cohort 1965 of Swiss students, collected by a research team at the preparatory stage of the University for Educational Sciences at Aarau, Switzerland, two variants of the model have been analyzed, one for the flow between subjects and one for the flow between universities. Both variants basically consist of several sets of transition matrices, each one of these matrices being divided into two parts: One set of rates for the internal movement between the states of the system and a second set of rates for the leaving of the system. In constructing the model, several adjustments had to be made to overcome the shortcomings of the data.

The model presented is intended to serve three purposes: Description, prediction and simulation. With respect to description, both, the flows between universities and those between subjects have been shown to be rather limited in scope. Secondly, the description of the four alternative ways of leaving the system yielded a variety of results: Overall rates of leaving at the different levels, average durations for the attainment of degrees at the respective levels and time series for the variation of leaving over time. The model of immanent change has proved to fit a majority of these time series.

Assuming that the transition rates are stable over time, the model was then shown to be useful for the prediction of the total number of students as well as of dropouts and graduates in future semesters. The predictive power of the model was tested by employing the model for the prediction of the total number of students in the WS73/74, the most recent semester, for which official figures exist at the time of this writing. The predictions turned out to be quite satisfactory.

Finally, applications in the field of simulation have been demonstrated. The model allows to work out the implications of several assumptions about the flows within the system and about specific patterns of leaving.

In the perspective of educational planning the model is, of course, rather limited in scope. It is concerned only with flows within the subsystem of higher education and the leaving therefrom. As has been pointed out in the context of prediction, this constitutes a major shortcoming. The usefulness of the model depends on the existence of similar models for the other subsystems within the whole educational system.

Moreover, even within the framework of higher education the model is not all-encompassing. The model is only concerned with the flow of students and other important aspects of the system of higher education, above all the different resources (faculty, finances, facilities, technical and administrative personnel etc.) do not enter into the model.

Finally, it has not been attempted to consider the output of the students from the point of view of the societal system receiving them. Questions concerning the demand of manpower by the economic system are not discussed.

Although the limited scope of the model presupposes the existence of related sets of models for the unfolding of its full power, it is good policy to construct models of this type, as pointed out by STONE (1966, p. 104):

> "In trying to model economic and social activity, it seems better to think in terms of a set of limited models operating by means of exchange of information rather than in terms of a monolithic structure."

Following such a piecemeal strategy, the incorporation of the present model into a larger model for the whole subsystem of higher aducation has been taken into consideration by the Swiss National Science Council (1). Similarly, the out-

1) This larger model TUSS has been developped by a team at the University of Utrecht, the Netherlands, and is presently adapted to Switzerland by a group of researchers at Genève (see CUENDET 1975).

put of the present model, the number of graduates at different levels, could be introduced as input into manpower planning studies.

In constructing a model of flow, I have not invented anything new. PERRENOUD (1973) and PERRENOUD/HUTMACHER (1973) mention several European countries that have developed models of flow in the years 1960 to 1970, in particular the Netherlands, Norway, Sweden and England. One might add Germany, where a group around von WEIZSAECKER (see von WEIZSAECKER/FREYTAG (1967) and KLING (1969)) has been working on a model for the whole educational system for years. More limited models are also being constructed in Germany, such as the one for the flow of students in medicine (FOERST et al. 1975). PERRENOUD in his own work uses similar techniques to construct a model of flow for the educational system of the canton of Genève. The statistics for the higher secondary school system of the canton of Zürich also work with transition matrices.

In constructing the model I have been influenced by COLEMANs System of Social Accounts (197?), as I have pointed out in section 2.2. above. A synthesis of similar ideas is also provided by two volumes edited by the OECD (see OECD 1969 and 1970). A textbook for educational planners is also relevant in this respect (CORREA 1969).

Models of flow are, in particular, related to two quite distinct research traditions. The first one of these traditions is input-output analysis originally developed for the analysis of economic systems by LEONTIEF (1941). The application of this strain of thought to demographic flows in general and to flows in the educational system in particular is primarily associated with the name of the English mathematical economist Richard STONE (1966, 1971). STONE (1971) stresses the fact that input-output models of the demographic flows "provide an analogy with the input-output analysis of economic systems which contributes to our understanding of both systems; and they provide a direct link with it, in that an economic input-output matrix for the educational system is simply a part of the input-output matrix for the whole productive system."

The second research tradition which is of particular relevance in the context of models of flow is the study of social mobility. A particular 'puzzle' arising in a particular paradigm within that framework (see KUHN 1962, 1971) will serve as point of departure for the second part of this study, where I will try to introduce independent determinants of flow into the model presented in this first part.

PART II:

INTRODUCING INDEPENDENT DETERMINANTS OF FLOW INTO THE MODEL

5. INTRODUCTION: THE RELATIONSHIP OF THE PRESENT MODEL TO MARKOVIAN MODELS

The flow of students through the Swiss University System is, of course, a special case of intragenerational social mobility, more specifically even, it is a special case of the flow through a graded system (see BARTHOLOMEW 1967, Chapt. 3). Individual students enter the University System, advance within the system, move between the units of the system and eventually leave it at a specific level.

Social mobility processes, both of the intergenerational and of the intragenerational type, have usually been modeled by Markov chains or extensions thereof. The first author to apply Markov chain theory to social mobility has, according to BOUDON (1973), probably been PRAIS (1955a, 1955b). Since this first attempt the use of Markov models has led to numerous substantive contributions in the field of social mobility (see McFARLAND 1974).

The limitations of the simple Markov chain model for social mobility processes have, however, soon become apparent (see McFARLAND 1970): On the one hand, the notion of treating intergenerational mobility as a simple Markov chain involves some serious logical difficulties, as pointed out by DUNCAN (1966). Although these logical difficulties do, on the other hand, not apply to the intragenerational case, the applications of Markov chain models to intragenerational mobility have run into difficulties on empirical grounds: The Markov chain assumptions imply that the probability of remaining in a state for two successive time periods is equal to the square of the probability for remaining in the same state for a single time period. Or, more generally, the Markov chain assumptions imply that the k-step transition matrix - the matrix of transition probabilities during a period of k time units - is equal to the k-th power of the one-step transition matrix. Now, this implication is empirically false. As has been shown repeatedly - first by BLUMEN, KOGAN and McCAR-

THY (1955), but also by HODGE (1966) and others - the amount of movements is overestimated by taking the k-th power of the one-step transition matrix. Or, stated differently: The estimation of the transition probabilities under the Markov chain assumption leads to a deficient main diagonal when compared with the observed k-step transition matrix.

The problem of the deficient main diagonal has given rise to several extensions of the Markov chain model. BOUDON (1973) gives a systematic overview of these extensions. For a somewhat different classification see COLEMAN (1973).

In this second part, I will start out by considering the problem of the deficient main diagonal for the model introduced in Part I, testing the performance of the model under the Markovian assumption. The discussion of alternative explanations of the outcome of this test, which correspond to different extensions of the simple Markov chain model, will, then, lead to the explicit consideration of independent determinants of flow.

In this second part, the discussion will be limited - without restriction of generality - to the university model. The case of the subject model could be treated in a manner entirely analogous to the analysis described below.

5.1. Performance of the Model under the Markovian Assumption

The simple Markov chain model operates under three assumptions (McFARLAND 1970):

- Stationarity: The transition probabilities remain constant over time.
- Markovian: The probability of transition to a given state during the next time unit depends only on the current state of the process and not on its history of previous moves from state to state.
- Homogeneity: The various members of the population are subject to identical sets of transition probabilities. This third assumption is not always made explicitly.

In comparing the 21 transition matrices of the set 1 in APPENDIX A, it immediately becomes apparent that the stationarity assumption does not hold: The transition matrices vary greatly from one semester to the next. The movement to the different degree levels - to mention but one aspect of variation - while being completely absent during the first few semesters later on gradually grows over time. This lack of stationarity already rules out the applicability of simple Makov chain theory to the present model.

It is, however, still possible that the flow of students through the Swiss University System is Markovian. In terms of the university model of Part I, this assumption states that the k-step transition matrix (which is a type 2 matrix) is equal to the product of the first k transition matrices of type 1. Or, stated alternatively:

$$M(t,t+k) = M(t,t+1) \cdot M(t+1,t+2) \cdot \ldots\ldots\ldots M(t+k-1,t+k)$$
$$= \prod_{i=1}^{k} M(t+i-1,t+i)$$

For example, the transition matrix W65/S67 should be equal to the product of the matrices W65/S66 times S66/W66 times W66/S67.

Therefore, in order to test the Markovian assumption for the present model, one has:

- first, to compute the set of products of the k first matrices of type 1, with k varying from 1 to 21 (1)
- then, to compare the resulting products with the corresponding k-step transition matrices of type 2.

An appropriate measure for comparing the two sets of matrices is the widely used <u>Index of Dissimilarity</u> (2). To define this index, let p_{ij} be the observed

---

1) For the computation of these products, the type 1 matrices, which originally are of rectangular (12x16) form, have to be supplemented by 4 additional rows to make them square. The 4 additional rows refer to the 4 absorbing states and consist of 1's in the main diagonal and 0's otherwise, indicating that the corresponding state, once entered, cannot be left anymore.

2) See, for example: TAUEBER/TAUEBER 1965, BLAU/DUNCAN 1967, LIEBERSON/FUGUITT 1967, DUNCAN 1968, SPILERMAN 1972.

k-step probability of transition from state i to state j, and let $\hat{p}_{ij}$ equal the corresponding prediction under the Markovian assumption. Then, the Index of Dissimilarity is given by:

$$ID = 50 \cdot \sum_j | p_{ij} - \hat{p}_{ij} | .$$

According to that definition, for each origin state (i.e. for each row of the matrices considered) the Index of Diessimilarity gives the percentage of persons who would have to change their destination states to bring the observed distribution into accord with the predicted one. The problem with this index is that there is no benchmark against which to check its empirical values. For all practical purposes, I suggest to accept the Markovian assumption for indices smaller than 5, and to certainly reject it for values larger than 10. Values between these two cut-off points would fall into a 'region of indeterminacy', where the decision could go either way, depending on additional information.

On the basis of these cut-off points, Table 32 shows that the Markovian assumption does not hold in general for the present data. It has some justification for the three first and the three last semesters (3), as can be seen from Table 33. Providing a mean index, the table shows a curvilinear distribution, steadily rising up to the 8th semester, stabilizing on a high level till the 14th one and then gradually dropping off again.

With respect to the different universities, the assumption does not hold either. In Table 34 the universities are ranked according to their mean index of dissimilarity, means taken over all 21 semesters. The assumption is most justified for the university of FRIBOURG, the mean index of which lies well within the 'region of indeterminacy'. The mean indices for NEUCHATEL and GENEVE are also relatively low.

---

3) Instead of speaking of semesters, it would be more precise to refer to the transition matrices between semester x and semester y. For the sake of brevity the shorthand term is used.

Table 32

Indices of Dissimilarity between Observed Matrices and Matrices Predicted under the Markovian Assumption by Semester and University

Semester	Basel	Bern	ETHL	ETHZ	Frib	HSG	Geneve	Laus	Neuch	Zuric	Inter
1	0.0	0.0	0.0	0.0	0.0	0.0	0.0	0.0	0.0	0.0	0.0
2	2.3	3.2	7.8	2.7	2.1	9.8	2.5	3.1	3.2	3.2	11.7
3	4.6	5.2	9.0	5.2	2.6	18.1	4.5	6.7	4.5	5.9	15.1
4	6.4	8.6	10.0	9.5	4.3	26.3	7.5	11.5	10.8	9.4	15.1
5	7.9	10.5	9.8	11.6	5.5	30.0	8.6	14.3	11.2	11.9	15.7
6	11.1	13.1	12.9	15.5	8.3	33.6	11.9	16.2	11.8	14.8	13.4
7	12.6	16.1	14.9	17.1	9.0	34.7	13.2	18.3	14.8	15.9	17.1
8	15.1	18.6	16.8	19.3	10.4	34.4	14.4	20.9	14.2	17.3	15.9
9	15.8	19.6	18.4	18.8	9.7	29.5	15.2	20.8	13.7	17.9	20.9
10	15.5	20.8	16.5	17.9	9.1	27.0	14.3	20.2	11.9	18.4	26.6
11	17.3	21.9	14.7	18.0	9.5	25.8	15.6	19.7	13.0	18.4	22.6
12	18.0	20.9	16.1	17.0	10.6	25.8	15.0	20.4	13.9	18.6	23.3
13	18.3	20.2	14.7	16.2	10.9	26.4	15.0	20.0	13.0	18.0	17.6
14	16.3	16.2	12.1	14.3	11.4	20.7	12.7	17.6	11.0	15.7	15.1
15	14.1	12.4	12.4	12.1	10.3	17.2	11.2	15.8	9.4	12.8	17.5
16	13.5	11.5	11.4	11.3	9.7	16.5	10.5	15.3	8.7	12.2	48.6
17	11.6	7.4	13.4	8.0	7.2	12.8	7.5	11.8	7.4	10.0	48.0
18	10.9	5.8	13.4	6.7	6.2	12.4	6.6	11.0	5.6	8.5	48.1
19	9.7	5.2	11.2	5.2	4.9	9.3	5.6	9.6	5.4	7.0	48.0
20	9.6	4.2	11.1	4.6	4.4	8.4	5.3	8.5	5.5	6.6	48.1
21	8.7	3.9	11.0	4.3	3.7	8.2	4.6	8.3	4.9	5.5	48.0

Table 33

Index of Dissimilarity by Semester: Means and Standard Deviations Taken over all 10 Universities

Semester	Mean	Standard Deviation
1	0	0
2	4.0	2.6
3	6.6	4.4
4	10.4	6.0
5	12.1	6.7
6	14.9	7.0
7	16.7	6.9
8	18.1	6.5
9	17.9	5.2
10	17.1	5.0
11	17.4	4.6
12	17.6	4.2
13	17.3	4.4
14	14.8	3.1
15	12.8	2.4
16	12.1	2.4
17	9.7	2.5
18	8.7	2.9
19	7.3	2.4
20	6.8	2.4
21	6.3	2.5

Table 34

Index of Dissimilarity by University: Means and Standard Deviations Taken over all 21 Semesters

University	Mean	Standard Deviation
Fribourg	7.1	3.4
Neuchâtel	9.5	4.2
Genève	9.6	4.7
ETHZ	11.2	6.0
Basel	11.4	5.1
Zürich	11.8	5.7
Bern	12.1	6.9
ETHL	12.3	3.9
Lausanne	13.8	6.1
HSG	20.3	10.1

Now, the question is whether this lack of congruence between the observed matrices and those predicted on the basis of the Markovian assumption is actually due to deficient main diagonals or whether it is a result of discrepancies in other parts of the matrices. In order to answer this question percentage differencies (predicted figures minus observed ones) have been computed for three separate sections of the matrices

- the main diagonal
- the movement between universities
- the leaving of the system

Table 35 gives the results for the main diagonal by universities and semesters. Virtually all the differences are negative, which is clear evidence of the suspected deficiency of the main diagonal. In the first part of Table 36 these results are summarized. On the average, the deficiency is comparatively small for the ETHL, FRIBOURG and NEUCHATEL. This was to be expected for the latter two universities in view of their low mean index of dissimilarity (see Table 34). For the ETHL, the comparative absence of deficiency in the main diagonal is compensated by relatively large differencies with respect to the other two subsections of the matrices, which can be seen from the other two parts of Table 36.

In both of these parts, absolute and signed values of the respective differences are presented in order to give an idea not only of the absolute amount of the incongruencies, but also of the degree to which they cancel.

With respect to the movement between universities, Table 36 shows that the Markovian assumption leads to a considerable exaggeration of the flows in practically all instances. With the exception of NEUCHATEL, absolute and signed values are in close agreement indicating that there is little cancellation in this subsection of the matrices.

With respect to leaving the system the results are less clearcut: Once again for FRIBOURG and NEUCHATEL, but also for BERN and GENEVE, the absolute

Table 35

The Deficient Main Diagonal: Percentage Differences between Observed Matrices and Matrices Predicted under the Markovian Assumption in the Main Diagonal, by Semester and University

	BASEL	BERN	ETHL	ETHZ	FRIB	FSG	GENEVE	LAUS	NEUCH	ZURICH
1	0.0	0.0	0.0	0.0	0.0	0.0	0.0	0.0	0.0	0.0
2	-1.9	-3.2	-7.7	-1.7	-1.3	-8.8	-2.1	-2.9	-2.7	-3.0
3	-3.8	-5.0	-8.9	-3.6	-2.1	-16.5	-4.2	-6.2	-3.5	-5.8
4	-6.0	-4.0	-8.8	-9.2	-3.3	-26.3	-6.2	-11.5	-5.7	-8.5
5	-7.4	-9.4	-3.5	-10.4	-4.1	-29.3	-7.2	-14.3	-8.0	-10.6
6	-10.9	-11.5	-10.6	-15.6	-6.6	-33.7	-9.1	-13.3	-7.8	-12.8
7	-11.9	-13.9	-12.0	-17.5	-7.4	-28.4	-10.4	-14.2	-9.3	-13.6
8	-12.6	-16.3	-12.9	-19.6	-7.6	-21.2	-10.9	-14.8	-7.6	-15.6
9	-12.9	-13.1	-2.9	-17.6	-7.9	-13.6	-12.9	-14.7	-5.8	-15.3
10	-13.7	-19.2	-6.7	-13.5	-8.4	-11.0	-12.6	-14.4	-4.2	-16.1
11	-14.7	-20.2	-2.2	-10.3	-7.7	-9.3	-13.9	-13.1	-5.4	-15.8
12	-15.1	-19.4	-3.9	-8.1	-6.3	-11.3	-13.5	-12.2	-5.9	-16.5
13	-16.4	-18.7	-0.7	-7.6	-6.4	-9.1	-13.7	-12.3	-5.4	-15.1
14	-13.1	-14.7	-0.7	-7.4	-7.1	-9.9	-10.9	-8.9	-5.1	-12.6
15	-9.5	-10.7	0.6	-6.4	-6.2	-10.3	[illegible]	-7.0	-5.0	-9.8
16	-8.7	-9.7	0.6	-8.3	-5.9	-9.9	[illegible]	-6.4	-4.9	-8.7
17	-5.8	-4.9	-3.1	-3.0	-4.6	-7.7	-5.7	-2.4	-3.5	-5.3
18	-6.2	-3.4	-1.0	-2.2	-3.2	-6.2	-4.3	-2.7	-2.6	-3.2
19	-4.9	-2.6	0.0	-0.9	-1.8	-5.0	-3.1	-1.5	-2.3	-1.7
20	-4.9	-1.1	0.0	-0.6	-1.0	-4.8	-2.8	-0.7	[illegible]	-1.3
21	-3.7	-0.9	0.0	-0.1	-0.8	-3.7	-2.2	-0.7	-1.5	-0.3

Table 36

Percentage Differences between Observed Transition Probabilities and Probabilities Predicted under the Markovian Assumption for the 10 Swiss Universities: Means and Standard Deviations Taken over 21 Semesters for 3 Subsections of the Transition Matrices

University	Main Diagonal 1)		Movement between Univ. 2)				Leaving the System 3)			
			absolute val.		signed val.		absolute val.		signed val.	
	Mean	S.D.	Mean	S.D.	Mean	S.D.	Mean	S.D.	Mean	S.D.
Basel	-8.8	4.8	7.8	4.9	7.5	4.7	6.3	4.8	1.2	1.4
Bern	-10.1	6.9	8.7	5.3	8.6	5.2	4.7	2.8	1.0	2.6
ETHL	-4.3	4.5	9.2	4.6	8.4	4.0	11.0	6.9	-4.6	5.2
ETHZ	-7.7	6.2	8.5	5.5	8.4	5.5	6.3	3.9	-2.2	3.0
Fribourg	-4.7	2.7	6.2	3.5	5.6	3.4	3.3	2.3	-.8	.9
HSG	-13.1	9.3	15.5	9.9	13.8	10.5	12.1	7.2	-1.0	5.1
Genève	-7.8	4.4	7.8	5.0	7.2	4.7	3.7	2.0	0	1.3
Lausanne	-8.3	5.6	10.3	6.3	10.0	6.3	9.1	5.3	-2.8	3.0
Neuchâtel	-4.7	2.4	10.0	6.4	2.6	2.2	3.9	1.9	1.2	1.7
Zürich	-9.1	5.9	8.5	5.3	8.2	5.1	6.0	3.7	0	1.9

1) Since virtually all the differences in the main diagonal are negative, a comparison of absolute values and signed ones is meaningless in this category
2) including movements to foreign universities
3) including interruptions

size of the difference is rather limited. The deviations are, on the other hand, quite large for the HSG, the ETHL and for LAUSANNE. Moreover, the deviations cancel in most cases indicating that the overestimation of moves to certain absorbing states is compensated by the underestimation of moves to some other ones.

In sum: The large indices of dissimilarity are due to

- a deficient main diagonal
- an overestimation of the movement between universities
- (to a lesser extent) an incorrect prediction of the leaving of the system, overestimating certain moves and underestimating some other ones.

I shall now consider two approaches that try to take into account these results by modifying the simple Markov chain.

5.2. Pseudo-Markovian Models and Their Explanation of the Deficient Main Diagonal

The general strategy in meeting the problem raised by the underestimation of the main diagonal cells has been to, first, relax some assumption of the ordinary Markov chain approach and, then, to reformulate the process so as to bring it back into the Markovian framework. The different attempts can be distinguished according to whether they focus on the non-stationarity of the process or the heterogeneity of the population involved.

5.2.1. The Phenomenon of Cumulative Inertia

The Cornell Mobility Model (McGINNIS 1968, GINSBERG 1971, HENRY 1971, HENRY et al. 1971) introduces non-stationarity on the basis of an axiom of cumulative inertia. This axiom states that the longer an individual has remained in a given state, the less likely he is to leave it: Attachements grow over time.

In order to account for cumulative inertia, the model of McGINNIS and associates introduces changing probabilities of remaining in an origin state as a function of duration of prior residence in that state. Aredefinition of the states brings the resulting non-stationary process back into the Markovian framework: States are partitioned into sub-states that are characterized by particular duration statutes.

In the present context, it is not the specifics of the Cornell Mobility Model that should concern us, but its underlying reasoning. The question is, whether some kind of cumulative inertia is operating in the process of the flow of Swiss students through the university system. The presence of cumulative inertia could explain the deficient main diagonal and the overestimation of inter-university moves noted above. With respect to the leaving of the system, the applicability of the concept of cumulative inertia has to be ruled out, not only, because the leaving occurs but once in the life of a student, but also because it is mainly governed by structural determinants, as we shall see later on. In this subsection, therefore, only the moves between universities are considered.

MORRISON (1967) and LAND (1969) have supplied evidence for the presence of the phenomenon of cumulative inertia in the case of migration between geographic regions. Applying a formula similar to the one used by MORRISON, it is possible to test for cumulative inertia in the context of the Swiss University System:

$$P(i,j) = \frac{M(i,j)}{Y(i,j)}$$

In this formula, the subscript j refers to the duration status, i.e. to the number of semesters one has stayed at a particular university between moves (4). The subscript i refers to a specific university; in the MORRISON case it referred

---

4) This definition of duration status differs from the one used in the first part, where duration status was defined as the amount of time spent in the university system in general. For the remainder of this study, the term duration status will always refer to the number of semesters stayed at a particular university.

to age. Y(i, j) refers to the duration specific number of intervals lived at a certain university i and the M(i, j)'s are the duration specific number of moves originating at university i. Y(i, j) standardizes the number of moves made by the total amount of risk, i.e. the total number of intervals during which one could have moved (5).

---

5) The following example may serve as an illustration of how this formula works: Let us consider a population of, say, five individuals and let us assume, for simplicity, that we are only interested in the overall number of moves (and not in university specific ones). These five individuals exhibit the following pattern of moves:

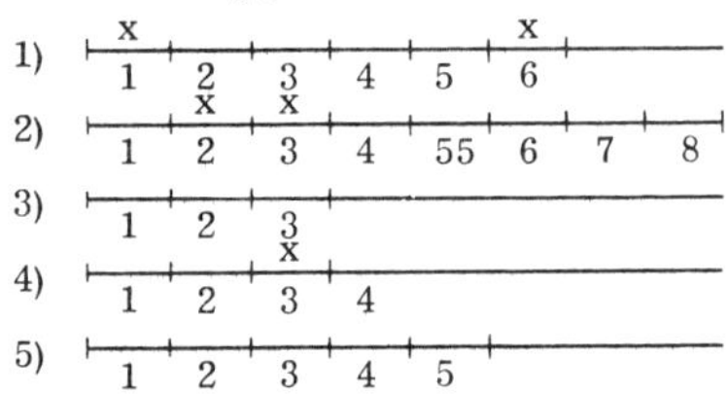

For the computation, an additional convention has to be adopted: If an individual moves in the i-th semester after entry into the system or after his last move, he will be assigned a duration status of i semesters for that particular move. The table below gives a schematic outline of the computational procedures involved:

Duration Status	Computation of Semester Probabilities of Moving		
	M(j)	Y(j)	P(j)=M(j)/Y(j)
1	2	1 + 3 + 1 + 2 + 1 = 8	.25
2	1	1 + 2 + 1 + 1 + 1 = 6	.166
3	1	1 + 1 + 1 + 1 + 1 = 5	.2
4	0	1 + 1 + 0 + 0 + 1 = 3	0
5	1	1 + 1 + 0 + 0 + 1 = 3	.333

Take, for example, row 2: There is one move with a duration status of two semesters, i.e. M(2) is equal to 1. With the exception of individual 2, who has lived in the system for two periods of 2 semesters, all the individuals have lived only for one two-semester interval, which amounts to 6 two-semester intervals lived in the system alltogether: Y(2) is equal to 6. P(2), finally is nothing but the ratio of the two quantities.

The computation of the required quantities M(i, j) and Y(i, j) is rather tedious, because there are a great number of possibilities to effectuate duration-specific moves over a period of 16 semesters, which is the period for which reliable data on inter-university moves are available (6).

The result of these computations are presented in the Tables 37 to 40. The first two of these tables give the duration specific number of moves M(i, j) and the duration specific number of intervals lived Y(i, j) for each university. Based on these figures, the duration specific semester probabilities of inter-university moves P(i, j) are calculated. Table 39 shows the P(i, j)'s for the ten

---

6) To get an idea of that number, consider the following table which shows the relationship between the duration status in a particular semester and moves in semesters prior to that semester.

Semester of the last move	Duration Status in: S65	W65/66	S66	W66/67	S67	W67/68	......	W72/73
no move	1	2	3	4	5	6	......	16
S65		1	2	3	4	5	......	15
W65/66			1	2	3	4	......	14
S66				1	2	3	......	13
W66/67					1	2	......	12
S67						1	......	11
.								
.								
.								
W72/73								1

This table refers to the spring starters. An analogous table could be drawn for the fall-starters.
Now, take, for example, a spring-starter who is still in the system during the WS67/68 and who has made no prior move. The table tells us that he will be assigned a duration status of 6 semesters. Similarly, someone who has moved in the third semester, but not since, will be assigned a duration status of 4 in the WS67/68. Students who are not in the system anymore at the time in question will get a missing value.
Based on such considerations, it is possible to create 17 variables for the duration status in each semester for every student. Appropriate cross-tabulations of these variables with the university of origin will yield the required quantities M(i, j) and Y(i, j).

Table 37

Number of Moves between Universities by University of Origin and Duration Status ($M_{i,j}$)

University	Duration Status															
	1	2	3	4	5	6	7	8	9	10	11	12	13	14	15	16
Basel	12	9	16	16	6	14	9	8	2	2	5	3	1	1	1	2
Bern	5	8	9	17	17	17	20	8	9	9	2	-	1	3	-	-
ETHL	5	4	6	-	6	1	1	-	-	-	1	-	-	1	1	-
ETHZ	1	24	19	12	14	7	8	4	6	5	3	3	4	-	4	-
Fribourg	2	12	14	6	13	29	17	5	5	4	5	1	2	-	-	-
HSG	3	7	7	1	1	-	1	4	2	2	1	-	1	-	-	-
Genève	21	22	9	10	7	5	2	2	4	2	2	2	2	-	1	-
Lausanne	28	18	15	7	10	8	4	3	4	1	4	1	1	1	-	-
Neuchâtel	6	7	20	11	3	1	2	1	1	2	4	-	-	-	-	-
Zürich	30	39	36	31	24	19	22	12	21	4	12	3	4	1	4	1
Foreign Univ.	118	71	15	6	1	2	2	1	-	-	2	-	-	-	-	-
Interruption	7	3	3	6	2	2	2	1	-	2	-	1	-	1	-	-
Total	238	224	169	123	104	105	90	49	54	33	41	14	16	8	11	3

Table 38

Duration Specific Number of Intervals Lived in the System by University ($Y_{i,j}$)

University	Duration Status															
	1	2	3	4	5	6	7	8	9	10	11	12	13	14	15	16[1)]
Basel	504	470	448	430	389	378	349	334	304	282	244	224	212	200	181	96
Bern	667	631	520	601	543	485	440	398	379	355	320	296	268	239	192	24
ETHL	105	98	97	93	91	83	81	80	71	33	35	16	16	9	9	0
ETHZ	873	830	729	709	691	671	675	680	633	513	384	280	212	175	159	1
Fribourg	391	373	354	327	289	255	216	175	140	116	97	85	66	55	51	4
HSG	170	147	135	130	134	126	125	102	72	50	40	35	33	27	28	1
Genève	503	464	423	405	360	332	302	282	235	221	199	197	175	167	126	2
Lausanne	430	383	345	313	303	276	233	209	182	163	141	120	99	93	65	2
Neuchâtel	192	178	171	144	127	115	98	93	74	59	42	36	37	33	29	2
Zürich	1190	1081	994	925	878	804	760	703	659	580	534	465	422	373	311	61
Foreign Univ.	297	168	87	68	53	43	35	32	22	14	12	4	0	0	0	0
Interruption	92	253	312	282	258	276	237	200	182	163	144	107	77	56	38	7
Total	5414	5076	4715	4427	4116	3844	3551	3288	2953	2549	2192	1865	1617	1427	1189	200

1) The corresponding numbers are so small, because only the spring-starters could have stayed at the same university for 16 consecutive semesters by the WS72/73, the last semester with reliable data on inter-university moves.

Table 39

Semester Probability of Moving between Universities by University of Origin and Duration Status, Including Confidence Limits

DURATION STATUS	UNIVERSITY OF ORIGIN BASEL	BERN	ETHL	ETHZ	FRIBOURG	HSG	GENEVE	LAUSANNE	NEUCHATEL	ZURICH
1	.024 (.013,.042)	.007 (.003,.018)	.048 (.018,.113)	.001 (.000,.007)	.005 (.001,.020)	.018 (.005,.055)	.041 (.026,.063)	.065 (.044,.094)	.031 (.013,.070)	.025 (.017,.036)
2	.019 (.009,.037)	.013 (.006,.026)	.041 (.013,.107)	.029 (.019,.043)	.032 (.018,.057)	.048 (.021,.099)	.047 (.031,.072)	.047 (.029,.075)	.039 (.017,.083)	.036 (.026,.049)
3	.036 (.021,.059)	.015 (.007,.028)	.062 (.025,.135)	.026 (.016,.041)	.040 (.023,.067)	.052 (.023,.108)	.021 (.010,.041)	.043 (.025,.072)	.117 (.075,.177)	.036 (.026,.050)
4	.037 (.022,.061)	.028 (.017,.046)	.0 (.001,.049)	.017 (.009,.030)	.018 (.007,.041)	.008 (.000,.048)	.025 (.013,.046)	.022 (.010,.048)	.076 (.041,.136)	.034 (.023,.048)
5	.015 (.006,.035)	.031 (.019,.051)	.066 (.027,.143)	.020 (.012,.035)	.045 (.025,.078)	.007 (.000,.047)	.019 (.009,.041)	.033 (.017,.062)	.024 (.006,.073)	.027 (.018,.041)
6	.037 (.021,.063)	.035 (.021,.057)	.012 (.001,.075)	.010 (.005,.022)	.114 (.079,.161)	.0 (.001,.037)	.015 (.006,.037)	.029 (.014,.058)	.009 (.000,.055)	.024 (.015,.037)
7	.026 (.013,.050)	.045 (.029,.071)	.012 (.001,.076)	.012 (.006,.024)	.079 (.048,.125)	.008 (.000,.050)	.007 (.001,.026)	.017 (.006,.046)	.020 (.004,.079)	.029 (.019,.044)
8	.024 (.011,.048)	.020 (.009,.041)	.0 (.001,.057)	.006 (.002,.016)	.029 (.011,.069)	.039 (.013,.103)	.007 (.001,.028)	.014 (.004,.045)	.011 (.001,.067)	.017 (.009,.030)
9	.007 (.001,.026)	.024 (.012,.046)	.0 (.001,.064)	.005 (.004,.022)	.036 (.013,.086)	.028 (.005,.106)	.017 (.005,.046)	.022 (.007,.059)	.014 (.001,.083)	.032 (.020,.049)
10	.007 (.001,.028)	.025 (.012,.049)	.0 (.003,.130)	.010 (.004,.024)	.034 (.011,.091)	.040 (.007,.149)	.009 (.002,.036)	.006 (.000,.039)	.034 (.006,.127)	.007 (.002,.019)
11	.020 (.008,.050)	.006 (.001,.025)	.029 (.001,.166)	.008 (.002,.025)	.052 (.019,.122)	.025 (.001,.147)	.010 (.002,.040)	.028 (.009,.076)	.095 (.031,.235)	.022 (.012,.040)
12	.013 (.003,.042)	.0 (.000,.016)	.0 (.006,.241)	.011 (.003,.034)	.012 (.001,.073)	.0 (.003,.123)	.010 (.002,.040)	.008 (.000,.052)	.0 (.003,.120)	.006 (.002,.020)
13	.005 (.000,.030)	.004 (.000,.024)	.0 (.006,.241)	.019 (.006,.051)	.030 (.005,.115)	.030 (.002,.175)	.011 (.002,.045)	.010 (.001,.063)	.0 (.002,.117)	.009 (.003,.026)
14	.005 (.000,.032)	.013 (.003,.039)	.111 (.006,.493)	.0 (.001,.027)	.0 (.002,.081)	.0 (.003,.155)	.0 (.001,.028)	.011 (.001,.067)	.0 (.003,.130)	.003 (.000,.017)

Table 40

Semester Probability of Moving between Universities by Duration Status for 3 Summary Categories, Including Confidence Limits

DURATION STATUS	FOREIGN U	SWISS U ONLY	TOTAL
1	.397 (.342,.456)	.022 (.019,.027)	.044 (.039,.050)
2	.423 (.348,.501)	.032 (.027,.038)	.044 (.039,.050)
3	.172 (.103,.272)	.035 (.030,.041)	.036 (.031,.042)
4	.088 (.036,.189)	.027 (.023,.033)	.028 (.023,.033)
5	.019 (.001,.114)	.027 (.022,.032)	.025 (.021,.031)
6	.047 (.008,.171)	.029 (.024,.035)	.027 (.022,.033)
7	.057 (.010,.205)	.026 (.021,.032)	.025 (.021,.031)
8	.031 (.002,.180)	.015 (.011,.021)	.015 (.011,.020)
9	.0 (.004,.185)	.020 (.015,.026)	.018 (.014,.024)
10	.0 (.007,.268)	.013 (.009,.019)	.013 (.009,.018)
11	.167 (.029,.491)	.019 (.014,.026)	.014 (.014,.026)
12	.0 (.024,.604)	.007 (.004,.013)	.008 (.004,.013)
13	.0 (****,****)	.010 (.006,.017)	.010 (.006,.016)
14	.0 (****,****)	.005 (.002,.011)	.006 (.003,.011)

Swiss universities, including confidence limits (7). The last table in this series, Table 40, contains the corresponding figures for the cohort as a whole and two of its most significant subsets, the group of all the Swiss universities taken as a whole and the group of the foreign universities.

Figure 18 and Figure 19 reveal the significance of the latter distinction. Figure 18, which provides a plot for the second and third column in Table 40, shows an overall trend toward cumulative inertia for the cohort taken as a whole. Clearly, the probability of moving between universities is decreasing with increasing duration status, although the relationship is by no means smooth or linear. Looking, however, at the curve for the Swiss universities alone, the trend is much less clear-cut: Up to a duration status of three semesters, there is a positive relationship between the probability of moving and duration status, and only afterwards does the curve for all the Swiss universities follow the expected general trend.

Responsible for this discrepancy are the moves to and from foreign universities. As can be seen from Fig. 19, most of the Swiss students who ever study at a foreign university do so for only a short period of time: 40% return after only one semester abroad and 42% of those remaining for more than one semester return after the second one. These comparatively short stays abroad are primarily due to structural constraints of the flow: Some courses of study, mostly in the field of the humanities, require the students to go abroad for a lmited period of time to get first hand acquaintance of the culture of the country and language they are studying. This sort of constraint, however, is incompatible with the phenomenon of cumulative inertia, which implies that the student be free to establish stronger ties to a given state.

---

7) For the computation of these confidence limits what is known as the quadratic normal approximation to the binomial distribution has been used (see BROWNLEE 1965, p. 150). Although this approximation presupposes large n, it gives quite accurate results for n as small as 30. With the exception of a few cases, most of the Y(i, j)'s are considerably larger than that.

Figure 18

Semester Probability of Moving between Universities by Duration Status for all the Universities and for Swiss Universities only

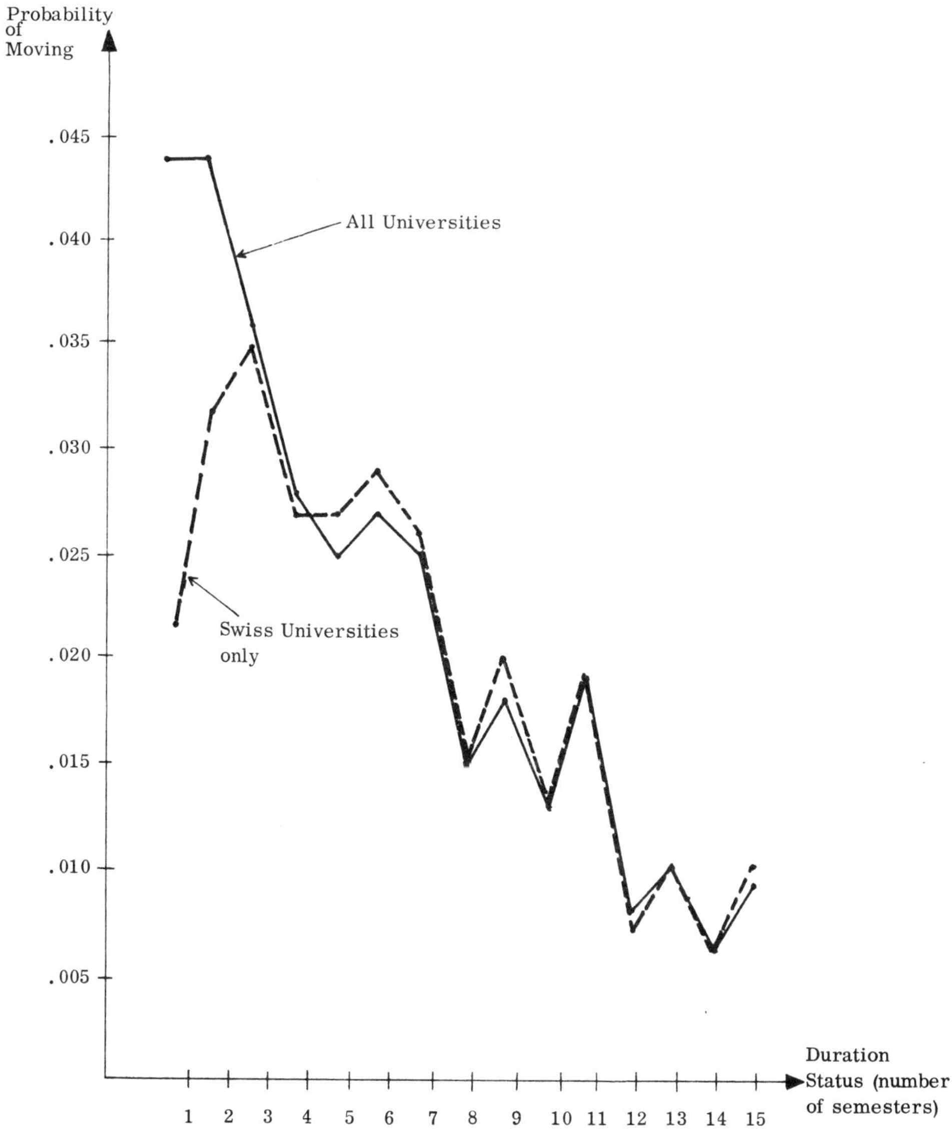

Figure 19

Semester Probability of Moving between University by Duration Status for Foreign Universities as Compared to all the Universities and the Swiss Universities only

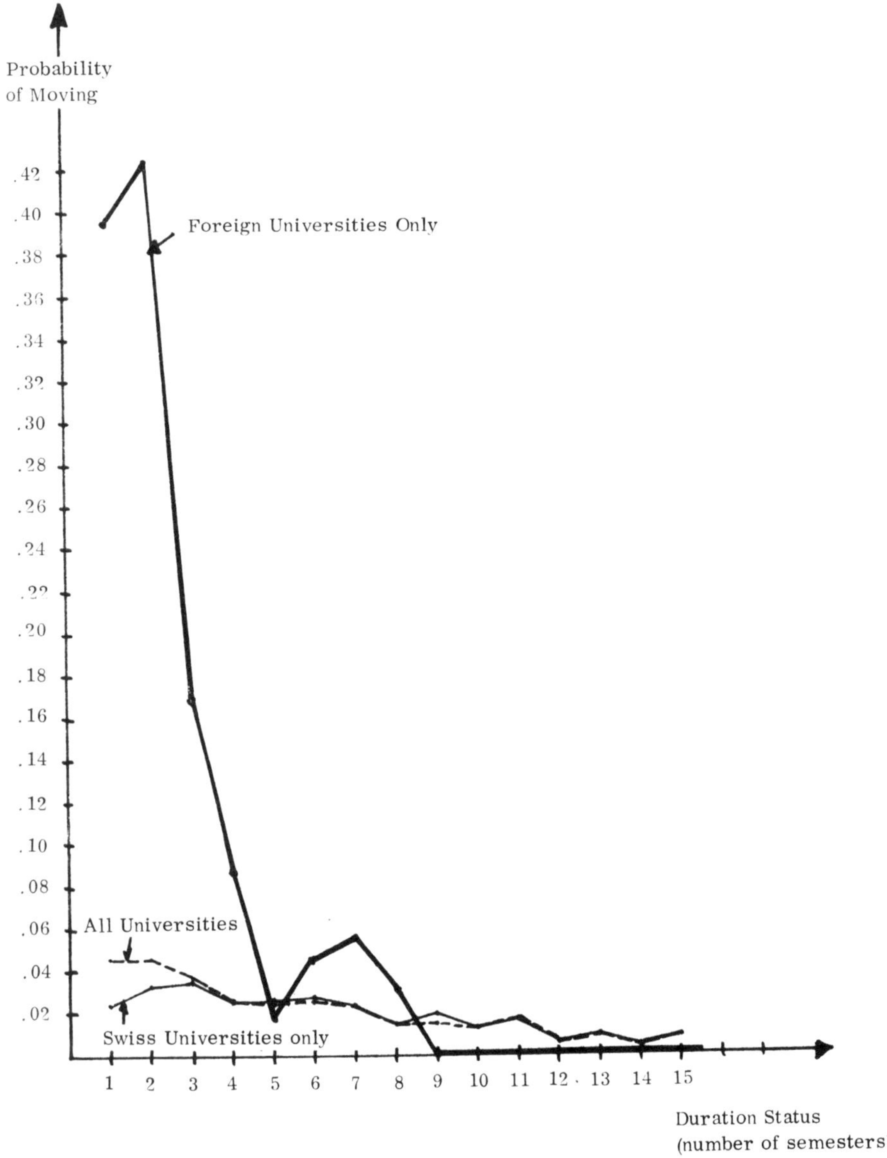

Thus, although the relationship between duration status and the probability of moving for foreign universities is as predicted by the axiom of cumulative inertia, it is no evidence for the presence of the said phenomenon.

This finding also disqualifies the clear-out overall trend noted in Fig. 18 as hard evidence for cumulative inertia.

The evidence becomes even more inconclusive, when we inspect the duration specific probability of moving at the level of a particular university. Fig. 20 provides the relevant plots for eight different universities. Only three of these can be said to roughly follow the hypothesized general trend toward cumulative inertia: LAUSANNE, GENEVE and ZUERICH (8). At the ETHZ, there is a trend towards increasing moves with high duration status. A similar increase, although much more dramatic and for lower duration statuses can be noted for the HSG. I suspect that these late increases are due to students who change the university after having received their medium level degrees. Since the ETHZ and the HSG offer only a limited group of subjects, a student who wants to continue his studies in another field after having received his degree can often do so only by changing the university. Thus, a student at the HSG who wants to get his doctorate in law, a not infrequent case, will have to leave the HSG for another university. The earlier starting point of the increasing trend at the HSG can be attributed to the lower average number of semesters necessary to obtain a medium level degree at the HSG (see Table 14).

The most spectacular deviations from the expected trend toward cumulative inertia are the cases of FRIBOURG and NEUCHATEL: More than 10% of the students with a duration status of 3 semesters (NEUCHATEL) or 6 semesters (FRIBOURG) leave these universities. This massive exit is due to the fact that medical students cannot complete their studies at these universities, but can

8) The plot for BASEL is rather erratic, although it bears some resemblance with the one for ZUERICH. This plot is not given in Fig. 20. Similarly, the plot for the ETHL is missing, because the figures for this university are too small.

Figure 20

Semester Probability of Moving between Swiss Universities by Duration Status for 8 Swiss Universities

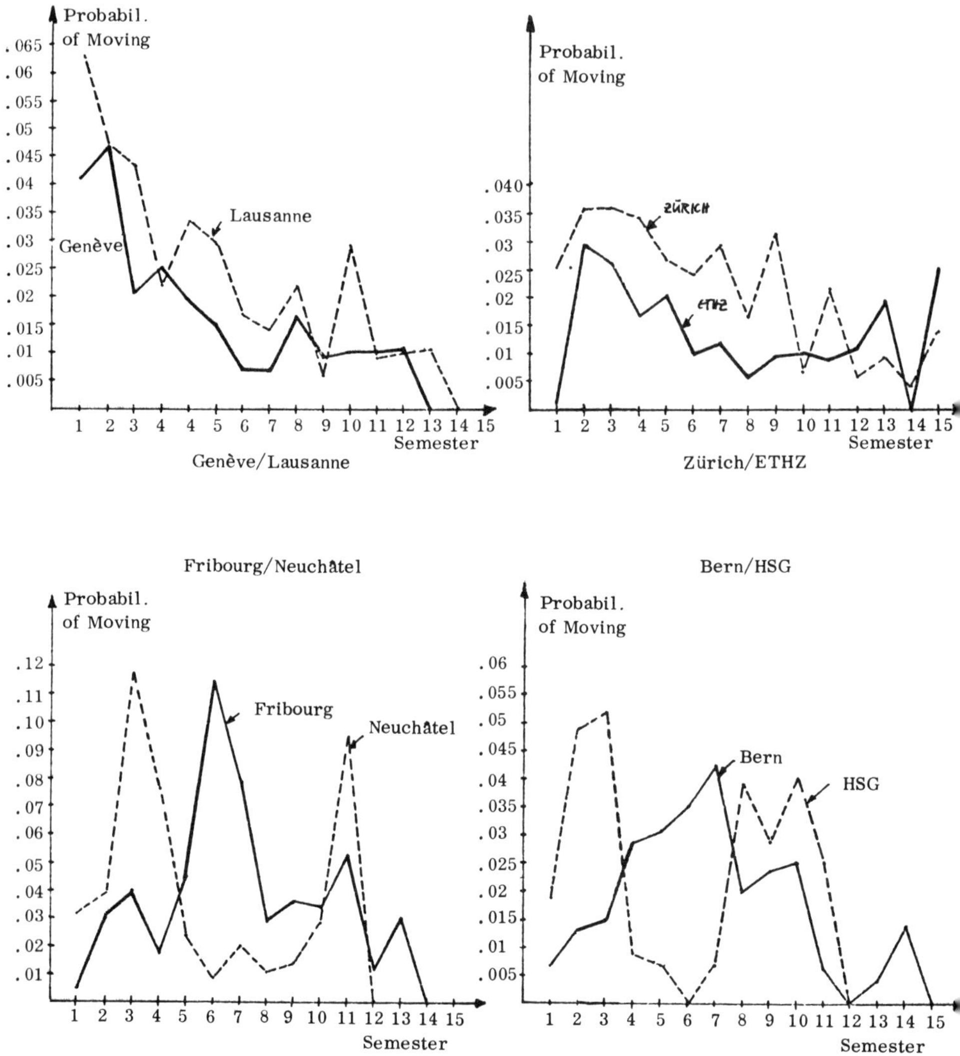

only pass the first preparatory exam at NEUCHATEL and the first two preparatory exams at FRIBOURG (9). It is again structural constraints which prevent the development of lasting ties to a particular university.

It is interesting to note, however, that for both, FRIBOURG and NEUCHATEL, the Markovian assumption had some plausibility (see Table 34), which leads one to speculate that it is just the observed lack of cumulative inertia, or, more precisely, the observed structural conditions preventing its development, that are responsible for the earlier finding.

Finally, the marked deviations from the trend at BERN as well as some minor deviations in the other curves are most likely a result of the moves to foreign universities discussed above.

It can be concluded that the principle of cumulative inertia is operating only to a limited extent within the Swiss University System due to structural considerations: Students are often constrained to move. They have to leave the system at some point or another and even their movements within the system may be prescribed, preventing them to develop lasting ties to a particular university. For those flows that are not constrained, however, the results reported give some indications for the presence of the phenomenon.

5.2.2. Heterogeneity

McFARLAND (1970, 1974) focuses on the heterogeneity in his attempt to explain the excess stability in the main diagonal. He points out that "even had non-stationarity been completely absent, the relationship between mobility and duration would still be negative, and thus the observed negative relationship is not necessarily due to non-stationarity" (1974, p. 884). McFARLANDs rationale is that those who leave a certain state tend to be different from the people who stay. The former tend to have a higher probability of moving. Therefore, the

---

9) See p. 52 above.

group remaining will have a higher average probability of staying than did the initial group.

This model attributes causality in a direction exactly opposite to the previous one: "In the Cornell Model one's probability of movement becomes low, be-cause of his long duration status; in this model one has a long duration in his current status, because his probability of movement is (and always was) low" (1970, p.471).

In McFARLANDs model, each individual is characterized by his own transition matrix. The model is brought back into the Markovian framework by the assumption that these individual matrices each follow a Markov chain. The aggregated population matrix is shown to have properties corresponding to ordinary Markov chain theory.

McFARLANDs model is, as BOUDON (1973) points out, intellectually appealing, because it explains the trend toward stability by a factor which has apparently no connection with stability, i.e. the heterogeneity of the population, which is in sharp contrast to the Cornell Model and other models introducing non-stationarity in the form of collective or individual history. Moreover, as BOUDON also stresses (p.99), McFARLANDs idea of explaining the excess stability by the heterogeneity of the population is also important, because it advances a step in the direction of sociological realism: As many mobility students have shown (e.g. GIROD (1971) or BLAU/DUNCAN (1967)), the moves of an individual depend not only on his current status or on the history of his moves, but also on other variables such as his social origin, his race, his educational status etc.

Above all, however, BOUDON sees in McFARLANDs model an indication of how a difficult dilemma confronting studies in social mobility could be resolved. This dilemma he describes as follows (1973, p.75):

"Either the models include a sufficient number of intervening variables,

but use general statistical instruments, the syntax of which is necessarily poor (for instance, the syntax of causal analysis where the only possible type of statement has the form: the variable x has an influence on the variable y*), or they use more sophisticated mathematical models but exclude a number of intervening variables which are essential for the explanation of the mobility processes.
This dilemma is generated by the fact that inclusion of a new intervening variable does not generally raise any problem, while the introduction of a new variable into a stochastic model generally amounts to constructing a new model."

The generally negative results of the preliminary tests of the Markovian assumption and of the phenomenon of cumulative inertia suggest that the flow of students through the Swiss University System is probably determined by a rather large number of intervening variables. Or, to put it in terms of heterogeneity, the population under consideration is most likely heterogenous in a large number of dimensions. To find out, what these dimensions are will be the main goal of the rest of this study.

The presumably large number of intervening variables also suggests the use of <u>general statistical techniques</u> in solving this problem. In the face of the dilemma invoked above, BOUDON opts for syntactically sophisticated models embodying only those variables that definitely play a role in the generation of mobility processes. While this option certainly is advantageous from the point of view of theory construction, it does not seem to be feasible in the present context, where we are primarily concerned with description, prediction and simulation, and only indirectly, if at all, with theory construction.

In resorting to general statistical techniques, I step outside of the Markovian framework, which has, however, been important in raising the issue to be discussed.

6. INTRODUCING INDEPENDENT DETERMINANTS OF FLOW INTO THE MODEL

6.1. Description of the General Technique

To use general statis tical techniques to get at the determinants of the individual differences in transition probabilities is to do precisely what SPILERMAN (1972) has done.

6.1.1. SPILERMAN's Technique

SPILERMAN uses regression analysis to get at the determinants of heterogeneity. Instead of disaggregating the population into separate transition matrices, he constructs m^2 regression equations, where m equals the number of states. Thereby, the crucial point is his definition of the dependent variables:

> "For each origin state S_i at t_o define an indicator variable y_{ij} which equals one for a person, if he moved to S_j at t_1, and zero otherwise. Corresponding to each origin state S_i, m such variables can be defined $(y_{i1}, y_{i2}, \ldots\ldots, y_{im})$. Exactly one element of this vector will equal one for an individual originating in S_i, specified by his destination state."

Now, using as observations all individuals in S_i at t_o, he regresses

$$y_{ij} = a_{ij} + \sum_{v}^{V} b_{ijv} X_v + e_{ij}, \text{ for } j = 1, \ldots\ldots, m \quad (1)$$

where the X_v's are the individual characteristics expected to explain the heterogeneity in the transition probabilities, and e_{ij} is the error term for the ij^{th} transition. This procedure yields m equations for each origin state and m^2 equations for the whole matrix.

He then goes on to show that the desired decomposition of the transition probabilities p_{ij} in terms of the individual characteristics $(X_1, \ldots\ldots, X_V)$ can be derived from these regression equations. The proof is straightforward and hinges on the fact that, for an individual c, with attribute profile $(\acute{X}_{1c}, \ldots\ldots, \acute{X}_{vc})$,

$$\hat{y}_{ijc} = \hat{a}_{ij} + \sum_v b_{ijv} X'_{vc} \qquad (2)$$

can be interpreted as the expected number of moves this individual made from i to j per unit time. This interpretation is a consequence of the dependent variable being coded 0-1. Since expected numbers can be summed over observations, the expected number of i to j transitions for the subpopulation at origin state i is given by the sum of the corresponding expected numbers of all the individuals at origin i.

But this sum is, in turn, an unbiased estimate of the number of persons who actually have made the move from i to j during the period in question. This can be stated in formal terms: Letting $n_{i\cdot}$ equal the number of persons at origin state i at time t_o and n_{ij} equal the number of persons who actually moved from i to j by time t_1, we have

$$\sum_c^{n_{i\cdot}} \hat{y}_{ijc} = \sum_c^{n_{i\cdot}} y_{ijc} = n_{ij} \qquad (3)$$

where the first equality derives from the least squares procedure of fitting a regression line, and the second from the definition of the y_{ij}.

Recalling that the transition probabilities p_{ij} can be defined as $n_{ij}/n_{i\cdot}$ (see p. 27), substituting from equations (2) and (3) completes the proof:

$$p_{ij} = \frac{1}{n_{i\cdot}} \sum_c^{n_{i\cdot}} \hat{y}_{ijc} = \hat{a}_{ij} + \sum_v^{V} \hat{b}_{ijv} \cdot \left(\frac{\sum_c X'_{vc}}{n_{i\cdot}}\right) \qquad (4)$$

The terms $\sum_c X'_{vc}/n_{i\cdot}$ for v=1, ..., V can be interpreted as the profile of a 'typical' individual in state i at t_o.

We, therefore, have a matrix of m^2 equations, in which the significant b's in an equation indicate the individual characteristics responsible for the heterogeneity with respect to that transition:

$$Y = \begin{bmatrix} y_{11} = \hat{a}_{11} + \sum \hat{b}_{11v} X_v, & \ldots\ldots, & y_{1m} = \hat{a}_{1m} + \sum \hat{b}_{1mv} X_v \\ \vdots & & \vdots \\ y_{m1} = \hat{a}_{m1} + \sum \hat{b}_{m1v} X_v, & \ldots\ldots, & y_{mm} = \hat{a}_{mm} + \sum \hat{b}_{mmv} X_v \end{bmatrix}$$

6.1.2. An Extension Incorporating both, Individual and Semester Heterogeneity*

SPILERMAN analyzes geographical migration data with this technique. He estimates the b-coefficients on the basis of the transitions made during an initial time period by a cohort with a given distribution of individual characteristics. Then he investigates the extent to which changes in subsequent one-step transition matrices of this cohort can be attributed to demographic shifts in the population.

Changes in subsequent transition matrices can be attributed to either one of two general processes, structural change or demographic change:

- Structural change, in the present context, refers to a change in the influence of particular attributes on an individual's probability of making a given transition. Structural change means that the b-coefficients in the regression equations change.
- Demographic change refers to a shift in the distribution of individuals with certain attribute profiles among the states of the system. In this case, the regression coefficients remain constant over time.

SPILERMAN is able to show that the major proportion of the change in the migration behavior of the cohort he studied is due to shifts in the population distribution and not to alterations in underlying relationships. This is probably not the case for the flow of Swiss students through the Swiss University System. The preliminary results reported in the introduction to this second part make it reasonable to assume that structural changes are much more influential in the present context than they were in SPILERMAN's migration data.

The most conspicuous example of structural change from one semester to the next is given by the fact that it is only possible to graduate after having been in the system for a certain number of semesters. Therefore, if we attempted to estimate the transition probabilities of, say, the 10th semester by evaluating its population profiles on the Y equations constructed for the first semester,

---

* This extension was suggested to me by Prof. James S. Coleman.

we would get very inaccurate estimates, simply because of the fact that there are a lot of students graduating in the 10th semester, but none in the first one.

Does this mean that we have to reestimate the Y equations for each consecutive one-step transition matrix of type 1? The answer is no, because this rather unpleasant prospect can be avoided by the introduction of the concept of semester heterogeneity. What this concept states is that not only individuals, but also semesters are characterized by specific profiles of attributes: Some semesters are different from others, because one can graduate at a particular level at that point in time. Some other semester is special, because it is the interval, when some students have to go abroad to get acquainted with the foreign culture they are studying. In still another semester, a student studying a particular subject at a particular university will have to change to another university.

The transition probabilities in semester t can be viewed as a function of both, individual characteristics and semester characteristics. More specifically, the transition probabilities for semester t, i.e. the probabilities p_{ijt}, can now be written in terms of the conditioning variables ($\acute{X}_{1t}$, ..., X_{vt}; $\acute{X}_{v+1,t}$, ..., $\acute{X}_{qt}$) as follows:

$$p_{ijt} = \frac{1}{n_{i\cdot t}} \sum_{c}^{n_{i\cdot t}} y_{ijct} = \hat{a}_{ij} + \sum_{v}^{V} \hat{b}_{ijv} \left(\frac{\sum_{c} \acute{X}_{vct}}{n_{i\cdot t}}\right)$$

The introduction of the concept of semester heterogeneity implies that we do not have to estimate n sets of Y equations - one for each one-step transition matrix, but just one unique set for all the one-step matrices of the cohort. The significant b's in this unique set indicate the individual characteristics and the semester characteristics responsible for heterogeneity with respect to all transitions of the cohort considered.

Changes in successive one-step transition matrices of the cohort can now be attributed either to demographic shifts as before, or to specific characteristics of particular semesters. Structural change from one transition matrix to another within a single cohort has thus been incorporated into the model.

The estimation of the unique set of Y equations presupposes the construction of a synthetic population, consisting of the original cohort for the first semester plus the remainder of that cohort for each subsequent semester. This implies that every individual enters that synthetic population once for every semester he has been in the university system. The set of attributes characterizing him will, however, not be identical for every semester, because some of his attributes are time-dependent. Among these time-dependent characteristics we find individual ones, such as age, but also the characteristics of the semester for which he is entered.

The construction of such a synthetic population is tedious. In addition to making the incorporation of semester heterogeneity possible, it has, however, an additional technical advantage: It increases the number of cases, which means that the estimation procedure will be more accurate.

6.2. Finding the Independent Determinants of the Rates of Flow

The procedure to find the independent determinants of the rates of flow breaks down into two broad steps: the construction of the synthetic population and the running of the regressions. Within the first step, we can further distinguish between the construction of the synthetic population file and the creation of the dependent and independent variables.

6.2.1. Construction of the Synthetic Population

6.2.1.1. The Synthetic Population File

Starting from the data-file for the cohort 1965, it is in principle not difficult to create a synthetic population file by means of SPSS. For every semester, starting with the first one and proceeding consecutively to each of the following ones, we select a case and write it on a new file, if the corresponding student has still been in the system in that semester (1). The resulting synthetic file will contain the cases of successive semesters in sequential order, as indicated by the following graph (2):

Case 1(1)	Case 2(1)	Case 3(1)	....	Case x(1)	Case 1(2)
Case 2(2)	Case 3(2)	....	Case y(2)	....	Case z(15)

It takes 22 semesters (23 for the minority of the spring-starters) for the whole cohort to pass through the system (see p. 34). As indicated at the outset, we have, however, reliable data on only 16 (or 17) semesters (see p. 15). To find the independent determinants of flow, moreover, only 15 of these 17 semesters will be used. The first semester will be dropped, because only spring-starters have been in the system during that period; the 17th semester has to be dropped, because, unfortunately, we do not have data on inter-university flows for that semester.

This means that the independent determinants to be found will have the serious shortcoming of explaining the transitions for only about three-quarters of the semesters a cohort takes to pass through the system. Although the semesters covered contain the most important ones with respect to the number of students enrolled, this limitation has to be kept in mind when interpreting the results. The other major shortcoming of the data, the underrepresentation of the drop-outs, will also have to be considered in this second part.

According to the semester for which they are entered, the cases in the synthetic population file are distributed as shown in Table 41.

At this point, each case in the synthetic file consists of two kinds of variables:

---

1) This is accomplished by a combination of the *SELECT IF and the WRITE CASES features of SPSS (see NIE et al. 1975).

2) The numbers in brackets refer to the semesters.

Table 41

Distribution of Cases in the Synthetic File by Semester

Number of the Semester	Semester	Number of Cases
1	WS65	4189
2	SS66	4140
3	WS66	4078
4	SS67	4013
5	WS67	3885
6	SS68	3796
7	WS68	3621
8	SS69	3479
9	WS69	3246
10	SS70	2917
11	WS70	2610
12	SS71	2328
13	WS71	2079
14	SS72	1870
15	WS72	1523
Total		47771

Auxiliary variables to be used for the construction of the dependent variables and a set of independent variables to be discussed below.

6.2.1.2. Dependent Variables

There are 192 possible transitions to be estimated in the 12 x 16 type 1 transition matrices. However, from Tables 8 and 13, we know that a sizable portion of these transitions is so rare or even non-existant that it is not advisable to estimate independent determinants for them. In order to limit the estimation to important transitions, the flows involving less than 10 students in the entire period covered, or less than 2% of the students at the university of origin, will not be estimated. Moreover, only the determinants of the flows to foreign universities and to the interruption state will be included. Since most of the students going abroad or interrupting their studies return to their university of origin, it is reasonable to assume that the determinants of the movements in both directions will be rather similar in nature. On the basis of these exclusion criteria, the number of transitions significant for the estimation process is reduced to 83, as indicated by Table 42.

Therefore, 83 indicator variables, one for every significant transition, have to be constructed. According to SPILERMAN's definition (6.1.1.), each one of these variables has the form of a dummy variable: It takes the value of 1, if the student makes the transition in question during a particular semester, and 0 otherwise. For every case, however, there will be 0 and 1 entries in only one row of the table and at most one entry in that row will equal 1 (3). This row is the one pertaining to the origin state of the student in the corresponding semester, i.e. the university at time t. All the dependent variables in the other rows will have a missing value for that case.

Although all the cases will thus have an entry for missing value on most of the 83 dependent variables, all of these variables have to be carried along for all the cases in order to preserve an exactly equivalent variable structure for all the cases in the file.

On the basis of the auxiliary variables which include the number t of the corresponding semester, the university attended in semester t, the university attended in semester t+1, the degree level of leaving and the semester of leaving, it is possible to construct the 83 variables using the data-modification features of SPSS (4).

---

3) This difference to SPILERMAN's definition, which states that exactly 1 entry will be equal to 1, is a result of the exclusion of insignificant transitions. Students effectuating an insignificant transition have zeros in all the significant indicator variables of the corresponding row.

4) The large number of cases and the large number of required IF and COMPUTE statements make this task rather tedious and very costly in terms of computer-time.

Table 42

Transitions for which Independent Determinants are Estimated

From/To	Transitions between Universities											Leaving the System				
	Basel 1	Bern 2	ETHL 3	ETHZ 4	Frib 5	HSG 6	Genève 7	Lausanne 8	Neuchât 9	Zürich 10	Forgn U 11	Interrp 12	Drop out 13	Low Level 14	Medium Level 15	High Level 16
1 Basel	Y11	Y12			Y15					Y110	Y111	Y112	Y113	Y114	Y115	Y116
2 Bern		Y22						Y28		Y210	Y211	Y212	Y213	Y214	Y215	Y216
3 ETHL			Y33					Y38				Y312			Y315	
4 ETHZ				Y44						Y410	Y411	Y412	Y413	Y414	Y415	Y416
5 Fribourg	Y51	Y52			Y55		Y57	Y58		Y510	Y511	Y512	Y513	Y514	Y515	Y516
6 HSG						Y66					Y611	Y612	Y613		Y615	
7 Genève	Y71						Y77			Y710	Y711	Y712	Y713	Y714	Y715	
8 Lausanne		Y82					Y87	Y88		Y810	Y811	Y812	Y813	Y814	Y815	
9 Neuchâtel							Y97	Y98	Y99			Y912	Y913	Y914	Y915	
10 Zürich		Y102			Y105		Y107	Y108		Y1010	Y1011	Y1012	Y1013	Y1014	Y1015	Y1016

6.2.1.3. Independent Variables

The set of the independent variables is partitioned into two subsets, the variables characterizing the individual and those characterizing the semester. The first subset is copied from the data file for the cohort 1965 onto the synthetic file in the process of the creation of the latter, the second subset has to be constructed within the synthetic file.

6.2.1.3.1. Individual Characteristics

Which are the individual characteristics relevant for the leaving of the system and for inter-university flows? From among the large number of characteristics provided by the researchers at AARAU, I have selected a limited number of variables which, on the basis of earlier results, seem to be the most likely candidates.

The most prominent one of these candidates is the group of subject studied by the individual. As noted by RIES (1975), the subject studied is the most important determinant of leaving the system. Results reported in Part I show, how the students studying different subjects are differentially distributed over the four levels of leaving (Fig. 11) and how they differ with respect to the average number of semesters needed to obtain a degree of a specific level (Table 17). Moreover, the discussion of cumulative inertia has provided evidence for the importance of certain subjects for the inter-university flows.

Now, the type of subject studied is a nominal variable to start out with. As such it cannot be entered into the regression. It will have to be transformed into a group of dummy variables (5). A dummy variable will be included for each one of the following 10 groups of subjects:
- medicine
- law

5) For the use of dummy variables in regression see SUITS (1957) and DRAPER/SMITH (1966, p.134ff.).

- economics/business
- social sciences
- humanities
- historical sciences
- natural sciences I
- natural sciences II
- engineering
- architecture

Theology and the miscellaneous group of subjects not categorized in any of the above groups will form the base category.

'Group of subjects' is a time-dependent variable, because a student can change his subject from one semester to the next. The important time-dependent individual characteristics also include the duration status of a student and his age. Although the discussion of the phenomenon of cumulative inertia has shown that there is no general trend for a decrease in the probability of moving with an increase in duration status, it still was indicated that duration status plays a non-negligible role for flows not constrained by structural considerations.

Age is only partly time-dependent. On the one hand, students grow older over the period of 15 semesters, but, on the other hand, they already differed in age when they entered the university system. Thus, although everybody grows older, the original age differences are preserved. These original differences are quite small: Three-quarters of the students entering the system are between 19 and 21 years old, only 4.8% are younger than that and only 3.6% are 26 or older. In spite of their limited range, age differences still appear to have a decisive impact on the level of moving (see RIES/KRIESI 1974d).

The time-dependent characteristics may vary according to the semester for which an individual case is written onto the synthetic file. The constant individual characteristics, on the other hand, are identical every time an individual enters the synthetic file. These characteristics include:

- sex
- SES
- language region of origin

- type of admission certificate
- work load

There are good reasons for including each one of these:

Sex presumably is the second most important determinant of the pattern of leaving. Female students drop out much more often than males (18.5% vs. 10.9%) and, if they get a degree, they usually seem to get lower ones than their male counterparts (see RIES/KRIESI 1974d and KRIESI/RIES 1974). A dummy variable is introduced for this determinant, which takes a value of 1 for female students and zero otherwise.

SES is selected on the basis of its estimated impact on inter-university flows. Assuming that the categories of SES are equally spaced one notes a roughly linear increase in percentage change of university with increasing socio-economic status from 13.3% in the lowest category to 23.9% in the highest one (RIES/KRIESI 1974c). SPILERMAN introduces 4 dummy variables for SES. Assuming interval quality allows to consider just one variable for this characteristic.

Language region of origin: The earlier findings suggest that Swiss German speaking students are more likely to visit universities of the other language region for a couple of semesters than French speaking students (see RIES/KRIESI 1974b and p. 44 above). A dummy variable is introduced which takes the value of 1 for students with French speaking origin, and zero otherwise. The bulk of the remainder is made up of students with Swiss German origin, but the 3% with Italian mother tongue and an even smaller group with Romantsch origin is also included in that category.

Type of admission certificate: The Federal Administration of Education has been showing considerable interest in the differential success of students with various admission certificates at the university level. The question of whether the committee type of the Federal Admission Certificate is a valid selection criterion or not has actually touched off the research project at AARAU (RIES 1975). Against this background, it will be interesting to find out, whether particular types of certificates are important determinants of leaving the system at specific universities.

Moreover, the different admission certificates do not allow equal freedom of movement between the universities. At least so it seems. While the two types of the Federal Admission Certificates impose no restrictions on movements, the special entrance exam usually allows to study at the university where the exam was taken only (see RIES/KRIESI 1974c).

There are four intermediary types of certificates regarding freedom of inter-university moves. These are: The cantonal admission certificate, the teacher's certificate, the business certificate and the foreign certificate.

To conceptualize this dimension, five dummy variables will be considered, one apiece for:

- federal certificates, committee type
- cantonal certificates
- entrance exams
- teacher's certificates
- business certificates

Students holding a normal federal certificate (64.5%) or a foreign certificate (3.4%) will form the base category.

Work Load: This characteristic refers to the number of hours of work for one's own living. The work load has proved to be important with respect to the duration of studies and to the level of leaving (RIES/KRIESI 1974 e). Among the three variables used to measure this characteristic - regular and irregular work during the semesters and work during vacations, the average regular work load during semesters per semester studied has been shown to have the strongest impact on the flow. This variable, which is measured on a ratio scale, is the last individual characteristic to be included among the set of independent variables. It will serve as a general indicator of the financial situation of the student (6).

6) There are strong correlations between the regular semester work load and the way the students finance their studies in general (see RIES/KRIESI 1974e, p. 73ff.).

6.2.1.3.2. Semester Characteristics

The most obvious characteristic of the semester is its number. Thus, every case is to receive a variable recording the number of the semester, as given in Table 41, for which it is entered into the synthetic file. It is also obvious that semesters differ from one another with respect to the possibility of leaving the system at a particular level: As indicated in Fig. 7, no student can graduate at a low level in the first semester. Similarly, Fig. 8 shows that, except for a couple of strange exceptions, nobody graduates at a medium level before the 7th semester. Finally, it takes at least 9 semesters to get a doctorate (see Fig. 9). This suggests the introduction of three dummy variables, one for each degree level, indicating whether a given semester falls above or below the minimal amount of time required to obtain the degree in question. These variables, called LOW, MEDIUM and HIGH according to the degree level which they refer to, take on the following values:

LOW	= 1 for semester 2, ..., 15
	= 0 for semester 1
MEDIUM	= 1 for semester 7, ..., 15
	= 0 for semester 1, ..., 6
HIGH	= 1 for semester 9, ..., 15
	= 0 for semester 1, ..., 8

The additional semester characteristics used are less obvious (7). They are suggested by various plots picturing probabilities of specific transitions as a function of the number of the semester. These plots can be derived from the type 1 transition matrices for the university model in the APPENDIX (8). Fig. 21 provides three examples. The first plot on the left hand side of that figure shows

7) In the first trial regressions run, I actually made use of only the four obvious semester characteristics, in addition to the individual characteristics described above. The results were quite satisfactory, but they could be improved by the introduction of the less obvious semester characteristics described below. In 6.3., I will provide a rationale for testing the accuracy of the estimated regression coefficients.

8) DRAPER/SMITH (1966) note that "suitable transformations of the independent variable are also sometimes suggested by plotting the data in various ways." (p. 132)

Figure 21

Derivation of the 3 Semester Characteristics XN Drop, XN Low, XN Medium

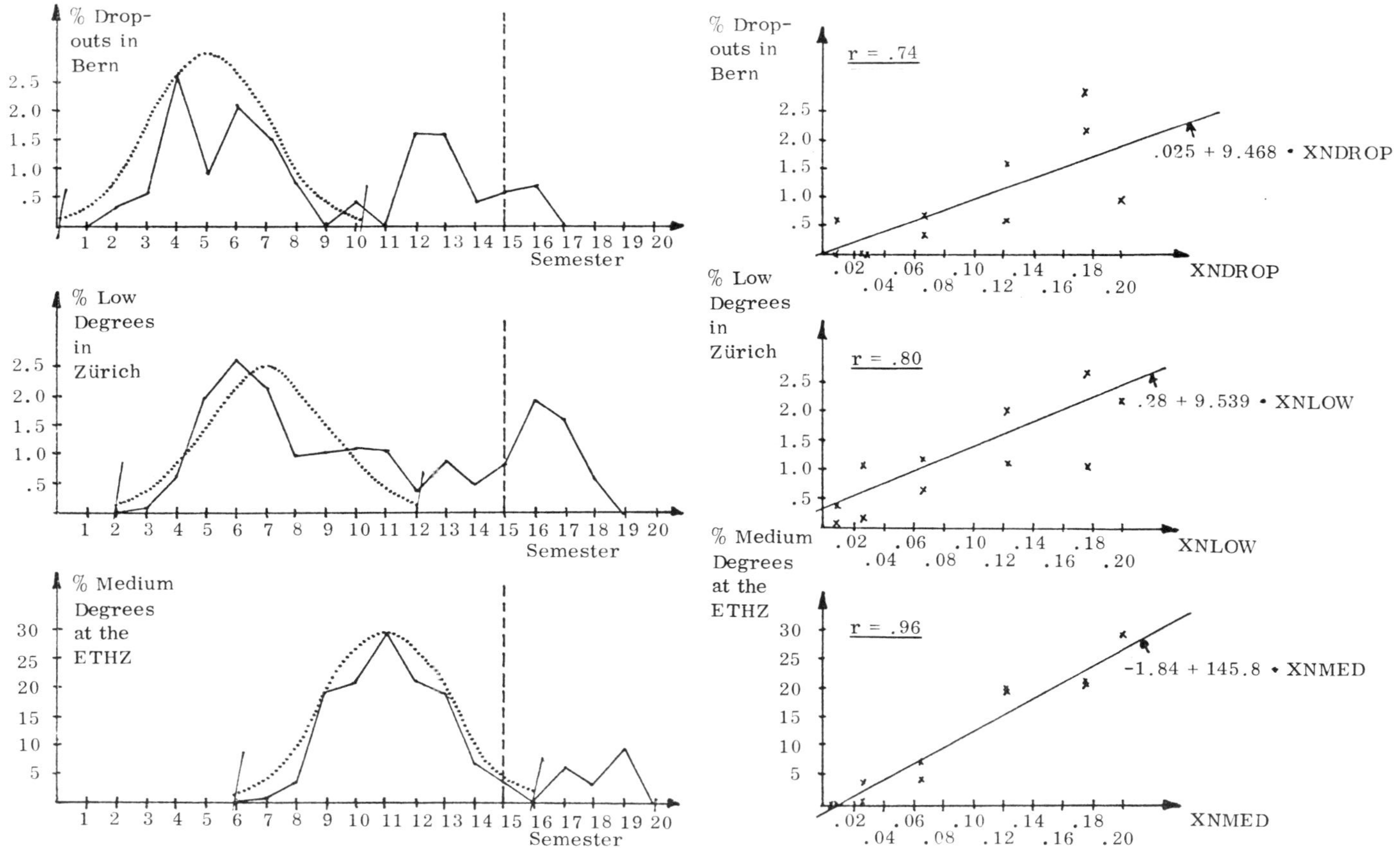

how the percentage of dropouts at BERN develops over successive semesters. Although the resulting graph is rather erratic, it can nevertheless be approximated between the first and the tenth semester by a multiple of the normal distribution with a mean of 5 and a standard deviation of 2 semesters. Similarly, the graph for the percentage of low level degrees at ZUERICH can be approximated between the second and the twelfth semester by a multiple of the normal distribution with mean 7 and standard deviation of again 2 semesters. Finally, the most accurate approximation is the one for the medium level degree curve at the ETHZ, which is given by a multiple of the normal distribution with mean 11 and, once again, a standard deviation of 2 semesters.

A transformation of the variable on the horizontal axis, the number of the semester, according to the formula for the corresponding normal distribution will yield an approximately linear relationship between the transformed variable and the dependent variable in every one of the three cases. This is shown on the right hand side of Fig. 21.

The transformed variables are defined as follows:

$$\text{XNDROP} = \frac{1}{2\sqrt{2\pi}} \cdot e^{-(x-5)^2/8}$$

$$\text{XNLOW} = \frac{1}{2\sqrt{2\pi}} \cdot e^{-(x-7)^2/8}$$

$$\text{XNMEDIUM} = \frac{1}{2\sqrt{2\pi}} \cdot e^{-(x-11)^2/8}$$

Since the dependent variables in question are linear in these transformed variables, they can be entered into the regressions (9). For the interpretation of the resulting semester characteristics - XNDROP, XNLOW and XNMEDIUM, note that in each case the mean is equal to the overall average number of semesters required to obtain a degree at the respective level (10).

9) The linearity assumption of the standard linear model does not require that the expectation of the dependent variable be linear in the independent variable, but linear in known functions of the independent variable (see THEIL 1971, p. 156).

10) These overall averages are computed by taking the respective averages of the averages for the Swiss universities given in Table 14.

Following an analogous rationale, 8 additional semester characteristics are defined. Two variables derive from the 'number of the semester' variable:

Log of SEMNUM = the natural logarithm of the number of the semester
SQ of SEMNUM = the square of the number of the semester

Two variables are provided for linear trends on the medium and high level of graduation respectively:

Lin Medium = (SEMNUM - 7), for semester numbers greater than 7, zero otherwise
Lin High = (SEMNUM - 11) for semester numbers greater than 11, zero otherwise

Three more variables are introduced, because there are indications of logarithmic trends at each level of graduation:

Log Low = log (SEMNUM - 4) (11)
Log Medium = log (SEMNUM - 7)
Log High = log (SEMNUM - 11)

Finally, a variable is provided for quadratic trends at the high level:

SQ High = $(SEMNUM - 11)^2$

6.2.1.3.3. A Summary List of the Independent Variables

There are alltogether 36 independent variables, 21 in the subset of the individual characteristics and 15 in the one of the semester characteristics:

Individual Characteristics

1. SEX = 1, if sex is female, 0 otherwise
2. AGE = age in the corresponding semester - 17
3. SES = ranging from 1 (low) to 5 (high)
4. ORIGIN = 1, if region of o. is French, 0 otherwise
5. WORK LOAD = average number of hours worked regularly during the semester (multiples of 10 h)
6. DURATION STATUS = number of semesters lived consecutively at the university in question

11) (SEMNUM -4) is defined analogously to Lin Medium and Lin High.

7.	CANTONAL ADM CERT	= 1, if admitted on the basis of a cantonal certificate, 0 otherwise
8.	COMMITTEE CERT	= 1, if admitted on the basis of a federal certificate, committee type, 0 otherwise
9.	ENTRANCE EXAM	= 1, if admitted on the basis of an entrance exam, 0 otherwise
10.	TEACHER'S CERT	= 1, if admitted on the basis of a teacher's certificate, 0 otherwise
11.	BUSINESS CERT	= 1, if admitted on the basis of a business certificate, 0 otherwise
12.	MEDICINE	= 1, if subject is medicine, 0 otherwise
13.	LAW	= 1, if subject is law, 0 otherwise
14.	ECON/BUS	= 1, if subject is economics or business, 0 otherwise
15.	SOC SCI	= 1, if subject is social sciences, 0 otherwise
16.	HUMANITIES	= 1, if subject is any sort of language studies, 0 otherwise
17.	HIST SCI	= 1, if subject is a historical science, 0 otherwise
18.	NATURAL SCI I	= 1, if subject is either mathematics or physics, 0 otherwise
19.	NATURAL SCI II	= 1, if subject is any other natural science, 0 otherwise
20.	ENGINEERING	= 1, if subject is engineering, 0 otherwise
21.	ARCHITECTURE	= 1, if subject is architecture, 0 otherwise

<u>Semester Characteristics</u>

22.	SEMNUM	= Number of the Semester
23.	Log of SEMNUM	= Log of the number of the semester
24.	SQ of SEMNUM	= Square of the number of the semester
25.	LOW	= 1, if graduation at the low level is possible, 0 otherwise
26.	MEDIUM	= 1, if graduation at the medium level is possible, 0 otherwise
27.	HIGH	= 1, if graduation at the high level is possible, 0 otherwise
28.	Lin MEDIUM	= (SEMNUM - 7)
29.	Lin HIGH	= (SEMNUM - 11)
30.	Log LOW	= Log (SEMNUM - 4)
31.	Log MEDIUM	= Log (SEMNUM - 7)
32.	Log HIGH	= Log (SEMNUM - 11)
33.	XN DROP	= $1/2\ \sqrt{2\pi} \cdot \exp(-(\text{SEMNUM} - 5)^2/8)$
34.	XN LOW	= $1/2\ \sqrt{2\pi} \cdot \exp(-(\text{SEMNUM} - 7)^2/8)$
35.	XN MEDIUM	= $1/2\ \sqrt{2\pi} \cdot \exp(-(\text{SEMNUM} - 11)^2/8)$
36.	SQ HIGH	= $(\text{SEMNUM} - 11)^2$

All relations between these variables and the dependent variables are assumed to be either linear or dichotomous.

6.2.2. Regression

The independent variables selected in the first step of the construction of the synthetic population consist of those variables which, on the basis of earlier analyses, were judged to be the most likely sources of heterogeneity. But not all of these variables are important for every transition. In a second step, the most important ones from among these variables as judged by a partial F criterion will be selected for each equation by means of a *stepwise* regression procedure. This is, of course, an extremely *inductive* procedure, not suited for the test of theories about particular variables. As SPILERMAN (1972) notes, it is, however, justified, if one wishes to obtain an efficient set of predictors by deleting superfluous variables. It is also well suited for descriptive and simulation purposes. But these are the uses to which I intend to put the method in the present context.

From among the several stepwise regression strategies, the *forward inclusion* option is used, because this is the only one supplied by SPSS (see NIE et al. 1975). In the opinion of DRAPER/SMITH (1966), this strategy is "basically a good idea", although they do not believe it is the best method (12).

For the partial F, a *significance level of .05* is used for cutoff. Two additional statistical criteria are specified: To avoid extreme *multicollinearity* a tolerance parameter of .33 is introduced. As defined by NIE et al. (1975), the tolerance

---

12) They explain: "One of its drawbacks is that it makes no effort to explore the effect that the introduction of a new variable may have on the role played by a variable which entered at an earlier stage (p. 171)." This deficiency is overcome by the so called 'stepwise procedure', as discussed by DRAPER/SMITH. According to THEIL (1971, p. 605), regression strategies, of this nature raise formidable unsolved problems at the theoretical level, which involve the sampling distribution of the coefficients appearing in the strategy. As noted by THEIL, this very complicated problem is not considered by DRAPER/SMITH.

of an independent variable being considered for inclusion is the proportion of variance of that variable not explained by the independent variables already in the regression equation. A tolerance of .33 means that a potential variable will not be entered, if less than one third of its variance is still unexplained by the predictors already entered. This criterion is rather important in the present context, because some of the semester characteristics are highly correlated in several instances. If the criterion were not specified and highly correlated predictors were jointly entered into the equation, the regression coefficients could not be estimated with any precision anymore (THEIL 1971) (13).

Finally, a third criterion limits the maximum number of independent variables to 10. This criterion prevents variables from entering that are indeed significant, but of no practical importance. This criterion did not prove to be very restrictive, because only in a few instances the maximum step was reached.

Tables 43 to 50 present the resulting regression equations using unstandardized coefficients. It is evident from the R^2 values in the bottom row that the 36 independent characteristics do in general not explain very much of the variation among individuals in their movements between Swiss universities and their exit therefrom. There are a few exceptions: 32% of the variation in the leaving of the ETHL at the medium level is explained by this set. For four more transitions the R^2 value exceeds .10: Y_{33}, Y_{38}, Y_{415} and Y_{98}. But in most cases, R^2 is lower than .05.

This result parallels the findings of SPILERMAN. But, as noted by that author, it must not disconcert us, because it is still possible that some of the most im-

13) THEIL makes this intuitively obvious: "For when the first explanatory variable takes values that are almost exactly equal to some multiple of those of the second, the data do not really enable the analyst to distinguish the effects of these variables on the dependent variable; at least, these data do not enable him to do so with any real precision, and the standard errors of the coefficients will therefore be large." And: "The situation of multicollinearity ... implies for the analyst that he is asking more than his data are able to answer." (p.153/154)

Table 43
Regression Equations for Transitions Originating at BASEL: Unstandardized Regression Coefficients

Independent Variables	Y11	Y12	Y15	Y110	Y111	Y112	Y113	Y114	Y115	Y116
Constant	.94393	.00123	.00143	-.00097	.00605	.02658	.00382	-.00347	.01800	.00125
Sex	-.03875						.01536	.01241	.01462	
Age				.00142			.00215			
SES										
Origin		.03546	.03703							
Work Load	.00192							-.00041	-.00070	-.00041
Duration Status				-.00103						
Cantonal A.C.								.01769		
Committee A.C.										
Entrance Exam										
Teacher's C.							-.02481			
Business C.	-.04731			.00873				.02186		
Medicine	.02714			.00440			-.00945			
Law							-.00127		.01960	
Econ/Bus										
Social Sci				.03435			.06405			
Humanities			.00449		.01565			.01764	-.03002	
Hist Sci					.01786				-.02325	
Natural Sci I										
Natural Sci II	.02891							.00953	-.02564	
SEMNUM										
Log of Semnum	-.02310									
SQ of Semnum									.00022	
Low										
Medium										
High										
Lin Medium										
Lin High										
Log Low							-.00764			
Log Medium										
Log High						-.02163				
XN Drop		.02572						-.09427	-.16635	
XN Low								.15321	.09499	
XN Medium										
SQ High	-.01068								.00504	.00793
R^2	.03154	.00302	.00437	.00928	.00422	.00282	.01590	.02770	.05691	.08213

Table 44

Regression Equations for Transitions Originating at BERN: Unstandardized Regression Coeff.

Independent Variables	Y22	Y28	Y210	Y211	Y212	Y213	Y214	Y215	Y216
Constant	.95470	-.00165	.00321	.00264	.06437	-.01323	-.02878	.00829	-.00183
Sex	-.02461						.00975		
Age				-.00173	-.00546	.00194	.00276		
SES									
Origin		.00751							
Work Load	.00199						-.00088		
Duration Status									
Cantonal A.C.									
Committee A.C.									
Entrance Exam									
Teacher's C.	-.03430			-.00622			.00556		.00976
Business C.							.01217	.01146	
Medicine		.00549							.00513
Law									
Econ/Bus			-.00635		.01361		-.00955		
Social Sci						.01060		-.02504	
Humanities	-.05126			.03413			.05640	-.01535	
Hist Sci				.01179				-.03322	
Natural Sci I			-.00604				.02352	-.02837	
Natural Sci II								-.03028	
SEMNUM									
Log of Semnum	-.07842								
SQ of Semnum									
Low	.07313								
Medium	.06426								
High				.01339					
Lin Medium								.01914	
Lin High									
Log Low									
Log Medium									
Log High	-.07420								
XN Drop		.02438				.07147	.07290		
XN Low			.04160	.12520			.10843		
XN Medium								-.13945	
SQ High									.00376
R^2	.03014	.00560	.00319	.02170	.00802	.00641	.06015	.07578	.03579

Table 45
Regression Equations for Transitions Originating at the ETHL and ETHZ: Unstandardized Regression Coeff.

Independent Variables	Y33	Y38	Y312	Y315	Y44	Y410	Y411	Y412	Y413	Y414	Y415	Y416
Constant	.90183	.01423	.09546	.04472	.93067	.00496	-.00139	.06186	.00563	-.00177	.01118	-.00057
Sex	-1.0259	.98748	-.00972				.00976			.01302		
Age								-.00653				
SES												
Origin												
Work Load	.00368	.00115				.00046				-.00029	-.00130	
Duration Status							.00040					
Cantonal A.C.												
Committee A.C.				-.05841					.01308			
Entrance Exam					.03490				.01113			
Teacher's C.					-.14313					.19749		
Business C.												
Medicine							.01182					
Law												
Econ/Bus							.01047		.02520			
Social Sci					-.77471						.84771	
Natural Sci I		.01815			.05005			-.01986			-.02205	
Natural Sci II	.12342			-.07840	.08186				-.00583		-.06144	.00381
Engineering						-.00530					.01730	
Architecture			.07311	-.07208	-.05059	-.00562		.06293		-.00331		
SEMNUM		-.00199										
Log of Semnum												
SQ of Semnum												
Low			-.07264		-.07331	.02136		.04716		.00357		
Medium									.01741			
High	.54012			-.54392								
Lin Medium												
Lin High												
Log Low						-.01059						
Log Medium									-.01159			
Log High	-.38979			.37048								
XN Drop						-.07426						
XN Low	.73249		.21311	-.68595	.34445						-.17414	
XN Medium	-3.9395			4.25203	-1.0259						1.38564	
SQ High												.00362
R^2	.18073	.10529	.03917	.32449	.06568	.00698	.00621	.02251	.00719	.08616	.16659	.03927

Table 46
Regression Equations for Transitions Originating at FRIBOURG: Unstandardized Regression Coefficients

Independent Variables	Y51	Y52	Y55	Y57	Y58	Y510	Y511	Y512	Y513	Y514	Y515	Y516
Constant	-.00214	.00117	1.00482	-.00117	-.00078	-.00060	.00689	.02182	.00417	-.05500	-.00843	-.00479
Sex			-.08233							.03129	.02478	
Age			-.00712							.00581		
SES			.01064								-.00531	
Origin						-.01132	-.00744					
Work Load												
Duration Status												.00147
Cantonal A.C.		.01584	.05884									
Committee A.C.												
Entrance Exam												
Teacher's C.								.01808		.02170		
Business C.											.03599	
Medicine	.02103	.03231	-.09865	.01830	.01677	.05828						
Law											.06991	
Econ/Bus												
Social Sci										.04043		
Humanities			-.04921				.01259		.01480	.05738		
Hist Sci		.01654					.01726					
Natural Sci I												
Natural Sci II			.06315									
SEMNUM												
Log of Semnum			-.02890									
SQ of Semnum												
Low												
Medium			-.05929								.07483	
High												
Lin Medium												
Lin High												.01311
Log Low					.00295							
Log Medium												
Log High												
XN Drop										.17483		
XN Low	.04504		-.50679	.03189		.13082			.06087		.19864	
XN Medium												
SQ High												
R^2	.01561	.01725	.06062	.01219	.00910	.04365	.00788	.00157	.00397	.05667	.06204	.03204

Table 47
Regression Equations for Transitions Originating at the HSG and NEUCHÂTEL: Unstandardized Regr. Coeff.

Independent Variables	Y66	Y612	Y613	Y615	Y97	Y98	Y99	Y912	Y913	Y914	Y915
Constant	.79032	.19575	.01438	-.00517	-.00417	-.02264	.99606	.00610	.01050	-.03335	-.00418
Sex											
Age		-.01330				-.00331			.00398	.00502	
SES								.00727	-.00549		
Origin											
Work Load		.00322	.00135					.00121			
Duration Status											
Cantonal A.C.											
Committee A.C.											
Entrance Exam											
Teacher's C.										.10693	
Business C.											
Medicine					.03786	.15449	-.20672				
Law											
Econ/Bus										-.04033	.06251
Social Sci								.04328			
Humanities											
Hist Sci							.09657				
Natural Sci I											
Natural Sci II					.01594					-.02667	
SEMNUM			-.00151								
Log of Semnum		-.02676					-.07930	-.01099			
SQ of Semnum											
Low	-.08504					.05184					
Medium											.11550
High	.16897			-.15592							
Lin Medium											
Lin High				-.02704							
Log Low				.13479							
Log Medium											
Log High											
XN Drop	1.53072	-.21621								.23832	
XN Low	-.39129									.19067	
XN Medium					.12364				-.14020		
SQ High											
R^2	.04245	.04986	.01270	.08824	.02527	.13668	.05341	.01466	.01356	.04993	.08475

Table 48

Regression Equations for Transitions Originating at GENEVE: Unstandardized Regression Coefficients

Independent Variables	Y71	Y77	Y710	Y711	Y712	Y713	Y714	Y715
Constant	.01476	.90779	.01684	.00742	.02841	-.00913	-.00142	.00413
Sex		-.03939		.00553			.01509	.01964
Age		-.01558				.00314		.00837
SES		.00971					-.00342	-.00713
Origin	-.01121	.03134	-.01914	-.00872				
Work Load		.00129			.00060			-.00121
Duration Status	-.00185	.00548	-.00334					
Cantonal A.C.								
Committee A.C.								
Entrance Exam								
Teacher's C.								
Business C.				.01371				
Medicine		.03849						
Law	.01295		.01156					
Econ/Bus						.02905		.03070
Social Sci			-.00667				.04249	
Humanities			.01462	.02119				
Hist Sci					.03277	.02140		
Natural Sci I						.02332		
Natural Sci II		.10565						.03955
SEMNUM								
Log of Semnum								
SQ of Semnum								.00052
Low			.01546				.01322	
Medium			.01355					
High			.00918			-.01768		
Lin Medium								
Lin High		-.02721						
Log Low								
Log Medium	.00955				-.01818			
Log High								
XN Drop						.02235		.23499
XN Lew	.03854	-.22150						.13059
XN Medium								.24844
SQ High								
R^2	.01596	.03765	.03542	.01374	.01110	.01327	.03375	.06642

Table 49

Regression Equations for Transitions Originating at LAUSANNE: Unstandardized Regression Coefficients

Independent Variables	Y82	Y87	Y88	Y810	Y811	Y812	Y813	Y814	Y815
Constant	.01510	.00094	.92652	.05163	.01018	.02104	.02363	.00638	-.00796
Sex			-.04690						.03043
Age			-.01537						
SES									
Origin	-.02093		.09045	-.06017	-.00867	.01706		-.01240	
Work Load								.00039	
Duration Status	-.00108	-.00240							
Cantonal A.C.		.03769							
Committee A.C.						.02128			
Entrance Exam				.02690		.06142			
Teacher's C.			-.52894	.22391				.26983	
Business C.			-.05396	.01785			.01844		.06207
Medicine	.01517	.00975					-.01947		-.04916
Law				.01494			-.01758		-.05097
Econ/Bus				-.02754					
Social Sci			-.05501	-.01987					.02661
Humanities				.01704	.01323			.01165	-.02859
Hist Sci									
Natural Sci I			-.09214						
Natural Sci II		.00740	.04496			-.01865	-.01865		-.03811
SEMNUM									
Log of Semnum									
SQ of Semnum									.00095
Low	.01141								
Medium		.02184							
High									
Lin Medium									
Lin High									
Log Low									
Log Medium						-.01311			
Log High									
XN Drop				.08324					
XN Low		.07075	-.26955					.04533	.20073
XN Medium									
SQ High			-.01086					.00106	.00420
R^2	.02120	.02203	.06113	.09152	.00657	.00922	.00954	.09263	.08629

Table 50
Regression Equations for Transitions Originating at ZUERICH: Unstandardized Regression Coefficients

Independent Variables	Y102	Y105	Y107	Y108	Y1010	Y1011	Y1012	Y1013	Y1014	Y1015	Y1016
Constant	.00040	.00074	-.00013	.00081	.94220	-.00110	.06597	.00260	-.00959	-.01709	.00079
Sex		.00281	.00290		-.02888		-.01678	.01135	.00961	.01368	
Age					-.00561		-.00273	.00225	.00406		
SES											
Origin		.02608									
Work Load				-.00015	.00091			.00025	-.00066	-.00050	
Duration Status	-.00116								-.00173	.00152	
Cantonal A.C.								.01431			
Committee A.C.											
Entrance Exam					-.14601					.06895	.03613
Teacher's C.									.02239		
Business C.											
Medicine					.02970		-.01672	-.00668		.02245	
Law			.00357				.01317	-.01060		.01491	
Econ/Bus										.03168	
Social Sciences		.00560				.01442			.01866		
Humanities			.00642	.00683	-.05552	.02995	.01883		.01508		
Hist Sci	.00839					.01257					
Natural Sci I										.02312	
Natural Sci II	.00572				.03674		-.00825	-.00817	.01107		
SEMNUM											
Log of Semnum	.00408						-.00977				
SQ of Semnum											
Low									-.01190		
Medium											
High											
Lin Medium										.00964	
Lin High											.01539
Log Low								-.00730			
Log Medium											
Log High											
XN Drop				.03941							
XN Low				-.02203		.06440			.10139		
XN Medium	.02261										
SQ High					-.00743					.00294	
R^2	.00589	.00418	.00387	.00519	.02465	.01441	.01478	.00948	.02822	.06073	.03357

portant determinants of the flow of students are included among the set of the selected predictors. The reason lies in the fact that we are dealing with individual-level data in estimating the regression equations, i.e. with data involving very large idiosyncratic components. The value of R^2 can always be raised by aggregating the data, because aggregation tends to reduce the relative importance of neglected variables (see THEIL (1971), STINCHCOMBE (1968) and BLALOCK (1961)). On the basis of the selected predictors we may therefore very well be able to predict the flows of aggregates, such as the students at a particular university, accurately. As SPILERMAN points out, it is the regression coefficients and not the R^2 values which must concern us here.

The resulting regression model is a very simple one in that it assumes the effects of the independent determinants to be entirely additive. No provision is made for interactions among the determinants.

6.3. Two Tests of the Independent Determinants of Flow

6.3.1. Reproduction of the Rates of Flow

In order to test the accuracy of the selected independent determinants, we can predict the rates of flow for the cohort 1965 on their basis. More specifically, it should be possible to reproduce the first 15 type 1 matrices for the cohort 1965 on the basis of the determinants found.

The program provided in APPENDIX B-6 allows the computation of the predicted transition matrices on the basis of the independent determinants (14). By

14) This program was written especially for this purpose. Some useful hints were provided by the program of DICKENS (1972), which SPILERMAN used in his analysis. For each individual in every semester, the program computes the vector of the expected values of his transitions to the 16 destination states, sums over individuals and divides the result by the total number of individuals in the respective origin states. The resulting probabilities are forced to lie in the [0,1] interval.

making use of the indices of dissimilarity again, these predicted matrices can be compared with the set of the observed ones. Tables 51 and 52 give a detailed account of the results of these comparisons. Table 53 summarizes Table 51.

Turning first to the overall averages reported in the latter table, we notice a fairly close overall correspondance between the two sets of matrices, which is to say that the selected determinants are on the whole accurate predictors indeed. With the exception of the ETHL, all the indices are markedly lower than the ones resulting under the Markovian assumption (see Table 34). Moreover, Table 53 also reveals a clear-cut pattern in the rank-order of the universities: In estimating transition probabilities on the basis of independent predictors we do consistently better for Swiss German institutions (including the Swiss German universities, the ETHZ and the HSG) than for French speaking ones (including LAUSANNE, GENEVE, NEUCHATEL and the ETHL) and FRIBOURG.

The result might be a reflection of the fact that I have better first hand knowledge of the situation at the Swiss German schools of higher education.

Turning to the two detailed tables, we notice that figures in corresponding cells differ slightly. The reason for this difference lies in the fact that the indices of dissimilarity reported in the first table take into account only those cells for which a transition rate has actually been computed, whereas those in the second one are based on all the cells of the 16 x 10 one-step transition matrices (15). A comparison of the two tables shows that the deletion of the rest of the cells was justified, because the excess discrepancy caused by the cells that were not considered in the estimation process is in general very small. In no case, this discrepancy amounts to more than 4%, in only two instances it is larger than 3%. The overall average discrepancy is as small as .5%.

---

15) The last 2 rows of the 16 x 12 matrices have not been considered in either table.

Table 51

Indices of Dissimilarity between Observed Matrices and Matrices Predicted on the Basis of Independent Determinants by Semester and University: Estimated Cells Included only

SEMESTER	BASEL	BERN	ETHL	ETHZ	FRIB	HSG	GENEVE	LAUS	NEUCH	ZURIC
1	1.2	0.6	1.6	0.6	1.8	1.3	1.4	1.4	0.8	0.8
2	1.9	1.1	2.7	3.5	1.8	2.4	2.2	1.0	5.3	0.7
3	2.0	1.4	1.4	4.7	2.3	1.6	2.1	2.3	5.1	1.6
4	2.9	1.4	2.6	4.1	2.8	2.9	3.4	4.6	3.8	2.5
5	2.3	2.6	4.1	3.0	4.9	1.6	2.0	3.9	4.6	2.3
6	1.8	3.4	3.8	2.5	5.1	7.8	1.7	5.3	4.2	1.8
7	2.3	2.6	2.9	4.8	4.0	7.4	5.2	4.4	7.2	1.4
8	1.3	2.6	13.2	4.1	3.4	3.9	6.6	2.9	5.0	3.0
9	4.7	0.7	5.1	5.6	6.1	2.1	4.1	2.4	1.1	2.4
10	2.4	1.8	17.7	2.7	9.6	8.0	3.5	3.7	14.6	1.4
11	1.3	1.3	22.3	2.4	5.9	3.8	3.9	3.3	7.8	3.8
12	3.6	1.2	14.7	4.2	7.0	3.5	4.0	5.9	12.0	1.6
13	4.8	2.4	9.3	3.1	4.4	3.5	8.6	12.8	8.0	3.3
14	4.1	3.9	15.0	3.0	4.6	3.6	12.4	8.[illegible]	3.2	2.5
15	1.6	2.9	5.4	1.4	6.8	2.3	8.2	4.2	3.9	1.4

Table 52

Indices of Dissimilarity between Observed Matrices and Matrices Predicted on the Basis of Independent Determinants by Semester and University: All Cells Included

SEMESTER	BASEL	BERN	ETHL	ETHZ	FRIB	HSG	GENEVE	LAUS	NEUCH	ZURIC
1	1.6	0.7	2.1	0.7	1.8	2.7	1.5	1.9	1.4	0.9
2	1.9	1.2	4.3	3.9	1.8	4.5	3.0	2.3	5.6	1.3
3	2.8	1.4	3.5	5.2	2.4	3.4	2.8	3.1	6.6	2.0
4	2.9	1.9	2.6	4.4	2.8	3.3	3.4	5.9	4.2	2.6
5	2.6	3.1	6.9	3.5	4.9	2.0	2.6	5.2	5.7	2.5
6	2.2	3.6	3.8	2.5	5.1	7.8	2.0	5.9	4.6	1.9
7	2.6	3.0	2.9	5.1	4.0	7.8	5.8	5.0	7.2	1.9
8	1.6	3.1	13.2	4.1	3.4	5.7	6.7	3.9	5.6	3.1
9	4.7	1.1	5.1	5.8	6.1	3.3	4.5	2.4	1.1	2.8
10	2.4	2.0	17.7	3.0	9.6	11.2	3.9	3.7	15.4	1.4
11	1.5	2.2	23.6	2.7	7.2	4.7	3.9	3.6	9.9	3.9
12	4.0	1.2	14.7	4.2	7.4	3.5	4.0	7.0	12.0	1.7
13	4.8	2.7	11.7	3.3	5.0	5.7	8.6	13.2	9.3	3.4
14	4.1	4.0	15.0	3.8	4.6	4.8	12.9	9.9	3.2	2.5
15	1.6	2.9	9.2	1.4	7.5	3.6	8.5	6.4	5.6	1.4

Table 53

Index of Dissimilarity by University: Means and Standard Deviations Taken over 15 Semesters

University	Mean	Standard Deviation
Zürich	2.0	.9
Bern	2.0	1.0
Basel	2.5	1.2
ETHZ	3.3	1.3
HSG	3.7	2.2
Lausanne	4.4	3.0
Genève	4.6	3.1
Fribourg	4.7	2.2
Neuchâtel	5.8	3.7
ETHL	8.1	6.7

The detailed tables also show that the estimation is somewhat more accurate for earlier semesters, although the trends are quite weak. All the very large entries in the two tables, however, pertain to semesters of the second half of the period covered. A closer inspection of these extreme discrepancies shows that they are invariably due to inaccurate estimations of the leaving at the medium level and the corresponding main diagonal cells:

Table 54

Comparison of Observed and Predicted Probabilities of Medium Level Leaving for Extremely High Indices of Dissimilarity in Tables 51 and 52

University	Semester	Medium Level Leaving		
		Observed	Predicted	Difference
ETHL	8	.050	.179	-.129
ETHL	10	0	.175	-.175
ETHL	11	.513	.300	.213
ETHL	12	.053	.183	-.130
ETHL	13	.300	.210	.110
ETHL	14	0	.112	-.112
Genève	14	.245	.139	.106
Lausanne	13	.031	.142	-.111
Neuchâtel	10	.212	.116	.096
Neuchâtel	12	.050	.114	-.064

The students at the ETHL seem to graduate at a medium level much more often in the winter-semester than in the corresponding summer-semesters. To account for this fact, an additional semester characteristic would have to be introduced which would allow to distinguish between these two types of semesters. An interactive term including this new variable and the XN MEDIUM characteristic would probably suffice to considerably improve the accuracy of the estimation procedure for the ETHL.

Apart from this notable deficiency, the outcome of the present test is quite satisfactory: On the basis of the selected set of predictors, the first 15 one-step transition matrices of the cohort 1965 can be reproduced with sufficient accuracy. Which is another way of stating that the main determinants of heterogeneity are included among the selected sets of predictors.

6.3.2. Reproduction of the Patterns of Leaving

On the basis of the transition matrices just obtained, it is also possible to reproduce the pattern of leaving after 15 semesters quite accurately. Making use of the program in APPENDIX B-7, we arrive at the pattern given in Table 55 (16). This pattern corresponds quite satisfactorily to the observed one (17), as can be seen from the discrepancy between the two reported in Table 56. There are, however, two serious exceptions. As was to be expected on the basis of the results of the previous test, graduation at the medium level at the ETHL is predicted less than accurately. It is seriously underestimated by the model using independent determinants. Unfortunately, the same is also true of me-

16) This program does exactly what is described in 3.2.2. under the title of prediction of drop-outs and graduates for a single cohort for the case of the university of exit. Since the last two rows of the one-step transition matrices have not been estimated (see previous footnote), the estimated transition matrices have been supplemented by the corresponding rows from the observed matrices for these computations.

17) The corresponding observed pattern is the one without adjustments for drop-out errors, which has not been reported before.

Table 55

Pattern of Leaving the System on the Basis of Estimated Transition Matrices after 15 Semesters by University of Exit

University	Level of Leaving			
	Drop Out	Low	Medium	High
Basel	.105	.076	.290	.105
Bern	.077	.160	.350	.064
ETHL	-	-	.515	-
ETHZ	.099	.019	.592	.024
Fribourg	.092	.182	.294	.056
HSG	.083	-	.495	-
Genève	.120	.095	.412	-
Lausanne	.155	.045	.564	-
Neuchâtel	.084	.187	.344	-
Zürich	.112	.109	.317	.092

Table 56

Discrepancy between Estimated and Observed Pattern of Leaving after 15 Semesters for Cells which have been Estimated (Obs-Est)

University	Level of Leaving			
	Drop out	Low	Medium	High
Basel	-.004	.008	.019	.008
Bern	-.003	.004	-.001	.003
ETHL	-	-	-.166	-
ETHZ	.013	0	-.003	0
Fribourg	-.003	.017	-.009	.004
HSG	.001	-	-.107	-
Genève	.019	.001	.012	-
Lausanne	.009	-.002	.007	-
Neuchâtel	.006	.025	-.035	-
Zürich	.005	.005	.006	-.004

dium level graduation at the HSG. One might also consider the discrepancies for low and medium level graduation at NEUCHATEL as minor ex eptions from the generally close correspondance.

All in all, it seems safe to conclude once more that the independent determinants selected represent the main determinants of heterogeneity in the population considered.

7. APPLICATIONS USING INDEPENDENT DETERMINANTS OF FLOW

The introduction of independent determinants of flow into the model presented in Part I allows some useful applications in the fields of description, prediction and simulation, which will again be taken up in that order.

7.1. Description

7.1.1. Description of the Independent Determinants of the Rates of Flow

The set of equations given in the Tables 43 to 50 allows a complete description of the determinants of the pertinent rates of flow for the first 15 semesters of the cohort 1965.

In describing these sets of determinants, four types of comparisons can be made:

- Comparisons within a single equation: What are the determinants of a specific transition and what is the relative importance of each one of them?
- Comparisons between different equations pertaining to a specific origin state: Do the determinants of flow for a certain origin state differ according to the destination states?
- Comparisons between different equations pertaining to a specific destination state: Do the determinants of flow and their relative importance for a certain destination state vary depending on the origin of the respective transition?
- Comparisons between equations for specific origin-destination combinations: The most interesting combination probably is the one, where origin and destination are identical. Thus, the question, for example, is whether the probability of staying in one's origin state is differentially affected from one university to the other.

From among these four alternatives, the second one seems to be less promising than the rest and it will therefore not be considered below (1).

---

1) By making use of more involved statistical instruments than are used here, comparisons of the second type might yield more interesting results, too. Thus canonical correlations may be a promising way to pursue in this context.

The description will concentrate on the determinants of the following types of transitions:

- staying at the university of origin
- moving between Swiss universities
- moving to foreign universities
- interrupting
- leaving the system

Rather than relying on the unstandardized regression coefficients, the discussion will make use of standardized coefficients, which provide the only sensible way to compare the relative weight of independent variables measured on different units such as the ones involved in the present context (2).

The results will be presented in a series of figures each one of which contains four kinds of information for every equation:

- a list of the determinants ranked according to their relative weight as given by the standardized regression coefficient
- an index measuring the relative weight of each determinant in terms of the standardized coefficient of the most important determinant in the equation in-question
- a bar-chart that allows not only a quick grasp of the relative weight of the determinants, but also provides an easy distinction between semester characteristics (white bars) and individual characteristics (black bars) as well as between positive influences (bars to the right of the center) and negative ones (bars to the left of the center)
- the rank ordered standardized regression coefficients

Unfortunately, the regression coefficients within a single equation cannot be compared by statistical tests, because SPSS does not provide covariance matrices. On the other hand, the unstandardized regression coefficients of different equations from different universities can be tested statistically using the usual two sample t-test with unequal variances (3). Since all the samples in-

---

2) see NIE et al. (1975, p.325)

3) I am indebted to Prof. Hampel from the ETHZ for pointing this out to me. A description of this so called Welch procedure is given by BROWNLEE (1965, p. 299-303).

volved are, however, comparatively large, virtually all the relevant tests of this nature yield significant differences. I will therefore omit all references to statistical tests for coefficients below.

Since it is easy to get lost in the details presented in the figures, they will be supplemented by several summary measures: First, the average number of predictors selected per university gives an idea about the complexity of the determination of the transition in question. The average number of times any single predictor is selected, on the other hand, allows some judgement about the degree to which this type of transition is governed by the same factors at all the universities. A very important part of description will be the assessment of the relative weight of semester and individual characteristics. Two additional summary measures try to get at this: First, the proportion of semester characteristics among the set of all the predictors selected for a certain type of transition is computed and tested against the actual proportion of semester characteristics among the predictors submitted to the regression procedure which equals 15/36 = .42. P - values will be provided for each test (4). Secondly, the proportion of times semester characteristics rank first among the predictors selected for a certain type of transition is also computed and tested against the actual proportion of semester characteristics (5).

Finally, the figures will be summarized by a list of the most important determinants for all the universities with respect to the type of transition in question. The rank of each determinant on this list is given by the sum of its index-values, summed over all universities for the type of transition in question.

---

4) For the test, the normal approximation to the binomial distribution is used, as suggested by LOETHER/McTAVISH (1974, p.109) and SIEGEL (1956, p.40). The P - values reported indicate the one-tailed probability of getting values at least as extreme as the ones observed.

5) For this test, the normal approximation to the binomial distribution cannot be used, because of the small numbers involved. Instead, the binomial test, as described by SIEGEL (1956) will be used.

For descriptive purposes, it is also useful to supplement the dominant distinction between semester characteristics and individual ones by some additional concepts. Thus, the individual characteristics can be divided into:

- socio-economic characteristics
- duration status
- admission certificates
- subject characteristics

Sex, age, SES, origin and work load belong to the first category. They provide a short-hand description of associated attitudes, aspirations, motivations, cognitions, abilities, financial resources and cultural environments (see PERRENOUD 1973). With respect to the subject characteristics, one more distinction which is related to the dominant one, although not identical with it, is relevant:

Determinants can be distinguished by whether or not they refer to the operation of structural forces within the university system or not. Now, the reference of the subject characteristics is primarily structural in nature.

It is true that students studying different subjects have been shown to have different social origins, different values and attitudes as well as different personality characteristics (NEWCOMB/FELDMAN 1970, SOFI 1973, RIES/KRIESI 1974b). But with respect to the flow of students, it seems that the differential structural characteristics of the various groups of subjects are the decisive aspects. Such characteristics are: The courses of study, the degrees offered, the financial resources, student-faculty ratios. Structural peculiarities of certain subjects are in turn partly a result of the specific situation characterizing any particular branch of science (JENNE et al. 1969), partly they depend on local idiosyncracies (RIES/KRIESI 1974a). As a result, every subject at a given university provides a distinct structural setting for the student studying there. The subject characteristics will therefore be interpreted in structural terms.

Similarly, all the semester characteristics will be interpreted as predominantly structural in nature. For some characteristics, such as LOW, MEDIUM and

HIGH, this is immediately clear. It is less apparent, for example, in the case of the normal transformations. The spread in the approximations in Fig. 21 is, of course, a result of individual idiosyncracies: Some student has worked harder than another one; some student has failed in a preliminary exam and had to take some courses all over again; someone else wanted to interrupt his studies for some reason or other and has thus lost time. But the spread is also a result of structural influences: The amount of time required to get a certain degree varies from subject to subject at a given university, as shown in Table 17. Moreover, the fact that the means correspond exactly to the overall averages required to obtain the respective levels is a clear-cut indication for the adequacy of the structural interpretation of these variables.

After all these preliminaries, it is now time to present the results.

7.1.1.1. Determinants of Staying at the University of Origin

Fig. 22 presents the results in detail. The determinant structure for staying at the university of origin is rather complex, as is indicated by the high average number of predictors selected for each university (7.6 out of a maximum of 10). In three instances, the maximum step is reached in the regression procedure. The complexity of the picture is increased by the fact that the average number of entries per predictor is only 2.9, which means that 26 out of 36 predictors are selected at least once.

In trying to assess the relative importance of each one of these, let us first consider the dominant distinction: Except for the French speaking universities (GENEVE, LAUSANNE, NEUCHATEL), the semester characteristics provide the most important determinant at every university, which is a significant departure from the expected (P = .071). The staying at the HSG is entirely determined by semester characteristics. They are very important at the technical schools and at BERN, too.

In contrast, the overall share of semester characteristics among the predictors

Figure 22

The Determinants of Staying at the University of Origin

	Variable	Index	standardized Regression Coefficient
BASEL	SQ High	100	-.12911
	Log Sem	46	-.06007
	Sex	43	-.05562
	Work Load	34	.04416
	Medicine	33	.04306
	Natural Sci II	30	.03810
	Business Cert	25	-.03353
BERN	Log Sem	100	-.20137
	Medium	53	.10696
	Log High	46	-.09309
	Low	33	.06631
	Work Load	25	.05017
	Humanities	25	-.04952
	Sex	16	-.03130
	Teacher's Cert	14	-.02836
ETHL	XN Medium	100	-.72223
	High	80	.57594
	Log High	37	-.26447
	XN Low	22	.15561
	Natural Sci II	18	.13160
	Sex	14	-.10049
	Work Load	11	-.07886
ETHZ	XN Medium	100	-.19785
	Natural Sci II	44	.08670
	XN Low	36	.07057
	Low	32	-.06323
	Natural Sci I	26	.05120
	Architecture	24	-.04777
	Teacher's Cert	13	-.02666
	Entrance Exam	13	.02498
	Social Sci	12	-.02444
HSG	XN Drop	100	.30597
	High	64	.19604
	XN Low	27	-.08182
	Low	24	-.07303

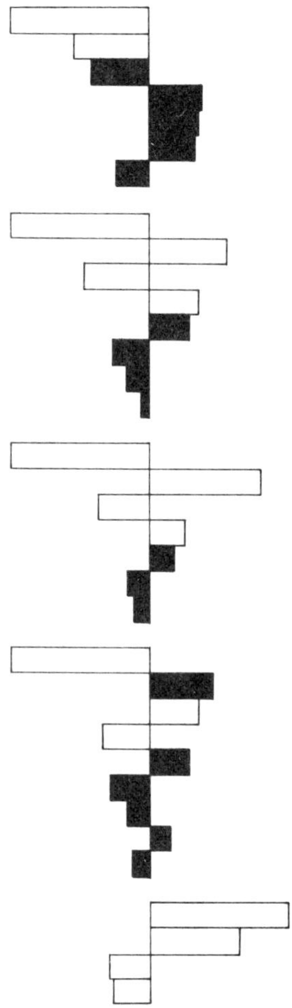

	Variables	Index	standardized Regression Coefficients
FRIBOURG	XN Low	100	-.10611
	Medizin	90	-.09573
	Sex	84	-.08861
	Medium	77	-.08175
	Logsem	63	-.06634
	Natural Sci II	58	.06121
	Age	53	-.05670
	Hist Sci	48	.05106
	Humanities	47	-.04975
	SES	35	.03742
GENEVE	Natural Sci II	100	.12882
	Age	98	-.12636
	Lin High	70	-.09010
	Duration Stat	57	.07298
	Sex	46	-.05910
	Medicine	42	.05442
	XN Low	40	-.05087
	Origin	34	.04406
	SES	30	.03898
	Work Load	26	.03394
LAUSANNE	Age	100	-.10822
	Teacher's Cert	93	-.10076
	SQ High	91	-.09827
	Origin	88	.09501
	Sex	60	-.06440
	Natural Sci I	53	-.05694
	XN Low	52	-.05623
	Natural Sci II	45	.04897
	Business Cert	42	-.04510
	Social Sci	37	-.03986
NEUCHATEL	Medicine	100	-.18463
	Log Sem	99	-.18209
	Hist Sci	38	.07076
ZUERICH	SQ High	100	-.08792
	Humanities	74	-.06448
	Age	54	-.04715
	Natural Sci II	48	.04179
	Medicine	47	.04173
	Sex	44	-.03843
	Entrance Exam	26	-.02325
	Work Load	25	.02165

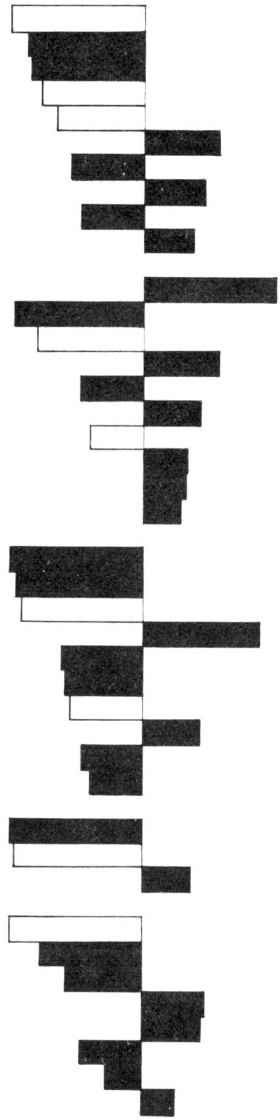

Table 57

The most Important Determinants of Staying at the University of Origin

Rank	Characteristic	Number of Entries	Index Points
1	Natural Sci II	7	343
2	Medicine	5	312
3	Log of Semester	4	308
4	Sex	7	307
5	Age	4	295
6	SQ High	3	291
7	XN Low	6	277
8	XN Medium	2	200
9	Humanities	3	146
10	High	2	144

selected (.342) is significantly smaller than expected (P = .076), indicating that semester characteristics are of only minor overall importance for staying at the university of origin after all. These mixed results are also brought out by the list of the 10 most important characteristics in Table 57 which is evenly divided between the two categories. The leading items on the list, however, are two subject characteristics, medicine and the natural sciences II. Students in the natural sciences II stay for a comparatively long time at their university of origin and this at all the universities, except at BERN and NEUCHATEL. This is so, because so many of them work for a degree at the highest level (see Fig. 11) and because theirs is a study of long duration anyway (see Table 31). The latter applies to the students in medicine, too, who stay comparatively long at BASEL, ZUERICH and GENEVE. BERN is again an exception, most likely because studies at this university are very long in general (see Table 14). At FRIBOURG and NEUCHATEL, on the other hand, medical students leave comparatively early for reasons discussed above (see 5.2.).

The third subject characteristic on the list of Table 57, the humanities, is negatively related to staying at ZUERICH, BERN and FRIBOURG, which is due to the large share of students graduating at low levels in these areas.

History students stay very long at NEUCHATEL and FRIBOURG; students in the natural sciences I remain longer at the ETHZ than the other students of this university, but they leave LAUSANNE earlier than the others.

The other two items on the list are the socio-economic categories of sex and age. Both, female students and older students, tend to leave the universities quite early, because either they graduate at low levels or drop out more often than the average student, as will be seen below. Regarding sex, this trend is prevalent at all universities with the exception of the ETHZ and the HSG, where there are only very few girls enrolled, and also with the exception of NEU-CHATEL. The impact of age shows at ZUERICH, FRIBOURG and, above all, at LAUSANNE and GENEVE. Of the other socio-economic characteristics, only work load plays an appreciable role, having some impact at five of the ten universities.

Duration status is generally unimportant and so are the admission certificates. Only at LAUSANNE, the students holding teacher's certificates leave earlier than others, for reasons to be discussed below.

Taking a second lock at the semester characteristics, we notice that the students at BERN and NEUCHATEL follow an overall logarithmic trend in leaving. The same trend, although to a lesser extent, is operating at BASEL and FRIBOURG. At BASEL, ZUERICH and LAUSANNE, students leave after the 11th semester at a geometrically increasing rate. XN LOW is positively related to staying at the technical schools, but inversely related to it at the HSG, LAUSANNE, GENEVE and, above all, FRIBOURG: Students at the technical schools predominantly leave around the 11th semester, as is indicated by the strong impact of XN MEDIUM; students at the HSG drop out or move to another university very early (impact of Low), graduate at the medium level in the 7th or 8th semester (impact of XN Drop, XN Low and High), or stay on to work for a high degree (impact of High); the importance of XN Low at FRIBOURG, finally, is a result of the large share of students graduating at a low level at this university.

Without being exhaustive, this discussion shows that the staying at the origin state is primarily regulated by structural determinants plus sex and age. The influence of structural determinants is most substantial in the cases of the HSG and the two technical schools, which are indeed characterized by the most rigidly structured courses of study in the whole Swiss University System (see 3.3.2.). Structural influences are also pervasive at BERN, making for generally long studies.

The discussion also provides evidence for the uniqueness of the setting at each university. Although there are some parallels, each university is characterized by a unique set of predictors. Moreover, the same variable may have opposite effects at different universities, such as the natural sciences I at the ETHZ and at LAUSANNE.

The discussion of the leaving of the system will, to a large extent, complement the present one.

7.1.1.2. Determinants of the Flow between Swiss Universities

As will be recalled, determinants of flow have been estimated for only a limited number of transitions between universities. These can be further divided into four groups: transitions between:

- Swiss German speaking universities
- French speaking universities
- universities of the two language groups
- Fribourg and the rest.

7.1.1.2.1. Transitions between Swiss German Universities

Fig. 23 gives the determinants for the five transitions in question. The average number of predictors per university is 4.2, making for a far less complicated picture than above. The determination of each transition is, however, even more unique than it was for the case of staying, because each predictor is entered just 1.1 times on the average. Only two characteristics are selected twice - duration status and XN Drop:

Figure 23

The Determinants of Transitions between Swiss German Universities

	Variables	Index	standardized Regression Coefficient
BASEL-BERN	Origin	100	.04515
	XN Drop	70	.03171
BASEL-ZUERICH	Duration Stat	100	-.07629
	Social Sci	83	.06370
	Age	83	.06315
	Medicine	49	.03710
	Business Cert	42	.03190
BERN-ZUERICH	XN Low	100	.04257
	Econ/Bus	74	-.03148
	Natural Sci I	59	-.02512
ZUERICH-BERN	Duration Stat	100	-.09386
	Log Sem	66	.06211
	Hist Sci	45	.04258
	Natural Sci II	42	.03932
	XN Medium	34	.03153
ETHZ-ZUERICH	Log Low	100	-.11443
	Low	72	.08200
	XN Drop	58	-.06586
	Engineering	29	-.03364
	Architecture	21	-.02364
	Work Load	19	.02217

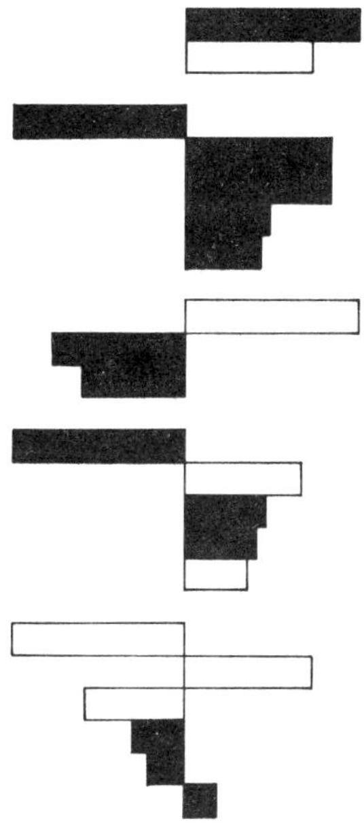

Cumulative Inertia seems to operate with respect to the moves from BASEL to ZUERICH and from ZUERICH to BERN. The longer someone has stayed at the respective university of origin, the less likely he is to move. This is an indication of the validity of the hypothesis put forward in 5.2.1. that the relative duration status is of some importance for moves unconstrained by structural determinants.

Semester characteristics are primarily important for the transitions between the ETHZ and the university of BERN, on the one hand, and the university of ZUERICH, on the other. As indicated by the selected semester characteristics, the transitions are generally made early in the course of study, which is another hint for the presence of cumulative inertia. Semester characteristics are neither over- nor underrepresented.

The subjects primarily involved in the moves are:

- social sciences and medicine for BASEL/ZUERICH
- economics/business and natural sciences I for BERN/ZUERICH
- historical sciences and natural sciences II for ZUERICH/BERN
- engineering and architecture for ETHZ/ZUERICH

While the specific attraction and push factors for the first three transitions are quite elusive phenomena and probably of changing nature, the motives involved in the last transition are more easily guessed at: Since the two subjects involved are not offered at the university of ZUERICH, the students left engineering and architecture at the ETHZ probably to take up subjects not offered at the ETHZ, such as specific courses in medicine, economics/business and the social sciences (see Table 10).

Admission certificates are again unimportant. The same is true of socio-economic characteristics with the unexpected exception that students of French speaking origin are more likely to move from BASEL to BERN than those of Swiss German origin.

7.1.1.2.2. Transitions between French Speaking Universities

As the discussion of cumulative inertia has shown (see Fig. 20), the students of medicine move out of NEUCHATEL around the turn of the 3rd semester. Fig. 24 now indicates that it is to GENEVE and LAUSANNE that they move. This figure also sheds some light on the previously unexplained sudden spurt of moves away from NEUCHATEL around the time of the 11th semester: At that time, several natural scientists transfer to GENEVE for whatever reasons.

Figure 24

The Determinants of Transitions between French Speaking Universities

	Variables	Index	standardized Regression Coefficient
ETHL-LAUSANNE	Sex	100	.29939
	Work Load	25	.07631
	Natural Sci I	25	.07472
	Semester	21	-.06414
LAUSANNE-GENEVE	Medium	100	.14739
	Duration Stat	88	-.12976
	XN Low	48	.07034
	Cantonal Cert	42	.06172
	Medicine	42	.06171
	Natural Sci II	26	.03842
NEUCHATEL-GENEVE	Medicine	100	.13782
	XN Medium	66	.09061
	Natural Sci II	50	.06848
NEUCHATEL-LAUSANNE	Medicine	100	.35075
	Low	36	.12573
	Age	19	-.06818

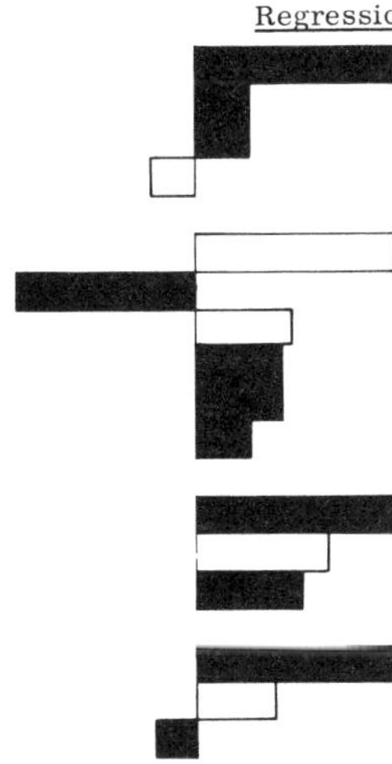

With respect to the transition LAUSANNE-GENEVE, this figure shows again some indication for the presence of cumulative inertia. Also, as in the case of NEUCHATEL, it is the medical students who move to GENEVE, mostly around the 7th semester (impact of Medium and XN Low).

Curiously enough, the main determinants of the fourth relevant transition, the one between the two institutions at Lausanne, is sex: This is a result of the fact that the only female student of the cohort 1965 enrolled at the ETHL left that university, too.

Admission certificates are, once more, unimportant. The only entry, moreover, has an effect contrary to expectations: Students holding cantonal certificates are more likely to move from LAUSANNE to GENEVE than holders of federal certificates.

7.1.1.2.3. Transitions between Universities of the two Language Groups

There are 7 transitions in this category, as presented by Fig. 25. 5.1 predictors are selected on the average per university and each predictor is entered an average number of two times. The resulting picture is moderately complex. Semester characteristics are underrepresented in the total set of predictors (P = .029), but not among the first ranks. The most important determinants are listed in Table 58:

Table 58

The most Important Determinants of Transitions between Universities of the two Language Groups

Rank	Characteristic	Number of Entries	Index Points
1	Origin	5	386
2	Duration Status	3	255
3	Humanities	4	244
4	XN Drop	3	197
5	Medicine	2	192
6	Law	4	147

As expected, origin is most prominent, especially in the direction from French speaking universities to German speaking ones: Students with Swiss German origin make these transitions more frequently. On the other hand, students with French speaking origin are more likely to move from BERN to LAUSANNE.

Figure 25

The Determinants of Transitions between Universities of the two Language Groups

	Variables	Index	standardized Regression Coefficient
BERN-LAUSANNE	Medicine	100	.05398
	XN Drop	69	.03739
	Origin	62	.03363
ZUERICH-GENEVE	Humanities	100	.05071
	Law	58	.02957
	Sex	51	.02585
ZUERICH-LAUSANNE	XN Drop	100	.05558
	Humanities	88	.04870
	XN Low	55	-.03081
	Work Load	39	-.02162
GENEVE-BASEL	Duration Stat	100	-.13192
	Log Medium	94	.12438
	Origin	64	-.08441
	Low	46	.06046
	XN Low	36	.04742
GENEVE-ZUERICH	Duration Stat	100	-.19486
	Origin	60	-.11783
	Medium	49	.09538
	Low	33	.06384
	High	31	.05989
	Humanities	29	.05694
	Law	23	.04410
	Social Sci	17	-.03285
LAUSANNE-BERN	Origin	100	-.09913
	Medicine	92	.09082
	Duration Stat	55	-.05501
	Low	44	.04390

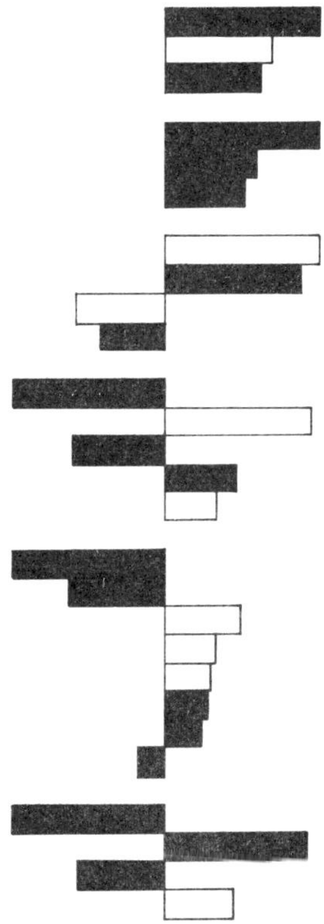

	Variables	Index	standardized Regression Coefficient
LAUSANNE-ZUERICH	Origin	100	-.22354
	Teacher's C.	67	.15087
	Econ/Bus	43	-.09545
	XN Drop	28	.06170
	Humanities	27	.05968
	Business Cert	24	.05275
	Social Sci	23	-.05093
	Law	20	.04468
	Entrance Exam	16	.03538

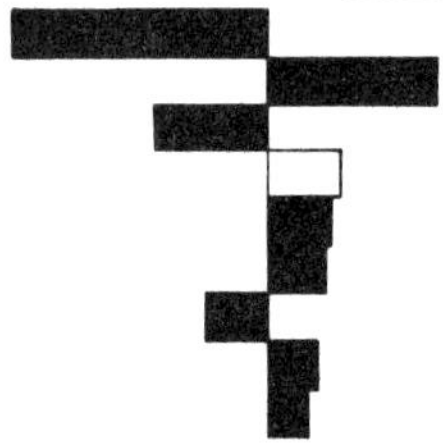

For the transitions from French to Swiss German speaking universities, duration status is almost as important as origin. Swiss German students who move back to their region of origin generally do so after only a short stay at the French speaking institution. Students returning from LAUSANNE to BERN and some students returning from GENEVE to ZUERICH originally enrolled at the French speaking institution and then moved back to their region of origin by the turn of the second semester (impact of Low).

Another reason for the effect of duration status has to do with the subject of the humanities, which is relevant for all the transitions between ZUERICH and the French speaking universities. As was pointed out in 5.2.2., the students in the humanities are required to get first hand acquaintance with the culture they are studying. This is the reason why they go abroad so often (see below). For students studying French at Swiss German universities and for those studying German at French speaking universities, however, it is not necessary to go abroad. Alternatively, they can visit a university in the respective language regions in Switzerland. This is precisely what some of the students of the universities in question seem to do.

Other subjects are also involved in the flows between the two language regions: The reciprocal moves between LAUSANNE and BERN are primarily made by medical students. Students in law move from GENEVE to BASEL and law is also an issue in the reciprocal moves between ZUERICH and GENEVE. Finally, students in the social sciences and in economics/business are unlikely to move from LAUSANNE to ZUERICH, and those in the social sciences only are similarly unlikely to move from GENEVE to ZUERICH.

The effect of admission certificates, which is restricted to LAUSANNE, is again contrary to expectation with holders of teacher's certificates moving more frequently than students with regular federal certificates.

7.1.1.2.4. Transitions between Fribourg and the Rest

Fig. 26 includes five transitions originating at FRIBOURG and two with FRIBOURG as destination. As was to be expected after the discussion in 5.2.2., medicine invariably is the dominant determinant of the leaving of FRIBOURG. Students in medicine have to leave this university after the second preparatory exam, which is around the 6th semester (see Fig. 20). The importance of XN Low reflects the peak of the movement out of FRIBOURG in the 6th semester. Except for XN Low, no other semester characteristic has any impact and in fact both of the summary proportions for semester characteristics are significantly underrepresented with P-values of .037 and .007 respectively. It is the factor medicine that is all powerful.

With respect to the movement to FRIBOURG, origin takes first place, indicating that French speaking students are more likely to transfer to FRIBOURG from BASEL and ZUERICH than those with Swiss German origin. Students of the humanities come from the former university and some students of the social sciences come from the latter.

Origin also plays a role in the determination of the destination of the medical students leaving FRIBOURG, although it has been selected just once, for the transition from FRIBOURG to ZUERICH. For three of the four remaining destinations, it would have been the next variable to be included by the stepwise procedure, if the F-criterion had been lowered. Only for the destination of

Figure 26

The Determinants of Transitions between Fribourg and the Rest

	Variables	Index	standardized Regression Coefficient
FRIBOURG-BASEL	Medicine	100	.11226
	XN Low	46	.05188
FRIBOURG-BERN	Medicine	100	.12220
	Hist Sci	46	.05594
	Cantonal Cert	34	.04169
FRIBOURG-GENEVE	Medicine	100	.10242
	XN Low	38	.03851
FRIBOURG-LAUSANNE	Medicine	100	.09385
	XN Low	46	.04302
FRIBOURG-ZUERICH	Medicine	100	.17766
	XN Low	48	.08605
	Origin	29	.05152
BASEL-FRIBOURG	Origin	100	.05961
	Humanities	51	.03022
ZUERICH-FRIBOURG	Origin	100	.05099
	Social Sci	59	.02983
	Sex	53	.02689

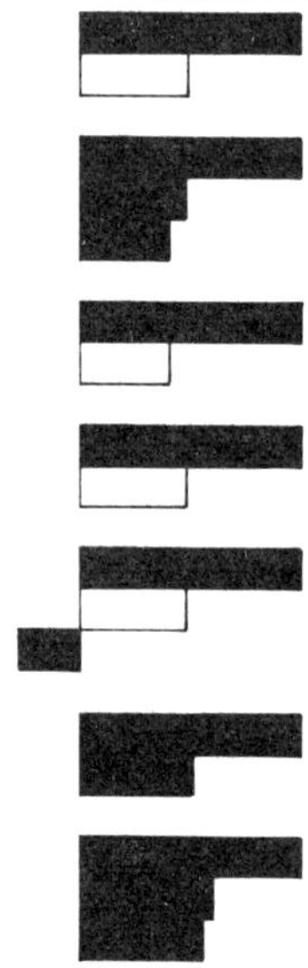

BERN it is irrelevant. In general, medical students of Swiss German origin are likely to leave FRIBOURG for Swiss German universities and those of French speaking origin are likely to go to either GENEVE or LAUSANNE. Fig. 2 above shows that the flows to the Swiss German universities are much more substantial than the ones to the two French speaking ones.

<u>Duration status</u> is no issue in the present context, because these moves are governed by structural constraints. <u>Admission certificates</u> are insignificant

once more and the only one selected has again an unexpected effect: Students holding cantonal certificates move particularly often from FRIBOURG to BERN.

In general, the hypothesis that certain types of admission certificates restrict the freedom of movement more severely than others could not be substantiated by this discussion of the flow between universities. On the other hand, it could be shown that the phenomenon of cumulative inertia is operating, if the flows are not restricted by structural considerations. The importance of the socio-economic category of origin for the flows between the two language regions could also be identified. Other socio-economic characteristics have not been shown to be significant; SES in particular has not been selected once, although it was hypothesized to have an impact on the basis of previous results.

The subject characteristic of medicine together with duration status and origin turn out to be the most important determinants of inter-university transitions.

7.1.1.3. Determinants of the Flow to Foreign Universities

If it was the subject of medicine that dominated the flows between Swiss universities, it is the humanities that dominate the transitions to the foreign universities. As pointed out, students in the humanities have to go abroad as part of their regular course of study. Fig. 27 gives the results in detail. Apart from the humanities, which is the first ranking determinant for all the universities except for the ETHZ, where it is not offered, there is no uniform pattern of determination. Each predictor is only entered an average number of 1.8 times.

There are 3.6 predictors per university on the average. The semester characteristics, however, are clearly underrepresented. Their proportion among the predictors is significantly smaller than expected (P = .000) and they never rank first (P = .007).

Table 59 gives the short list of the most important determinants:

Figure 27

The Determinants of the Flow to Foreign Universities

	Variables	Index	standardized Regression Coefficient
BASEL	Humanities	100	.05223
	Hist Sci	82	.04319
BERN	Humanities	100	.11877
	XN Low	89	.10592
	High	64	.07578
	Age	47	-.05611
	Hist Sci	26	.03086
	Teacher's Cert	16	-.01853
ETHZ	Sex	100	.04706
	Medicine	89	.03934
	Duration Stat	73	.03437
	Econ/Bus	50	.02375
FRIBOURG	Humanities	100	.05336
	Cantonal Cert	92	.04885
	Origin	85	-.04519
GENEVE	Humanities	100	.07887
	Origin	65	-.05132
	Business Cert	65	.05131
	Sex	44	.03471
LAUSANNE	Humanities	100	.06646
	Origin	70	-.04621
ZUERICH	Humanities	100	.10971
	XN Low	42	.04629
	Social Sci	30	.03313
	Hist Sci	30	.03277

Table 59

The most Important Determinants of the Flow to Foreign Universities

Rank	Characteristic	Number of Entries	Index Points
1	Humanities	6	600
2	Origin	3	220
3	Sex	2	144
4	Historical Sci	3	138
5	XN Low	2	131

The second most important characteristic, as shown, is origin, which is selected for the two French speaking universities GENEVE and LAUSANNE and for FRIBOURG. At these institutions, students of Swiss German origin are more likely to go abroad than French speaking ones.

Sex is the main determinant of flow abroad from the ETHZ, where it is closely followed by medicine. Most likely there is an interaction between these two characteristics, because the department of pharmacy, which makes up the group of subjects of medicine at the ETHZ, is the only one with a majority of female students at this institution (6).

In addition to the female pharmacists, students in economics/business at the ETHZ are particularly likely to go abroad.

At the Swiss German universities, not only students of the humanities, but also some students in the historical sciences go abroad, maybe because, as a subsidiary subject, they take some sort of language courses.

---

6) According to the Swiss Students Statistics (ESTA 1975), the proportion of female students in the different departments at the ETHZ in the WS 73/74 were as follows:
- pharmacy 53.2%
- gymnastics and sports 44.6%
- natural sciences II 15.6%
- architecture 13.5%
- agriculture 8.9%

The remaining 6 departments had proportions of less than 5% female students.

Duration status, admission certificates, socio-economic and subject characteristics other than the ones mentioned are generally insignificant for the moves to foreign universities. The determinants are primarily structural in nature, although sex and origin play a non-negligible role.

7.1.1.4. Determinants of Interruptions

For all the ten universities, the determinants of interruptions have been estimated, because there are a lot of interruptions at all of them (see Tables 8/9). Fig. 28 presents the results. Only 2.5 determinants are selected per university, in two instances, the selection procedure already stopped after the first step (BASEL and FRIBOURG). The patterns at the different universities are quite unique, the average item having been selected just 1.4 times. The proportion of semester characteristics is somewhat underrepresented among the set of predictors (P = .069), but semester characteristics rank first as often as expected.

The list of the eminent determinants is headed by age:

Table 60

The most Important Determinants of Interruptions

Rank	Characteristics	Number of Entries	Index Points
1	Age	4	367
2	Log of Semester	3	255
3	Work Load	3	204
4	Log Medium	2	200
5	Architecture	2	198
6	Low	2	168
7	SES	2	147
8	Natural Sci II	2	114
9	Log High	1	100
10	Social Sci	1	100

The younger someone is, the more likely he is to interrupt his studies at the HSG, the ETHZ and at the Swiss German universities of BERN and ZUERICH. This result is explained by the fact that younger students often have to inter-

Figure 28

The Determinants of Interruptions

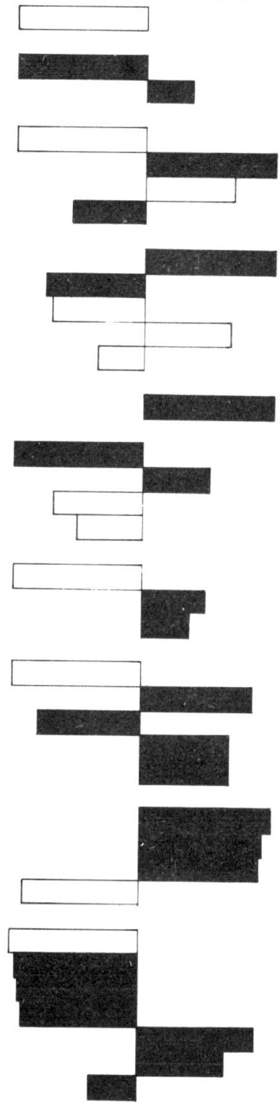

	Variables	Index	standardized Regression Coefficient
BASEL	Log High	100	-.05308
BERN	Age	100	-.08922
	Econ/Bus	32	.02836
ETHL	Low	100	-.14215
	Architecture	98	.13998
	XN Low	68	.09757
	SES	56	-.07969
ETHZ	Architecture	100	.09172
	Age	74	-.06785
	Lin Low	69	-.06328
	Low	68	.06278
	Natural Sci I	34	-.03956
FRIBOURG	Teacher's Cert	100	.03956
HSG	Age	100	-.13767
	Work Load	70	.09664
	Log Sem	66	-.09125
	XN Drop	49	-.06781
GENEVE	Log Medium	100	-.09039
	Hist Sci	49	.04420
	Work Load	36	.03245
LAUSANNE	Log Medium	100	-.05438
	Entrance Exam	85	.04637
	Natural Sci II	76	-.04125
	Origin	67	.03638
	Committee Cert	67	.03628
NEUCHATEL	Social Sci	100	.06622
	Work Load	98	.06488
	SES	91	.06026
	Log Sem	89	-.05883
ZUERICH	Log Sem	100	-.04708
	Medicine	95	-.04495
	Age	93	-.04399
	Sex	91	-.04270
	Humanities	90	.04249
	Law	66	.03118
	Natural Sci II	38	-.01794

rupt their studies to take the initial military training in the recruit school, while older students usually have completed this training before their entrance into the university system (7).

Although they do not have to do military service, female students interrupt their studies less frequently than their male counterparts only at ZUERICH. They apparently have other reasons for interruptions.

Other socio-economic characteristics are influential, too. Students with a high work load during the semester interrupt their studies more frequently at the HSG, at GENEVE and at NEUCHATEL. This is, I conjecture, an indication of a negative Matthew effect: Those who have to work a lot to make their living during the semester are also forced to interrupt their studies for financial reasons from time to time. SES also ranks among the important variables. Its impact, however, is ambiguous: At the ETHL, the lower classes interrupt more often than the upper ones; at NEUCHATEL it is just the other way around.

The semester characteristics rank second in importance after the socio-economic ones. The interruptions are generally decreasing in the course of the studies. Most often they decrease at a logarithmic rate: Log High (BASEL), Log Medium (LAUSANNE, GENEVE) and Log of Semester (ZUERICH, HSG, NEUCHATEL) are the indicators. At the ETHZ, the decrease follows a linear trend (Lin Low) which succeeds a particular spurt in interruptions around the turn of the second semester (impact of Low).

---

7) The most important reasons for interrupting the studies given by the cohort 1965 are (RIES/KRIESI 1974c):

- military service	35.9%
- practical work	27.3%
- financial reasons	9.3%
- accidents	7.2%
- travels abroad	6.1%
- family problems	5.1%

(see also p. 44 above)

At both of the technical schools, architects interrupt more often than other students. This is hardly news, but only replicates the results provided in Tables 10/11. With respect to the subject characteristics there is otherwise no distinct pattern. The subjects with extreme percentages of interruptions vary from one university to the next:

particularly few interruptions	particularly many interruptions
- natural sciences I (ETHZ)	- economics/business (BERN)
- natural sciences II (LAUSANNE, ZUERICH)	- historical sciences (GENEVE)
- medicine (ZUERICH)	- social sciences (NEUCHATEL)
	- humanities (ZUERICH)
	- law (ZUERICH)

The case of the social sciences at NEUCHATEL is notable, because it is an exception from the overall picture given in Table 10, which shows the social scientists to interrupt comparatively seldom.

While duration status is no issue in this context, admission certificates become relevant at the universities of FRIBOURG and LAUSANNE. At FRIBOURG, the holders of teacher's certificates interrupt particularly often, at LAUSANNE the same is true for the students with an entrance exam or a committee type certificate.

In general, it seems that, with respect to the determination of interruptions, structural characteristics of the university system are heavily supplemented by structural constraints from outside of this system. Students have to do their military duty or to replenish their financial resources etc. Socio-economic characteristics function as indicators of these outside constraints.

7.1.1.5. Determinants of Leaving the System

Each one of the four levels will be considered separately.

7.1.1.5.1. Dropping out

Fig. 29 presents the determinants of dropping out at 9 of the 10 Swiss universities; those for the ETHL could not be estimated, because of too few cases. The picture is quite complex again: 4.2 predictors are selected on the average per university, each one of them being entered an average number of 1.6 times. Semester characteristics are significantly underrepresented in the set selected (P = .014), but not among the first ranks. Among the ten most important determinants we find only two semester characteristics. As shown by Table 61, socio-economic and subject characteristics are the main determinants of dropping out.

Table 61

The most Important Determinants of Dropping out

Rank	Characteristic	Number of Entries	Index Points
1	Age	5	360
2	Medicine	3	207
3	XN Low	2	185
4	Law	3	174
5	Sex	2	165
6	Social Sci	2	146
7	Natural Sci II	3	140
8	Econ/Business	2	130
9	Work Load	2	127
10	XN Drop	2	119

Age is the most important single determinant, just as it was in the case of interruptions. It has an impact at the three Swiss German universities (BASEL, BERN, ZUERICH) and at the two French speaking universities of GENEVE and NEUCHATEL. Its impact is the same throughout: Older students are more likely to drop out. The reason for this result has probably to be sought in the very special situation of the older students, which is in many respects more difficult than the one of their younger colleagues, not the least of which being the financial one (see RIES/KRIESI 1974e). It is still worth mentioning, how-

Figure 29

The Determinants of Dropping out

	Variables	Index	standardized Regression Coefficient
BASEL	Log Low	100	-.07102
	Social Sci	96	.06805
	Sex	95	.06713
	Age	77	.05487
	Medicine	64	-.04565
	Law	57	-.04060
	Teacher's Cert	50	-.03585
BERN	Age	100	.06509
	XN Drop	99	.06418
	Social Sci	50	.03225
ETHZ	Medium	100	.08993
	Log Medium	89	-.07983
	Committee Cert	43	.03876
	Entrance Exam	33	.02970
	Econ/Bus	30	.02685
	Natural Sci II	26	-.02301
FRIBOURG	Humanities	100	.04922
	XN Low	85	.04191
HSG	Work Load	100	.09663
	Semester	63	-.06127
GENEVE	Econ/Bus	100	.07256
	High	99	-.07219
	Age	96	.06992
	Natural Sci I	77	.05555
	Hist Sci	53	.03868
	XN Drop	20	.01424
LAUSANNE	Medicine	100	-.07295
	Natural Sci II	71	-.05173
	Business Cert	60	.04348
	Law	57	-.04194
NEUCHATEL	Age	100	.10420
	XN Medium	78	-.08140
	SES	60	-.06298
ZUERICH	XN Low	100	-.06674
	Age	87	.05813
	Sex	70	.04646
	Law	60	-.04036
	Medicine	43	-.02888
	Natural Sci II	43	-.02858
	Cantonal Cert	41	.02767
	Work Load	27	.01826

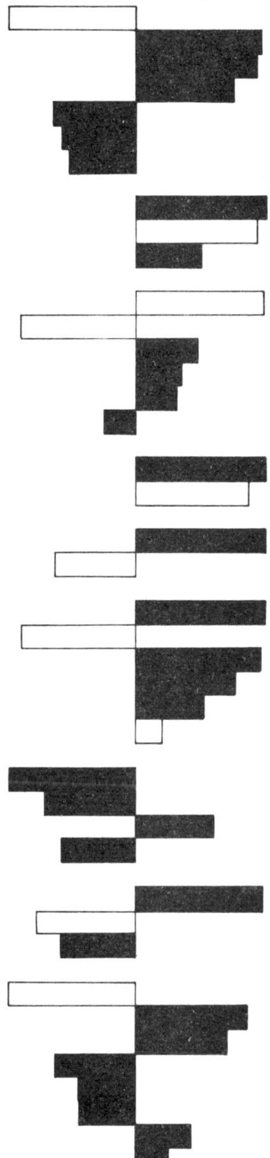

ever, that age is not an issue in dropping out at the other four universities, i.e. not at the ETHZ, the HSG, not at LAUSANNE and not at FRIBOURG, which has by far the oldest student body (see RIES/KRIESI 1974b).

The other two socio-economic characteristics of importance are sex and work load. Female students drop out significantly more often than males at BASEL and at ZUERICH, but not at the other universities. Students with a large work load are particularly likely to drop out again at ZUERICH and, above all, at the HSG.

Half of the ten characteristics listed in Table 61 are subject characteristics, indicating that a lot, if not most of the dropping out is determined by the particular context at the university the student is studying at. Both, at LAUSANNE and at ZUERICH, students in medicine, law and the natural sciences II drop out less frequently than students in other subjects. The same applies to students in medicine and law at BASEL, and to students in the natural sciences II at the ETHZ. The natural scientists II at BASEL also drop out comparatively infrequently; in the case of a lower partial F criterion, they would have been selected next. These four universities are, then, responsible for the low overall level of dropping out in law, medicine and the natural sciences II shown in the Tables 15/16. On the other hand, social science students at BASEL and BERN and economics/business students at GENEVE and the ETHZ drop out particularly often. At GENEVE, students in the natural sciences I and in the historical sciences also drop out rather often, making, together with those in economics/business, for the highest overall drop out rate in the country (see Tables 12/13).

At FRIBOURG, finally, it is the humanities students, who are particularly prone to drop out.

The structural interpretation of the impact of the subject characteristics is supported by these results which show that students in a particular subject do not uniformly drop out more or less frequently than students in some other subject, but that the impact of the same characteristic varies from one university to the next. This finding also corroborates RIES's (1975) proposition that the structural context of a particular subject at a particular university is the most important determinant of dropping out.

In his conclusions, RIES plays down the role of admission certificates and in the light of the present results he is correct in doing so. Admission certificates enter in five instances, but never very prominently. The committes certificate, which attracted so much interest, is associated with dropping out only at the ETHZ. Those who take the entrance exam at the ETHZ are, however, also more likely to drop out than holders of the regular federal certificate. Students with cantonal certificates at ZUERICH and with business certificates at LAUSANNE drop out comparatively often, those with teacher's certificates at BASEL comparatively seldom.

The semester characteristics indicate, when most of the dropping out at a particular university occurs. In the case of the ETHZ, they reflect the peculiarity mentioned in 3.1.4.2. that students in engineering can complete the coursework without having passed the prelims and leave after 8 semesters as 'engineers' (impact of Medium and Log Medium).

7.1.1.5.2. Graduating at the Low Level

The determination of graduation at the low level is as complex as the one of staying at the university of origin, as is documented by Fig. 30. On the average, 6.8 determinants are selected per university. Each predictor is entered an average number of three times, which is the highest average of all types of transitions, but still quite low. Semester characteristics are even more heavily underrepresented in the set of the predictors selected than in the case of dropping out ($P = .001$), but again their share among first places does not significantly deviate from the expected.

Figure 30

The Determinants of Graduating at the Low Level

	Variables	Index	standarized Regression Coefficients
BASEL	XN Low	100	.12908
	XN Drop	62	-.07989
	Humanities	50	.06461
	Sex	49	.06302
	Business Cert	41	.05317
	Natural Sci II	34	.04444
	Cantonal Cert	32	.04183
	Work Load	26	-.03339
BERN	Humanities	100	.14251
	Teacher's Cert	84	.12026
	XN Low	47	.06660
	Age	46	.06490
	Natural Sci I	41	.05875
	Work Load	41	-.05815
	XN Drop	32	.04589
	Business Cert	26	.03686
	Sex	23	.03243
	Econ/Bus	20	-.02845
ETHZ	Teacher's Cert	100	.28306
	Sex	22	.06090
	Work Load	9	-.02438
	Architecture	8	-.02402
	Low	8	.02369
FRIBOURG	Humanities	100	.13354
	Age	80	.10647
	Social Sci	72	.09559
	XN Drop	63	.08440
	Sex	58	.07753
	Teacher's Cert	37	.04963

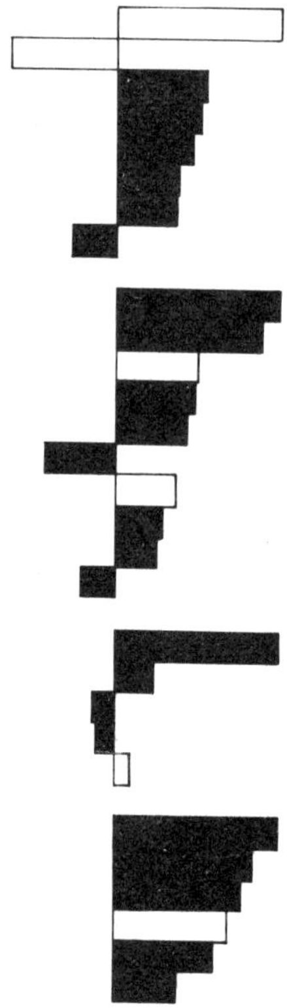

	Variables	Index	standardized Regression Coefficients
GENEVE	Social Sci	100	.15391
	Sex	47	.07289
	SES	29	-.04416
	Low	26	.04014
LAUSANNE	Teacher's Cert	100	.26983
	Origin	25	-.06836
	Humanities	22	.06059
	SQ High	19	.05051
	XN Low	18	.04964
	Work Load	15	.03995
NEUCHATEL	XN Drop	100	.10465
	Teacher's Cert	87	.09121
	Age	84	.08821
	XN Low	81	.08453
	Econ/Bus	80	-.08364
	Natural Sci II	58	-.06096
ZUERICH	Age	100	.10565
	XN Low	69	.07252
	Duration Status	67	-.07114
	Teacher's Cert	65	.06865
	Humanities	52	.05499
	Work Load	46	-.04868
	Social Sci	40	.04265
	Sex	37	.03949
	Natural Sci II	37	.03889
	Low	31	-.03314

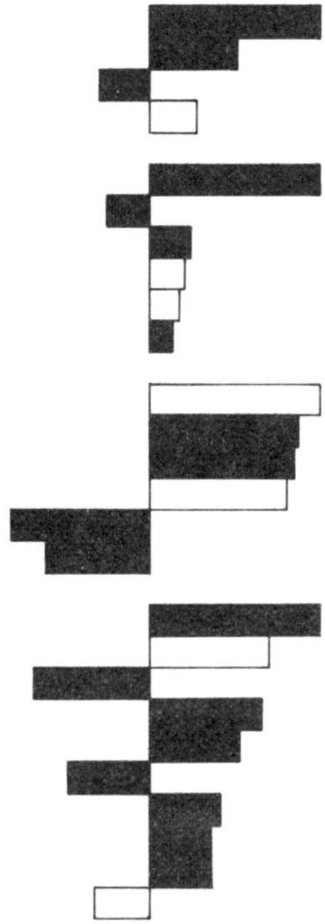

Graduation at a low level provides the only instance, where admission certificates prove to be very important: Teacher's certificates head the list in Table 62:

Table 62

The most Important Determinants of Graduation at the Low Level

Rank	Characteristic	Number of Entries	Index Points
1	Teacher's Cert.	6	473
2	Humanities	5	324
3	XN Low	5	315
4	Age	4	310
5	XN Drop	4	257
6	Sex	6	226
7	Social Sci	3	212
8	Work Load	5	137
9	Natural Sci II	3	129
10	Econ/Business	2	100

To understand the overall importance of the teacher's certificates in this context, recall that lower level degrees primarily consist of degrees for teaching at the lower level of the secondary schools, for teaching sports and for therapeutic education. As has been shown in previous publications, a majority of the students entering the university with a teacher's certificate wants to stay in the teaching profession, mostly at the lower level of the secondary school (see KRIESI/RIES 1974). From the present results, we see that they actually realize their intentions: At the universities of BERN, ZUERICH, LAUSANNE, NEUCHATEL, FRIBOURG and at the ETHZ, students holding teacher's certificates are particularly likely to get lower level degrees.

At the universities of FRIBOURG, BASEL, BERN, LAUSANNE and ZUERICH, these degrees are primarily obtained in the humanities. At the university of GENEVE, it is in the field of the social sciences that students are particularly prone to graduate at a low level. This is a result of the fact that all future lower level teachers at GENEVE are enrolled at the 'Ecole de Psychologie et de Science de l'Education', which has been classified among the social sciences. At FRIBOURG and ZUERICH, on the other hand, the branch of therapeutic education is responsible for the social sciences having particularly many low level graudates. At ZUERICH and BASEL, students in the natural sciences II, at BERN, those in the natural sciences I are also overrepresented among the low level graduates. At NEUCHATEL, it is just the other way around with respect to the former. Economics/Business students are unlikely to graduate at low levels in BERN and NEUCHATEL. At the ETHZ, finally, lower level graduates concentrate in the field of gymnastics and sports. This does not show, because this field is used as base category in the regressions.

Three socio-economic characteristics are important. As in the case of dropping out, they are age, sex and work load. Age specifies the impact of the teacher's certificates: Earlier analysis has shown that it is primarily the older students having a teacher's certificate who aspire to lower level degrees (KRIESI/RIES 1974). Although no interactive term is provided for this specification, it is still significant that at four universities (BERN, ZUERICH, NEUCHATEL and FRIBOURG) both characteristics are selected together. Thus, older students and among them primarily the ones holding teacher's certificates graduate particularly often at the low level.

Except at the two French speaking universities of LAUSANNE and NEUCHATEL, female students are more likely to graduate at the low level than males, as was to be expected.

The higher the work load, the less likely is a graduation at the low level at the Swiss universities of the German speaking part, including the ETHZ; at LAUSANNE, however, it is just the other way around, which makes this result difficult to interpret.

The semester characteristics, it has been pointed out, are underrepresented. Only XN Drop and XN Low are on the list of the ten most important characteristics, indicating that low level degrees are most frequently acquired around the 5th and/or around the 7th semester. Graduation at a low level is the one type of transition that is least governed by structural aspects of the university system. The main impact of the teacher's certificate indicates structural influences of institutions at lower levels in the educational system. These structural influences are probably mediated by interpersonal contacts (see CAMPBELL/ALEXANDER 1965).

Thus, students with teacher's certificates have taken advice from their teachers and peers in choosing their subject: A lot of them have also decided to go on studying only after a prolongued period of teaching on their own (see KRIESI/RIES 1974).

7.1.1.5.3. Graduating at the Medium Level

Fig. 31 presents again a rather complicated picture. For graduation at a medium level, 7.2 characteristics have been selected on the average, and each characteristic is only selected an average number of 2.5 times. In contrast to graduation at the low level, semester characteristics are all important now. Although they are not overrepresented with respect to the selected set of predictors, they take first place at all the ten universities, which is a highly unlikely event (P = .000). Their impact is reflected in the list of Table 63:

Table 63

The most Important Determinants of Graduating at the Medium Level

Rank	Characteristic	Number of Entries	Index Points
1	SQ Semester	3	279
2	XN Medium	4	271
3	Lin Medium	2	200
4	Medium	2	200
5	Natural Sci II	6	177
6	SQ High	3	166
7	XN Low	6	161
8	Law	4	147
9	Sex	5	135
10	High	2	130

Figure 31

The Determinants of Graduating at the Medium Level

	Variables	Index	standarized Regression Coeff.
BASEL	SQ High	100	.11181
	SQ Semester	79	.08819
	XN Drop	65	-.07312
	Natural Sci II	55	-.06200
	Humanities	51	-.5702
	XN Low	37	.04151
	Sex	34	.03851
	Law	34	.03747
	Hist Sci	29	-.03199
	Work Load	27	-.02979
BERN	Lin Medium	100	.29603
	XN Medium	20	-.05968
	Natural Sci II	18	-.05336
	Natural Sci I	16	-.04877
	Hist Sci	15	-.04346
	Social Sci	12	-.03678
	Humanities	9	-.02670
	Business Cert	8	.02388
ETHL	XN Medium	100	.98969
	High	74	-.73636
	Log High	32	.31913
	XN Low	19	-.18501
	Natural Sci II	11	-.10614
	Architecture	8	-.07912
	Committee Cert	6	-.06345
ETHZ	XN Medium	100	.38950
	Natural Sci II	24	-.09485
	XN Low	13	-.05200
	Social Sci	10	.03898
	Engineering	9	.03593
	Natural Sci I	8	-.03287
	Work Load	5	-.02041
HSG	Log Low	100	.46552
	High	56	-.25946
	Lin High	21	-.09855

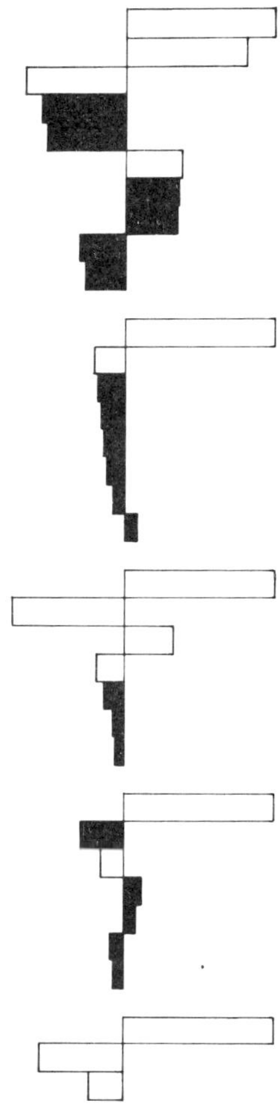

	Variables	Index	standardized Regression Coefficient
FRIBOURG	Medium	100	.20118
	Law	64	.12902
	XN Low	40	.08110
	Business Cert	35	.07025
	Sex	26	.05201
	SES	18	-.03639
GENEVE	SQ Semester	100	.17064
	Age	62	.10553
	XN Medium	51	-.08736
	XN Drop	50	-.08485
	Natural Sci II	44	-.07493
	Work	29	-.04937
	XN Low	27	.04660
	Sex	27	.04579
	SES	26	-.04445
	Econ/Bus	25	.04347
LAUSANNE	SQ Semester	100	.25121
	Medicine	39	-.09786
	Business Cert	31	.07776
	Law	26	-.06463
	XN Low	25	.06278
	Sex	25	.06265
	Natural Sci II	25	-.06223
	SQ High	23	.05691
	Humanities	17	-.04245
	Social Sci	12	.02891
NEUCHATEL	Medium	100	.27567
	Econ/Bus	39	-.10656
ZUERICH	Lin Medium	100	.14894
	SQ High	43	.06469
	Medicine	39	.05861
	Econ/Bus	35	.05206
	Duration Status	25	.03764
	Law	23	.03429
	Sex	23	.03382
	Natural Sci I	20	.03042
	Work Load	15	-.02217
	Entrance Exam	14	.02040

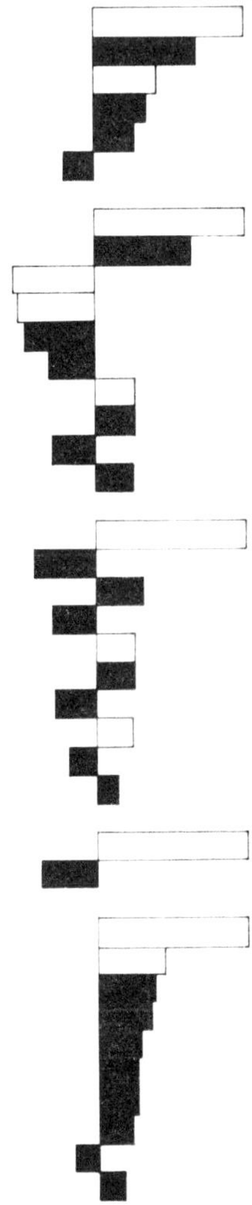

7 out of the ten most important determinants are semester characteristics: Medium level graduation is primarily a question of the fulfilment of given requirements which it takes a minimum amount of time to do. This duration aspect is represented by the importance of the semester characteristics.

The kind of semester characteristics selected varies from university to university, according to the courses offered and to the extent to which these courses are well organized or not (see 3.2.2.).

The detailed results reported in Fig. 31 replicate and specify the findings of Table 14 and Fig. 8. Thus, they indicate, for example, for the HSG, the courses of which are highly structured, that medium level graduation starts very early (impact of Log Low) and is concentrated before and around the 9th semester (impact of High and Lin High). Similarly, they show that at the technical schools as well as at FRIBOURG and NEUCHATEL, students are also likely to graduate comparatively early at the medium level, whereas at the three Swiss German universities, medium level graduation sets in comparatively late and keeps increasing over the semesters.

The distributive aspect of leaving not only affects the kind of semester characteristics selected, it is also expressed by the subject characteristics: Among the latter, the natural sciences II are most prominent: At BASEL, BERN, at the two technical schools as well as at LAUSANNE and at GENEVE, students in this group of subjects are less likely to graduate at a medium level than the average student. As was already noted in the context of the discussion of staying, students in the natural sciences II very often work for high degrees. Students in law at BASEL, FRIBOURG and ZUERICH, on the other hand, are particularly likely to get a medium level degree, although the share of high level degrees in law for the whole cohort is second only to the natural sciences II. In accordance with this overall trend for high degrees in law, students in this subject at LAUSANNE are unlikely to graduate at the medium level.

The relative likelihood of medium level graduation varies also within other subjects from one university to the other:

students in	are particularly <u>likely</u> to graduate at the medium level	are particularly <u>unlikely</u> to graduate at the medium level
- medicine	ZUERICH	LAUSANNE
- economics/bus.	GENEVE, ZUERICH	NEUCHATEL
- social sci	ETHZ, LAUSANNE	BERN
- humanities		BASEL, BERN, LAUSANNE
- historical sci		BASEL
- natural sci I	ZUERICH	BERN, ETHZ
- engineering	ETHZ	
- architecture		ETHL

This bewildering variety again stresses the significance of the particular situation at a specific subject within a given university.

Semester characteristics and subject characteristics make for an <u>almost complete structural</u> determination of this type of transition. <u>Sex</u> is the only non-structural determinant on the list in Table 63. Although its impact is less important at this level than at the previous one, its direction is the same: At BASEL, ZUERICH, FRIBOURG, LAUSANNE and GENEVE, it is the female students who are particularly likely to graduate at the medium level.

Other socio-economic characteristics are quite unimportant, except for <u>age</u> at GENEVE: Older students are graduating more often at a medium level than younger ones. <u>Work Load</u>, although entered four times (BASEL, ZUERICH, GENEVE, ETHZ), has only a weak, generally negative impact.

<u>Duration status</u> is of no significance and the same applies to <u>admission certificates</u>, which have proved to be so prominent in the case of low level graduation. Only business certificates have a moderate or small positive effect on the medium level at BERN, FRIBOURG and LAUSANNE.

7.1.1.5.4. <u>Graduating at the High Level</u>

In trying to get at the determinants of high level graduation, the limitations of the data become apparent: In the first 15 semesters covered, only a few students have completed work for the high level degree. At the French speaking universities, they are so few that it is not possible to estimate any determinants at all, although, eventually, sizable portions wind up getting a degree of the high level at these universities, too (see Tables 12/13). For the remaining five universities, determinants have been estimated and the results are given in Fig. 32 on this page. They are, however, to be taken with a grain of salt. If data had been available on all the semesters, a much more differentiated picture would probably have emerged.

Figure 32

The Determinants of Graduating at the High Level

	Variables	Index	standardized Regression Coefficient
BASEL	SQ High	100	.28579
	Work Load	10	-.02804
BERN	SQ High	100	.18496
	Teacher's Cert	18	.03259
	Medicine	17	.03194
ETHZ	SQ High	100	.19223
	Natural Sci II	15	.02792
FRIBOURG	Lin High	100	.13353
	Duration Status	49	.06539
ZUERICH	Lin High	100	.18241
	Entrance Exam	11	.01977

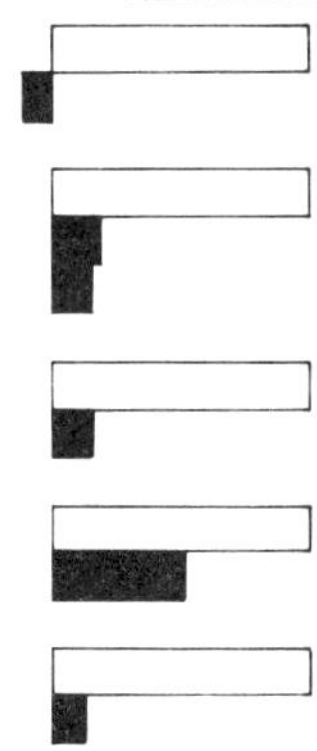

Fig. 32 shows the prime importance of semester characteristics for high level graduation. As in the case of the medium level, they take first rank in every case (P = .013), but in contrast to medium level graduation, this leading semester characteristic is virtually the only determinant in each one of the five cases. All the individual characteristics seem to be negligible in the face of the structural requirements for a high level degree. - As indicated, this result is most likely due to the few cases involved.

Table 64 summarizes the most important characteristics determining the leaving of the system at all four levels.

Table 64

The most Important Determinants of Leaving the System

Rank	Characteristic	Number of Entries	Index Points
1	Age	9	732
2	XN Low	13	661
3	Teacher's Cert.	8	541
4	Sex	13	526
5	Humanities	9	501
6	XN Drop	8	491
7	Natural Sci II	14	461
8	SQ High	7	450
9	Social Sci	8	392
10	XN Medium	5	349

Age turns out to be the most important determinant of leaving. The second socio-economic characteristic, sex, follows on place 4. The teacher's certificate, which is so important for low level graduation, is the only admission certificate included in the list. In addition, there are four semester characteristics among the most influential determinants of leaving, one for every level. And, finally, three subject characteristics complete the list.

7.1.1.6. Conclusion

Tables 65 and 66 summarize the findings of this section: Semester characteristics (Table 65) are of prime importance in the determination of staying at the university of origin and of graduation at the medium and high level. On the other hand, semester characteristics are underrepresented among the determinants of leaving at the two low levels and among those of the flows between universities of the two language regions. They are of no importance for the flows to

Table 65

Summary of the Summary Statistics for Semester Characteristics

Type of Transition	First Ranks		Charact. Selected	
	Proportion Semester Character.	P-value	Proportion Semester Character.	P-value
Staying	.7	.071	.342	.076
Swiss Germ./Swiss G	.4	ns	.333	ns
French/French	.25	ns	.313	ns
Swiss Germ./French	.143	ns	.278	.029
Fribourg/Rest	0	.007	.235	.037
Foreign University	0	.007	.120	.000
Interruption	.5	ns	.306	.069
Drop out	.33	ns	.263	.014
Low Level	.25	ns	.241	.001
Medium Level	1.0	.000	.375	ns
High Level	1.0	.013	.455	ns

Table 66

Summary of the most Important Determinants

Rank	Characteristic	Number of Entries	Index Points	Index
1	Medicine	23	1742	100
2	Humanities	24	1632	94
3	Age	21	1543	89
4	XN Low	29	1486	85
5	Sex	26	1232	71
6	Origin	16	1133	65
7	Natural Sci II	24	1036	59
8	XN Drop	15	965	55
9	Teacher's Certificate	14	808	46
10	Work Load	23	758	44
	Duration Status	11	755	43
	Social Sciences	16	753	43

foreign universities and for those between FRIBOURG and the rest.

The single most important determinant, as indicated by the sum of the index points (Table 66), is the subject of medicine. Medicine is, together with origin and duration status, the pivotal characteristic in the determination of the inter-university flows in Switzerland. In addition, it also takes a prominent place with respect to staying, which is primarily structurally determined by the joint impact of subject and semester characteristics.

The second most important determinant, the subject of humanities, accumulates most of its index points in the context of the transitions to foreign universities, although it is also the leading subject characteristic among the determinants of leaving the system.

Age and sex contribute to the determination of staying. Together with work load they also contribute to the substantial impact of socio-economic characteristics on dropping out and on low level graduation. Age is the most important determinant of leaving the system and of interrupting.

In the case of interruptions, it serves as an indicator of structural influences from outside of the university system, as does the only admission certificate of any relevance, the teacher's certificate, in the case of low level graduation. While the graduation at the low level is least determined by the structural context of the university system, this context assumes all-out importance for the graduation at the two higher levels, and it is also very important for dropping out.

The picture that results from this description is, of course, subject to several limitations. One has been stressed in discussing the determinants of the high level of leaving: The determinants selected do not cover the entire length of time the cohort 1965 has been in the system. A second and even more vital limitation stems from the changing nature of the determinant structure over time. The picture presented here is based on the transitions made by the cohort 1965. Although transition probabilities have a comparatively high stability (see Part I), their determinants are nevertheless subject to change. As an example, consider the most important determinant on the list, the subject of medicine. As a result of the impending numerus clausus in this subject, students are no longer free to choose the university they want to study at. If the university preferred by an applicant has no space left, he is transferred to another university. So far, the numerus clausus has been avoided by such measures. But it is apparent that transfers of this nature profoundly affect the flows between universities.

7.1.2. Homogenous Subpopulations: The Case of the Female Students

There is an alternative way to ascertain the relationship between social and semester attributes, on the one hand, and transition probabilities on the other: Instead of introducing independent determinants of flow, one could construct separate transition matrices for homogenous subpopulations. This procedure is, of course, highly inefficient, as SPILERMAN (1972) points out: First, one cannot control for several variables simultaneously and retain sufficient cases for estimating the transition probabilities. Moreover, it becomes impossible to attribute the heterogeneity to particular factors, if one has to compare matrices constructed for patterns of characteristics rather than for subpopulations characterized by a single attribute.

In spite of these shortcomings, the method has some advantages for descriptive purposes, because it allows a detailed description of the flows for a single variable controlled.

SPILERMAN shows how the set of regression equations can be used to construct transition matrices for subpopulations. On the basis of the synthetic population file, which has been used to find the regression equations in the present study, it is, however, possible to construct such matrices without having to rely on the regression equations: Instead of getting estimated transition matrices by using the regression equations, one can obtain observed transition matrices for homogenous subpopulations by making use of the indicator variables y_{ij} of the synthetic population file.

Recall that for a particular time period, the indicator variable y_{ij} is 1 for a person who has moved from i to j during that period, and zero otherwise. The sum of the y_{ij} over all individuals in the origin state i at the beginning of the period therefore is equal to the number of students who have moved from i to j in that period. Dividing this sum by the total number of students in the origin state at the beginning of the period gives the corresponding transition probability p_{ij} (8):

$$p_{ij} = \frac{n_{ij}}{n_{i\cdot}} = \frac{\sum_{c=1}^{n_{i\cdot}} y_{ijc}}{n_{i\cdot}}$$

In arriving at the observed transition matrices for a homogenous subpopulation, one thus follows a two-step procedure: First, the homogenous supopulation is constructed by selecting the relevant cases from the synthetic population file. Then, the transition matrices are computed in the way indicated. The two steps are accomplished by the two programs in APPENDIX B-8.

While the first one is simply an application of the SELECT IF and WRITE CASES features of SPSS, the second one is somewhat more involved, although still straightforward.

This procedure does not make use of the independent determinants, but it is still an indirect consequence of their introduction (9).

To illustrate the usefulness of matrices for homogenous subpopulations for descriptive purposes, I have computed such matrices for the female students. Sex has been shown to be one of the most important determinants of flow (see Table 64): Female students are generally less likely to stay at the university of origin

---

8) For the definition of the transition probability, see p. 27.

9) It shares the negative aspects of the independent determinants, too. Since these determinants have only been estimated for the most important transitions, the indicator variables have also been defined for these transitions only. The matrices of the subpopulations will therefore only have non-zero entries in the most important cells for the ten universities. The rows of transition probabilities pertaining to the foreign universities and to the interruption state as origins, will, moreover, always have to be the same as for the total population, since no independent determinants have been estimated for them. Finally, only the first 15 semesters are covered.

than males, because they drop out more often or leave at a lower level, or move to a foreign university more frequently. Comparing the first 15 observed type 1 transition matrices for female students with the ones for the total cohort, these results can be corroborated. For comparison purposes, the indices of dissimilarity are used again. Table 67 gives these indices by university and semester, Table 68 summarizes the results.

The enormous differences for the ETHL and the HSG are artificial artifacts, because of the very few female students enrolled at both of these institutions in the cohort 1965 (1 at the ETHL, 5 at the HSG). With the exception of FRIBOURG and the ETHZ, the rest of the discrepancies are, on the average, not as striking as we might have expected. This is, of course, also an artifact in part, because the female students are included among the total cohort (21.7% of it are female), too.

As was done in testing for the Markovian assumption, the sources of the discrepancies can be located more accurately by computing average percentage differences between the two sets of matrices for three subsections:

- the main diagonal, which corresponds to staying at the origin state
- the flow between universities
- the leaving of the system

Table 69 reports these results for the 8 universities with a sufficient number of cases of female students. Absolute and signed values are given to indicate not only the absolute amount of the differences, but also their net direction.

In comparing the absolute values of the three subsections, we note the largest average incongruencies for the leaving of the system, the smallest one for the inter-university flows. The discrepancies for staying lie in-between. In all three cases, the females at FRIBOURG deviate the most from their male counterparts, always followed by those at the ETHZ.

The signed values are generally quite smaller than the absolute ones, i.e. the discrepancies cancel to a considerable extent. This is an indication that there are by no means uniform trends in a given direction. The net result is, never-

Table 67

Indices of Dissimilarity between Observed Transition Matrices for Female Students and for the whole Cohort by Semester and University

SEMESTER	BASEL	BERN	ETHL	ETHZ	FRIB	HSG	GENEVE	LAUS	NEUCH	ZURIC
1	3.7	2.6	50.0	2.6	1.1	11.2	0.6	2.3	3.5	4.8
2	4.1	1.8	50.0	12.7	2.8	20.3	4.5	4.1	3.7	4.2
3	6.4	3.4	49.9	12.6	7.6	25.5	2.9	4.3	6.5	6.0
4	3.3	7.1	50.0	10.1	14.9	7.1	8.9	5.1	2.7	3.1
5	8.8	7.0	49.9	8.1	11.2	22.5	3.7	4.5	8.7	4.3
6	4.7	16.0	50.0	10.9	18.8	3.1	4.0	6.3	11.4	7.0
7	5.9	10.7	49.9	5.8	16.7	32.9	5.0	6.6	4.1	5.8
8	9.0	5.5	49.9	7.3	17.9	16.6	10.1	4.2	8.4	5.8
9	6.1	4.7	50.0	24.2	11.7	29.4	4.6	8.1	8.6	4.6
10	4.8	4.8	50.0	16.7	8.2	36.0	4.7	5.6	14.8	7.8
11	4.7	12.8	50.0	9.3	19.4	11.3	5.8	7.5	9.1	3.4
12	3.0	9.5	50.0	18.1	19.1	8.3	7.7	16.8	2.5	5.8
13	4.0	3.7	50.0	8.2	18.7	93.5	3.1	4.7	7.9	8.3
14	7.2	8.8	50.0	12.6	20.6	50.0	2.1	19.4	8.1	6.4
15	4.6	12.7	46.1	8.1	9.9	50.0	7.8	10.3	11.8	10.7

Table 68

Mean Indices of Dissimilarity between Observed Transition Matrices for Female Students and for the whole Cohort by University, Including Standard Deviations

University	Mean	Standard Deviation
ETHL	49.71	1.00
HSG	27.81	23.16
Fribourg	13.24	6.30
ETHZ	11.15	5.37
Neuchâtel	7.45	3.63
Bern	7.41	4.23
Lausanne	7.32	4.81
Zürich	5.90	1.98
Basel	5.35	1.85
Genève	5.03	2.62

Table 69

Percentage Differences between Observed Transition Probabilities for Female Students and for the whole Cohort at Swiss Universities[1]: Means and Standard Deviations Taken over 15 Semesters for 3 Subsections of the Transition Matrices

University	Main Diagonal				Movement between Universit.				Leaving the System			
	absolute Val.		signed val.		absolute val.		signed val.		absolute val.		signed val.	
	Mean	S.D.	Mean	S.D.	Mean	S.D.	Mean	S.D.	Mean	S.D.	Mean	S.D.
Basel	3.1	2.4	-2.6	3.0	2.3	1.4	-.1	1.3	5.3	2.3	2.2	2.6
Bern	5.4	4.8	-2.8	6.8	2.5	2.1	.7	1.6	7.0	3.9	1.5	6.3
ETHZ	8.1	4.7	-2.0	9.4	2.6	2.8	1.1	2.9	11.7	5.8	.8	10.8
Fribourg	9.0	6.2	-6.6	8.9	5.6	4.2	1.2	5.0	11.9	6.8	2.5	9.0
Genève	3.8	3.1	-1.6	4.7	2.1	1.3	.2	1.7	4.2	2.8	1.0	4.1
Lausanne	5.3	4.8	-3.7	6.2	2.4	1.4	-1.3	1.6	6.9	5.5	3.1	5.5
Neuchâtel	3.8	2.2	.6	4.4	2.2	2.5	-1.7	1.8	8.9	6.7	.2	5.1
Zürich	3.4	2.7	-3.3	3.3	2.3	1.4	.6	1.3	5.8	2.7	2.1	3.5

1) The Values for the ETHL and the HSG have been dropped, because of too few cases involved.

theless, in agreement with the previous findings. With respect to staying, the signed values show for all universities with the exception of NEUCHATEL that female students are more likely to leave their university of origin than male ones. This result replicates the one from 7.1.1.1., the only difference being that, there, sex was no issue in the determination of staying at the ETHZ.

With respect to leaving, the earlier results are also corroborated by the signed values. By the 15th semester, proportionally more female students have left the university system than male students at all the universities. Table 69, however, does not tell us at which level the female students have left more often than their male counterparts. To get an idea of these differences, turn to Tables 70 and 71.

Table 70

Pattern of Leaving the System after 15 Semesters for Female Students by University of Entrance

University	Level of Leaving the System			
	Drop Out	Low	Medium	High
Basel	.194	.201	.291	.067
Bern	.106	.354	.265	.009
ETHZ	.163	.143	.408	0
Fribourg	.141	.438	.234	.047
Genève	.125	.211	.375	.007
Lausanne	.155	.043	.638	.026
Neuchâtel	.100	.333	.383	.037
Zürich	.195	.183	.282	.062

Table 71

Discrepancy between the Pattern of Leaving of Female Students and the one of the Total Cohort by University of Entrance (Female-Total)

University	Level of Leaving the System			
	Drop Out	Low	Medium	High
Basel	.061	.094	-.009	-.059
Bern	.037	.186	-.088	-.058
ETHZ	.070	.106	-.172	-.030
Fribourg	.044	.246	-.111	-.041
Genève	-.017	.106	-.023	-.009
Lausanne	.005	.009	.075	-.002
Neuchâtel	.008	.128	-.033	.001
Zürich	.071	.066	-.011	-.031

Table 70 gives the pattern of leaving after 15 semesters for female students, by university of entrance. Unfortunately, similar tables for the university of exit cannot be provided (10). Table 71 shows how these patterns differ from the corresponding ones for the total cohort.

Because the results reported for the independent determinants relate to the university of exit and the present ones to the university of entrance, the two are not directly comparable. They support each other, though, because they point in the same general direction: Female students leave the university system not only earlier, but also at generally lower levels. From Table 71 we see that this trend is less pronounced for the French speaking universities than for the Swiss German ones. There are some differences in details between the two types of results; however, these seem to be attributable to the different reference points referred to above.

7.1.3. The Profiles of the 'Typical' Individual

7.1.3.1. The Profile of the 'Typical' Individual and the Semester Mean

As will be recalled, according to SPILERMAN the desired decomposition of the transition probabilities p_{ij} in terms of the conditioning variables (X'_1, ..., X'_v) is given by the following formula (see p. 146):

$$p_{ij} = \frac{1}{n_{i\cdot}} \sum_{c}^{n_{i\cdot}} \hat{y}_{ijc} = \hat{a}_{ij} + \sum_{v}^{V} \hat{b}_{ijv} \left(\frac{\sum_{c} X'_{vc}}{n_{i\cdot}}\right),$$

where the terms $\sum_c X'_{vc}/n_{i\cdot}$, for v=1, ..., v can be interpreted as the profile

10) The reason lies in the fact that the last 2 rows in the one-step transition matrices have not been estimated. An attempt to use the corresponding observed rows of the total cohort instead, as in the case of the total cohort (see footnote 16, p. 177) leads to erroneous results, because, apparently, the patterns of interruptions and movements to foreign universities for female students differ from the ones for males. If it had not been for this fact, the same procedure as the one applied in 6.3.2. could have been used to compute the pattern of leaving by university of exit.

of a 'typical' individual in state i at t_o in SPILERMAN's case. In the present case, with the introduction of semester heterogeneity, this formula has become (see p. 148):

$$p_{ijt} = \frac{1}{n_{i\cdot t}} \sum_{c}^{n_{i\cdot}} \hat{y}_{ijct} = \hat{a}_{ij} + \sum_{v}^{V} \hat{b}_{ijv} \left(\frac{\sum_{c} X'_{vct}}{n_{i\cdot t}}\right),$$

where the terms $\sum_{c} X'_{vct}/n_{i\cdot t}$, for v=1, ..., v can be interpreted as the profile of a 'typical' individual in state i at time t. These terms have, however, also a second interpretation, which is very important for the application of the model. They are nothing else but the semester means of the corresponding independent variables, i.e. the means of these variables taken over all individuals at a given university in a given semester.

In using the program B-6 to test the accuracy of the independent determinants in 6.3. above, I have computed the relevant transition probabilities on the basis of the individual probabilities, given by the individual profiles. That is, I have made use of the first equality sign in the above formula. Instead, I could also have used the second equality sign, which says that the relevant transition probabilities can also be obtained by multiplying the semester means of the selected predictor variables with the corresponding regression coefficients and summing over all the products plus the constant term in the regression. The program in APPENDIX B-9 does precisely that.

An example shall illustrate this procedure. Consider the graduation at the medium level in the 10th semester at the university of ZUERICH. Table 50 gives the relevant independent variables and the corresponding understandardized regression coefficients. The semester means for the selected independent variables can be obtained by applying the SELECT IF feature of SPSS in combination with the CONDESCRIPTIVE one to the synthetic population file. As indicated by the test in 6.3., the estimate of the probability in question is .039. Using alternatively the proposed procedure, we arrive at the same value:

Independent Determinant	Unstandardized Regression Coefficient		Semester Mean		Contribution to the Transition Probabilities
Constant					-.01709
Sex	.01368	x	.195	=	.00269
Duration Status	.00152	x	8.827	=	.01342
Work Load	-.00050	x	4.083	=	-.00204
Entrance Exam	.06895	x	.003	=	.00021
Medicine	.02245	x	.280	=	.00629
Law	.01491	x	.182	=	.00271
Economics/Business	.03168	x	.086	=	.00272
Natural Sciences I	.02312	x	.048	=	.00111
Lin Medium	.00964	x	3.000	=	.02892
SQ High	.00294	x	0	=	0
Transition Probability				=	.039

This new way of viewing the profile of the 'typical' individual has several advantages which will become apparent as soon as we are in possession of the semester means.

7.1.3.2. The Semester Means for the Cohort 1965

The same limitation of the data encountered throughout this paper also applies to the semester means: They are available only for the first 15 semesters. For these 15 semesters, however, the semester means for individual characteristics can easily be calculated from the synthetic population file using SPSS, as indicated above. The corresponding values for the semester chracteristics can be obtained without the help of the computer, because all the students in a particular semester have the same value on a specific semester characteristic, regardless of where they are studying.

Table 72 gives the semester means for the semester characteristics for all the universities, Tables 73 to 82 present the semester means for the individual characteristics for the ten different universities.

Table 72

Means of Semester Characteristics by Semester

Semester Characteristic	Semester														
	1	2	3	4	5	6	7	8	9	10	11	12	13	14	15
SEMNUM	1	2	3	4	5	6	7	8	9	10	11	12	13	14	15
Log of Semnum	0	.693	1.099	1.386	1.609	1.792	1.946	2.079	2.197	2.303	2.398	2.485	2.565	2.639	2.708
SQ of Semnum	1	4	9	16	25	36	49	64	81	100	121	144	169	196	225
Low	0	1	1	1	1	1	1	1	1	1	1	1	1	1	1
Medium	0	0	0	0	0	0	1	1	1	1	1	1	1	1	1
High	0	0	0	0	0	0	0	0	1	1	1	1	1	1	1
Lin Medium	0	0	0	0	0	0	0	1	2	3	4	5	6	7	8
Lin High	0	0	0	0	0	0	0	0	0	0	0	1	2	3	4
Log Low	0	0	0	0	0	.693	1.099	1.386	1.609	1.792	1.946	2.079	2.197	2.303	2.485
Log Medium	0	0	0	0	0	0	0	0	.693	1.099	1.386	1.609	1.792	1.946	2.079
Log High	0	0	0	0	0	0	0	0	0	0	0	0	.693	1.099	1.386
XN Drop	.027	.065	.121	.176	.200	.176	.121	.065	.027	.009	.002	0	0	0	0
XN Low	.002	.009	.027	.065	.121	.176	.200	.176	.121	.065	.027	.009	.002	0	0
XN Medium	0	0	0	0	.002	.009	.027	.065	.121	.176	.200	.176	.121	.065	.027
SQ High	0	0	0	0	0	0	0	0	0	0	0	1	4	9	16

Table 73

Means of Individual Characteristics by Semester for the University of BASEL

Individual Characteristic	Semester														
	1	2	3	4	5	6	7	8	9	10	11	12	13	14	15
Sex	.307	.305	.285	.278	.270	.241	.239	.224	.208	.192	.196	.193	.196	.191	.184
Age	3.335	4.321	4.275	5.254	5.232	6.224	6.233	7.167	7.157	8.168	8.100	9.039	9.045	9.966	10.054
SES	3.432	3.446	3.446	3.443	3.430	3.395	3.366	3.400	3.387	3.395	3.422	3.433	3.445	3.460	3.443
Origin	.005	.005	.005	.005	.005	.003	.006	.006	.006	.003	.004	.004	.008	.009	.011
Work Load	4.135	4.167	3.928	3.919	3.923	3.759	3.853	3.543	3.577	3.650	3.712	3.816	4.067	4.114	4.901
Duration Status	1.666	2.662	3.602	4.570	5.505	6.369	7.202	8.027	8.879	9.608	10.49	11.44	12.41	13.11	14.15
Cantonal A.C.	.057	.055	.048	.049	.049	.043	.046	.042	.042	.038	.030	.028	.029	.030	.038
Committee A.C.	.059	.055	.058	.057	.054	.054	.055	.048	.048	.056	.052	.047	.053	.060	.070
Entrance Exam	.010	.008	.008	.008	.008	.009	.009	.009	.010	.010	.011	.012	.008	.009	.011
Teacher's Cert.	.025	.023	.023	.022	.022	.017	.020	.018	.022	.024	.026	.016	.016	.013	.011
Business Cert.	.052	.048	.053	.054	.054	.051	.055	.045	.035	.035	.037	.031	.024	.026	.027
Medicine	.319	.320	.315	.324	.295	.287	.314	.342	.348	.395	.411	.417	.429	.426	.330
Law	.098	.101	.113	.116	.122	.125	.121	.133	.137	.105	.078	.083	.082	.085	.103
Econ/Business	.076	.073	.088	.086	.092	.094	.092	.094	.096	.094	.089	.083	.078	.064	.065
Social Sci	.015	.015	.013	.011	.014	.006	.009	.009	.010	.010	.007	.008	.008	.013	.016
Humanities	.152	.141	.116	.095	.105	.099	.107	.103	.083	.077	.100	.094	.102	.094	.119
Hist Sci	.052	.058	.058	.062	.059	.060	.049	.048	.045	.042	.041	.047	.049	.051	.065
Natural Sci I	.069	.065	.063	.062	.062	.068	.063	.067	.064	.063	.067	.059	.057	.055	.059
Natural Sci II	.162	.171	.179	.192	.205	.213	.207	.176	.195	.189	.185	.193	.188	.204	.232

Table 74

Means of Individual Characteristics by Semester for the University of BERN

Individual Characteristic	Semester														
	1	2	3	4	5	6	7	8	9	10	11	12	13	14	15
Sex	.219	.224	.218	.211	.193	.191	.159	.154	.155	.149	.149	.135	.126	.140	.140
Age	3.737	4.759	4.746	5.710	5.685	6.662	6.572	7.563	7.618	8.583	8.580	9.471	9.486	10.42	10.59
SES	3.379	3.373	3.366	3.387	3.377	3.409	3.425	3.416	3.400	3.369	3.419	3.422	3.415	3.398	3.358
Origin	.025	.029	.031	.040	.043	.050	.057	.059	.061	.060	.066	.053	.059	.051	.043
Work Load	3.786	3.872	3.856	3.751	3.900	4.037	4.242	4.122	4.260	4.122	4.168	4.339	4.239	4.196	4.713
Duration Status	1.158	2.153	3.084	4.004	4.870	5.761	6.461	7.323	8.155	8.759	9.525	10.23	11.05	12.25	13.08
Cantonal A.C.	.029	.031	.031	.030	.024	.028	.032	.031	.033	.034	.035	.037	.039	.038	.054
Committee A.C.	.049	.049	.049	.051	.049	.052	.045	.048	.052	.053	.056	.055	.059	.057	.066
Entrance Exam	.033	.033	.033	.034	.037	.039	.032	.033	.038	.039	.035	.026	.025	.025	.031
Teacher's Cert.	.102	.104	.099	.085	.063	.059	.045	.048	.049	.046	.045	.042	.050	.045	.054
Business Cert.	.153	.149	.142	.146	.159	.152	.155	.150	.134	.133	.129	.127	.112	.096	.097
Medicine	.270	.281	.283	.294	.289	.296	.316	.330	.322	.333	.336	.332	.333	.366	.280
Law	.135	.124	.129	.134	.149	.154	.157	.162	.160	.157	.162	.161	.154	.134	.128
Econ/Business	.125	.124	.131	.130	.141	.137	.143	.147	.155	.152	.139	.119	.115	.108	.117
Social Sci	.076	.079	.074	.067	.059	.057	.057	.052	.056	.055	.058	.055	.059	.061	.078
Humanities	.135	.134	.133	.128	.116	.113	.080	.069	.066	.067	.051	.063	.062	.070	.078
Hist Sci	.041	.041	.047	.043	.037	.043	.045	.048	.049	.053	.058	.061	.064	.057	.078
Natural Sci I	.104	.102	.099	.095	.094	.087	.084	.083	.078	.077	.076	.084	.078	.089	.097
Natural Sci II	.092	.090	.082	.085	.096	.093	.100	.097	.099	.092	.104	.106	.109	.102	.125

Table 75

Means of Individual Characteristics by Semester for the ETHL

Individual Characteristic	Semester														
	1	2	3	4	5	6	7	8	9	10	11	12	13	14	15
Sex	.01	0	0	0	0	0	0	0	0	0	0	0	0	0	0
Age	2.371	3.239	3.194	4.180	4.202	5.148	5.063	6.075	6.120	7.222	7.282	8.211	8.250	9.231	9.231
SES	3.115	3.286	3.152	3.148	3.148	3.150	3.152	3.190	3.162	3.167	3.205	3.684	3.600	3.923	3.923
Origin	.866	.870	.871	.876	.876	.877	.875	.850	.867	.833	.795	.684	.700	.769	.769
Work Load	3.804	3.047	3.207	3.049	3.030	2.826	2.585	2.356	2.017	3.496	3.385	2.029	1.949	2.427	2.471
Duration Status	1.010	1.989	2.978	3.921	4.921	5.914	6.850	7.737	8.653	9.389	10.44	10.53	11.60	11.85	12.85
Cantonal A.C.	.010	.011	.011	.011	.011	.012	.012	.012	.013	0	0	0	0	0	0
Committee A.C.	.082	.065	.075	.079	.079	.086	.087	.075	.080	.111	.128	.211	.200	.231	.231
Entrance Exam	.031	.033	.032	.034	.034	.037	.012	.025	.040	.056	.051	0	0	0	0
Natural Sci I	.340	.326	.312	.315	.281	.284	.275	.287	.320	.278	.256	.158	.100	.154	.154
Natural Sci II	.113	.130	.118	.135	.135	.136	.137	.137	.093	.194	.179	.368	.350	.538	.538
Engineering	.454	.467	.493	.472	.506	.519	.512	.487	.520	.417	.410	.211	.250	.077	.077
Architecture	.082	.076	.075	.079	.079	.062	.075	.087	.067	.111	.154	.263	.300	.231	.231

Table 76

Means of Individual Characteristics by Semester for the ETHZ

Individual Characteristic	Semester														
	1	2	3	4	5	6	7	8	9	10	11	12	13	14	15
Sex	.054	.055	.048	.044	.041	.037	.037	.042	.045	.052	.065	.063	.048	.042	.053
Age	3.301	4.333	4.401	5.397	5.332	6.320	6.326	7.306	7.198	8.093	8.080	9.136	9.190	10.23	10.23
SES	3.257	3.252	3.240	3.234	3.221	3.213	3.214	3.237	3.258	3.268	3.284	3.286	3.229	3.249	3.269
Origin	.098	.098	.109	.105	.095	.097	.096	.093	.095	.102	.085	.093	.078	.085	.082
Work Load	1.342	1.342	1.349	1.322	1.396	1.418	1.320	1.317	1.254	1.262	1.296	1.521	1.806	1.953	2.092
Duration Status	1.001	1.997	2.984	3.954	4.940	5.917	6.926	7.889	8.879	9.832	10.68	11.50	12.54	13.50	14.43
Cantonal A.C.	.006	.006	.006	.006	.007	.006	.006	.004	.003	.004	.005	.007	.004	.005	.006
Committe A.C.	.084	.089	.086	.085	.085	.083	.084	.086	.081	.071	.077	.093	.091	.095	.094
Entrance Exam	.067	.070	.076	.073	.073	.073	.072	.070	.053	.050	.060	.073	.069	.069	.076
Teacher's Cert	.010	.101	.004	.003	.003	.003	.004	.004	.003	.002	.002	0	0	0	0
Business Cert.	.002	.003	.003	.003	.003	.003	.003	.003	0	.002	.002	.003	.004	.005	.012
Medicine	.018	.015	.020	.020	.022	.019	.018	.022	.023	.021	.035	.050	.030	.026	.023
Econ/Bus	.006	.006	.006	.007	.009	.012	.013	.015	.013	.012	.010	.020	.017	.011	.006
Social Sci	0	0	0	0	0	0	0	0	.002	0	0	0	0	0	0
Natural Sci I	.151	.155	.160	.152	.151	.146	.143	.140	.144	.149	.144	.176	.169	.180	.187
Natural Sci II	.141	.146	.137	.135	.135	.142	.143	.143	.142	.166	.201	.252	.320	.381	.398
Engineering	.534	.534	.546	.553	.557	.559	.560	.560	.537	.492	.435	.359	.320	.286	.304
Architecture	.131	.123	.117	.125	.117	.112	.115	.111	.128	.153	.169	.140	.139	.111	.076

Table 77

Means of Individual Characteristics by Semester for the University of FRIBOURG

Individual Characteristic	Semester														
	1	2	3	4	5	6	7	8	9	10	11	12	13	14	15
Sex	.195	.194	.197	.189	.161	.152	.144	.116	.103	.115	.111	.085	.080	.092	.070
Age	4.229	5.215	5.142	6.112	6.047	7.036	6.936	7.979	8.019	8.986	9.026	10.08	10.05	10.97	10.97
SES	2.947	2.966	2.975	3.013	3.032	3.051	3.072	3.021	3.013	3.000	3.043	3.038	3.148	3.197	3.268
Origin	.399	.400	.401	.397	.394	.402	.423	.432	.429	.439	.427	.396	.364	.368	.423
Work Load	2.825	2.772	2.638	2.622	2.694	2.571	2.593	2.782	3.202	3.436	4.259	4.613	4.274	4.529	4.732
Duration Status	1.084	2.075	3.019	3.870	4.853	5.648	6.491	7.253	8.205	8.957	9.590	10.31	11.07	12.08	13.06
Cantonal A.C.	.053	.056	.051	.049	.054	.051	.045	.047	.064	.058	.060	.066	.068	.066	.056
Committee A.C.	.043	.047	.048	.049	.050	.047	.045	.037	.045	.050	.043	.047	.045	.039	.014
Entrance Exam	.003	.003	.003	.003	.004	.004	0	0	0	0	0	0	0	0	0
Teacher's Cert.	.158	.153	.153	.143	.118	.109	.090	.089	.090	.122	.145	.123	.091	.125	.127
Business Cert.	.124	.122	.127	.121	.118	.117	.144	.153	.147	.151	.137	.123	.125	.132	.127
Medicine	.170	.172	.156	.176	.190	.188	.122	.068	.064	.043	.009	.019	.011	0	0
Law	.105	.109	.118	.117	.129	.141	.153	.132	.103	.094	.094	.085	.080	.092	.099
Econ/Business	.127	.131	.131	.130	.136	.137	.167	.179	.179	.165	.154	.160	.159	.145	.155
Social Sci	.146	.144	.153	.140	.125	.129	.122	.132	.141	.173	.162	.132	.136	.132	.113
Humanities	.176	.169	.178	.166	.151	.129	.099	.095	.109	.108	.103	.094	.102	.092	.113
Hist Sci	.071	.072	.070	.075	.075	.078	.108	.132	.141	.122	.128	.160	.148	.132	.127
Natural Sci I	.056	.056	.057	.068	.061	.063	.068	.079	.071	.086	.111	.094	.091	.092	.085
Natural Sci II	.102	.100	.099	.091	.093	.098	.104	.116	.135	.151	.188	.189	.193	.224	.239

Table 78

Means of Individual Characteristics by Semester for the HSG

Individual Characteristic	Semester														
	1	2	3	4	5	6	7	8	9	10	11	12	13	14	15
Sex	.029	.025	.027	.027	.033	.023	.023	.028	.024	.031	.019	.021	.022	0	0
Age	3.810	4.803	4.917	5.910	5.803	6.794	6.794	7.679	7.530	8.516	8.500	9.511	9.489	10.48	10.50
SES	3.388	3.347	3.418	3.366	3.431	3.406	3.391	3.444	3.588	3.672	3.642	3.583	3.435	3.415	3.385
Origin	.050	.042	.045	.054	.049	.047	.053	.046	.059	.078	.075	.063	.043	.024	0
Work Load	3.116	3.387	3.615	2.850	2.898	2.696	2.865	3.250	3.353	3.407	3.556	3.795	4.048	3.814	4.170
Duration Status	1.043	2.025	3.009	3.795	4.724	5.500	6.444	7.139	7.941	8.422	9.170	10.08	11.28	11.76	13.10
Cantonal A.C.	.022	.025	.027	.018	.016	.016	.023	.028	.024	.016	.019	.021	.022	.024	.026
Committee A.C.	.029	.034	.027	.027	.033	.031	.038	.028	.024	.031	.038	.042	.043	.049	.051
Entrance Exam	.115	.102	.127	.134	.122	.117	.113	.093	.082	.094	.057	.063	.065	.073	.077
Business Cert.	.547	.551	.536	.527	.537	.523	.519	.528	.518	.500	.509	.563	.630	.537	.538
Law	.007	.017	.018	.018	.016	.016	.015	.019	.024	.031	.019	.021	.022	.024	.026
Economics/Bus	.935	.915	.918	.920	.911	.914	.925	.926	.929	.922	.925	.896	.891	.878	.897
Social Sci	.036	.042	.036	.027	.041	.031	.023	.028	.035	.016	.038	.063	.065	.073	.051

Table 79

Means of Individual Characteristics by Semester for the University of GENEVE

Individual Characteristic	Semester														
	1	2	3	4	5	6	7	8	9	10	11	12	13	14	15
Sex	.352	.354	.340	.350	.313	.318	.302	.306	.278	.281	.266	.247	.229	.240	.238
Age	3.257	4.270	4.296	5.281	5.185	6.151	6.188	7.148	7.113	8.092	8.089	9.086	9.065	10.09	10.05
SES	3.400	3.386	3.404	3.423	3.455	3.472	3.462	3.451	3.526	3.558	3.532	3.527	3.537	3.582	3.583
Origin	.751	.762	.755	.757	.753	.756	.751	.747	.756	.750	.768	.765	.757	.745	.742
Work Load	4.699	4.798	4.81	4.635	4.827	4.945	4.225	4.331	4.696	4.658	4.831	4.921	5.054	4.974	5.865
Duration Status	1.039	2.023	2.992	3.893	4.869	5.735	6.544	7.407	8.079	8.877	9.823	10.49	11.61	12.41	13.52
Cantonal A.C.	.020	.021	.022	.022	.021	.019	.016	.017	.015	.015	.013	.021	.028	.024	.033
Committee A.C.	.068	.072	.073	.068	.071	.065	.075	.081	.086	.077	.072	.078	.065	.063	.053
Entrance Exam	.015	.015	.016	.014	.012	.012	.010	.007	.008	.008	0	0	0	0	0
Teacher's Cert.	.012	.010	.005	.005	.006	.006	.010	.010	.004	.008	.004	.004	.005	.005	.007
Business Cert.	.108	.097	.111	.112	.098	.096	.075	.071	.064	.065	.055	.041	.033	.038	.033
Medicine	.203	.210	.208	.197	.217	.207	.236	.246	.305	.315	.346	.333	.341	.332	.245
Law	.078	.077	.078	.082	.080	.090	.095	.094	.090	.085	.076	.078	.075	.067	.086
Econ/Bus	.103	.103	.113	.117	.119	.111	.085	.081	.056	.058	.034	.041	.051	.048	.060
Social Sci	.181	.179	.175	.183	.155	.154	.141	.141	.102	.096	.089	.091	.079	.087	.093
Humanities	.088	.095	.086	.087	.086	.083	.085	.088	.090	.077	.072	.053	.047	.067	.073
Hist Sci	.049	.049	.043	.044	.045	.049	.043	.044	.041	.042	.046	.049	.037	.038	.040
Natural Sci I	.086	.087	.086	.077	.080	.080	.082	.077	.083	.081	.089	.091	.098	.082	.086
Natural Sci II	.125	.123	.129	.128	.137	.145	.154	.158	.173	.192	.211	.210	.229	.240	.278
Architecture	.022	.018	.022	.022	.024	.025	.023	.024	.026	.023	.017	.021	.009	.010	.013

Table 80

Means of Individual Characteristics by Semester for the University of LAUSANNE

Individual Characteristic	Semster														
	1	2	3	4	5	6	7	8	9	10	11	12	13	14	15
Sex	.353	.361	.369	.357	.349	.348	.354	.320	.333	.310	.309	.293	.250	.240	.187
Age	2.649	3.648	3.595	4.592	4.533	5.497	5.321	6.301	6.298	7.291	7.172	8.201	8.317	9.325	9.292
SES	3.476	3.480	3.505	3.503	3.474	3.538	3.591	3.548	3.599	3.587	3.606	3.633	3.711	3.712	3.692
Origin	.808	.818	.840	.830	.826	.834	.831	.859	.865	.875	.885	.880	.867	.848	.846
Work Load	3.104	3.376	3.081	2.938	3.162	3.059	2.727	3.003	3.110	3.051	2.911	3.194	3.309	3.462	4.215
Duration Status	1.038	2.017	2.959	3.718	4.516	5.424	6.161	6.929	8.014	8.962	9.885	10.66	11.58	12.34	12.74
Cantonal A.C,	.019	.017	.017	.014	.010	.014	.020	.025	.024	.033	.024	.020	.008	.016	.022
Committee A.C.	.095	.086	.092	.088	.089	.086	.102	.095	.111	.082	.073	.067	.078	.080	.088
Entrance Exam	.019	.017	.020	.017	.016	.017	.012	.012	.019	.022	.012	.013	.023	.024	.022
Teacher's Cert.	.013	.010	.003	.007	.010	.003	0	0	0	0	0	0	0	0	0
Business Cert.	.148	.152	.133	.129	.135	.117	.075	.050	.019	.022	.018	.013	.016	.016	.022
Medicine	.192	.195	.184	.204	.230	.259	.315	.344	.348	.375	.412	.467	.477	.480	.374
Law	.101	.099	.102	.102	.095	.076	.087	.095	.106	.114	.121	.093	.086	.080	.110
Econ/Bus	.174	.169	.171	.167	.174	.162	.114	.079	.072	.071	.079	.087	.078	.080	.099
Social Sci	.101	.093	.096	.092	.082	.086	.067	.046	.039	.038	.036	.027	.039	.040	.033
Humanities	.155	.159	.154	.146	.135	.134	.122	.129	.130	.130	.127	.107	.086	.088	.121
Hist Sci	.044	.040	.041	.034	.036	.038	.039	.041	.034	.033	.024	.033	.016	.016	.022
Natural Sci I	.060	.060	.061	.058	.056	.055	.047	.050	.043	.038	.024	.020	.016	.016	.011
Natural Sci II	.139	.149	.157	.163	.158	.159	.177	.183	.188	.174	.152	.147	.180	.176	.209

Table 81

Means of Individual Cahracteristics by Semester for the University of NEUCHÂTEL

Individual Characteristic	Semester														
	1	2	3	4	5	6	7	8	9	10	11	12	13	14	15
Sex	.333	.345	.355	.356	.362	.333	.320	.316	.312	.348	.298	.350	.325	.314	.310
Age	3.543	4.557	4.503	5.556	5.539	6.381	6.344	7.309	7.273	8.227	8.468	9.575	9.775	10.69	10.86
SES	3.376	3.365	3.327	3.276	3.333	3.289	3.299	3.379	3.299	3.288	3.213	3.225	3.300	3.343	3.310
Origin	.672	.685	.675	.678	.677	.684	.701	.705	.740	.727	.702	.675	.675	.686	.690
Work Load	4.735	4.575	4.475	4.642	4.892	4.771	5.072	5.250	5.660	5.664	5.259	5.329	6.503	5.868	6.654
Duration Status	1.029	2.024	2.988	3.897	4.900	5.868	6.887	7.800	8.636	9.455	10.23	11.60	12.60	13.71	14.79
Cantonal A.C.	.092	.095	.096	.110	.108	.114	.093	.063	.078	.076	.064	.075	.100	.057	.069
Committee A.C.	.034	.042	.042	.048	.054	.053	.062	.063	.052	.045	.064	.075	.075	.086	.034
Entrance Exam	.023	.024	.024	.021	.015	.018	.021	.021	.026	.030	.043	.050	.050	.057	.069
Teacher's Cert.	.017	.018	.018	.027	.015	.018	.010	.011	.013	.015	.021	.025	.025	.029	.034
Business Cert.	.121	.125	.133	.137	.138	.140	.144	.137	.104	.091	.064	.050	.075	.086	.069
Medicine	.218	.208	.193	.116	.108	.044	.041	.011	.013	0	0	0	0	0	0
Law	.029	.030	.030	.034	.038	.044	.041	.042	.039	.030	.043	.050	.050	.029	.034
Econ/Bus	.103	.113	.108	.137	.146	.175	.175	.179	.117	.076	.043	.050	.075	.086	.069
Social Sci	.046	.042	.042	.041	.046	.044	.041	.074	.065	.091	.085	.075	.075	.057	.069
Humanities	.230	.232	.241	.253	.262	.281	.268	.242	.260	.288	.255	.325	.325	.343	.345
Hist Sci	.034	.036	.036	.048	.054	.061	.072	.084	.104	.121	.128	.150	.125	.114	.138
Natural Sci I	.138	.131	.139	.130	.115	.088	.082	.095	.117	.121	.170	.125	.125	.143	.138
Natural Sci II	.115	.131	.133	.144	.138	.175	.196	.200	.234	.227	.255	.200	.200	.200	.207

Table 82

Means of Individual Characteristics by Semester for the University of ZUERICH

Individual Characteristic	Semester														
	1	2	3	4	5	6	7	8	9	10	11	12	13	14	15
Sex	.259	.251	.238	.230	.221	.220	.209	.198	.187	.195	.182	.190	.176	.170	.175
Age	3.729	4.682	4.688	5.647	5.573	6.475	6.432	7.376	7.348	8.308	8.344	9.348	9.300	10.28	10.31
SES	3.354	3.353	3.354	3.379	3.382	3.411	3.408	3.436	3.434	3.428	3.420	3.452	3.468	3.502	3.462
Origin	.003	.004	.004	.005	.004	.003	.003	.007	.011	.012	.011	.012	.013	.014	.010
Work Load	4.455	4.586	4.524	4.543	4.330	4.212	4.199	4.029	4.058	4.083	4.154	3.925	3.926	3.804	4.103
Duration Status	1.236	2.225	3.139	4.012	4.884	5.708	6.480	7.283	8.057	8.827	9.687	10.52	11.27	12.15	13.01
Cantonal A.C.	.040	.043	.039	.039	.036	.037	.033	.035	.037	.036	.039	.036	.045	.040	.048
Committee A.C.	.086	.085	.087	.091	.089	.094	.089	.089	.094	.092	.097	.096	.093	.091	.094
Entrance Exam	.001	.002	.002	.004	.004	.003	.001	.001	.004	.003	.003	.002	.004	0	0
Teacher's Cert.	.147	.140	.145	.136	.122	.104	.095	.083	.078	.077	.076	.080	.074	.071	.070
Business Cert.	.076	.082	.079	.076	.079	.076	.079	.072	.071	.071	.059	.060	.057	.061	.070
Medicine	.184	.197	.198	.206	.207	.226	.245	.264	.270	.280	.304	.315	.309	.323	.266
Law	.164	.164	.170	.173	.180	.181	.192	.201	.198	.182	.164	.150	.158	.164	.175
Econ/Bus	.075	.077	.081	.078	.084	.080	.082	.079	.086	.086	.082	.080	.075	.073	.074
Social Sci	.073	.057	.058	.050	.044	.050	.049	.051	.050	.050	.057	.053	.054	.057	.062
Humanities	.210	.201	.198	.190	.180	.167	.145	.130	.127	.118	.117	.120	.122	.115	.129
Hist Sci	.064	.061	.071	.072	.074	.074	.071	.069	.063	.070	.071	.075	.077	.077	.089
Natural Sci I	.065	.067	.052	.051	.053	.047	.049	.050	.046	.048	.045	.041	.039	.034	.031
Natural Sci II	.149	.159	.149	.153	.149	.146	.136	.129	.129	.136	.133	.142	.135	.133	.149

These tables provide yet a third descriptive tool, in addition to regression equations and homogenous subpopulations. They display the distributive shifts in the student population at a particular university in the course of the 15 semesters covered. A few remarks will have to suffice to indicate the usefulness of this additional instrument for description:

Regarding socio-economic characteristics, consider once again sex as an example: With the exception of NEUCHATEL, a pervasive trend toward decreasing shares of female students can be noted with increasing numbers of semesters. Similarly, a shift in the distribution of subjects takes place in the course of the semesters, which is in accord with previous findings: Thus, the natural sciences II are generally gaining in numbers, and so is medicine, with the exception of FRIBOURG and NEUCHATEL, where the share of medical students is gradually vanishing. Humanities, on the other hand, are generally losing in importance, again with the exception of NEUCHATEL. Finally, distributive shifts can also be found with respect to admission certificates. Here, the share of the teacher's certificate is generally decreasing in the course of the semesters, which is again an indication of particularly frequent low level graduation of the holders of this certificate.

7.1.3.3. Transition Matrices for the Last 6 Semesters

The first advantage of the new way to look at the typical individual is that, based on the procedure which derives from this view, it is possible to estimate the transition matrices for the last 6 semesters (11) using independent determinants. Two additional assumptions, however, have to be made:

- The semester means of the individual characteristics for all the remaining semesters will be the same as the ones for the last semester in the synthetic population file, i.e. the 15th semester.

11) The final semester, i.e. the 23rd one for spring-starters and the 22nd one for fall-starters, will not be considered, because the figures are negligible (see p. 81).

- The impact of the independent determinants on the transition probabilities in the last 6 semester will be the same as in the first 15 ones, i.e. the same coefficients that apply to the first 15 semesters can also be used for the remaining ones.

Whereas the first assumption has at least some plausibility, because distributive shifts are probably less prominent in later semesters, the second assumption is quite questionable, especially with respect to semester characteristics. Logarithmic, linear and, above all, geometric trends do not generally continue up to the last semester. This is especially true in the case of medium level graduation, where linear and/or geometric trends have been observed during the first 15 semesters for the following universities:

BASEL	: geometric trends (SQ High, SQ of SEMNUM)
BERN	: linear trend (Lin Medium)
GENEVE	: geometric trend (SQ of SEMNUM)
LAUSANNE	: geometric trend (SQ High, SQ of SEMNUM)
ZUERICH	: geometric and linear trends (SQ High, Lin Medium)

Fig. 33 shows that, at all of these universities, medium level graduation reaches its peak in the 17th semester and decreases for the rest of the semesters. This is to say that at these universities, medium level graduation will be greatly overestimated, if the trends are allowed to continue past the 17th semester.

One way to eliminate the influence of these trends after the 17th semester is to set the means for these semester characteristics equal to zero for the later semesters at the universities in question. This procedure will, of course, have side-effects for other transitions originating at these universities which also involve the corresponding semester characteristics. These side-effects are, however, generally desirable, because they result in an increased probability of staying at the respective universities.

I have estimated the transition probabilities for the last 6 semester twice, once without the modifications of the linear and geometric trends involved, and a second time including the modifications. Table 83 gives the unmodified semester characteristics for the last 6 semesters, Table 84 summarizes the results of

Figure 33

Observed Probability of Medium Level Graduation by Semester for Selected Universities

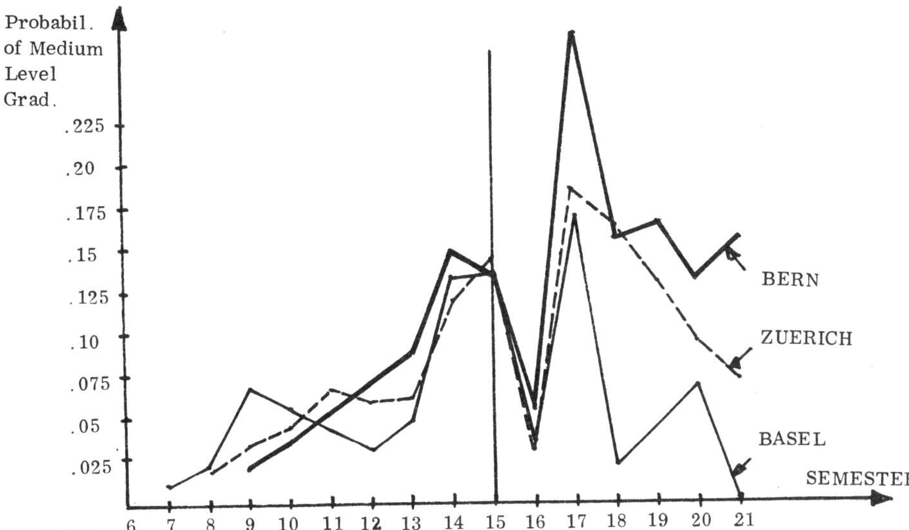

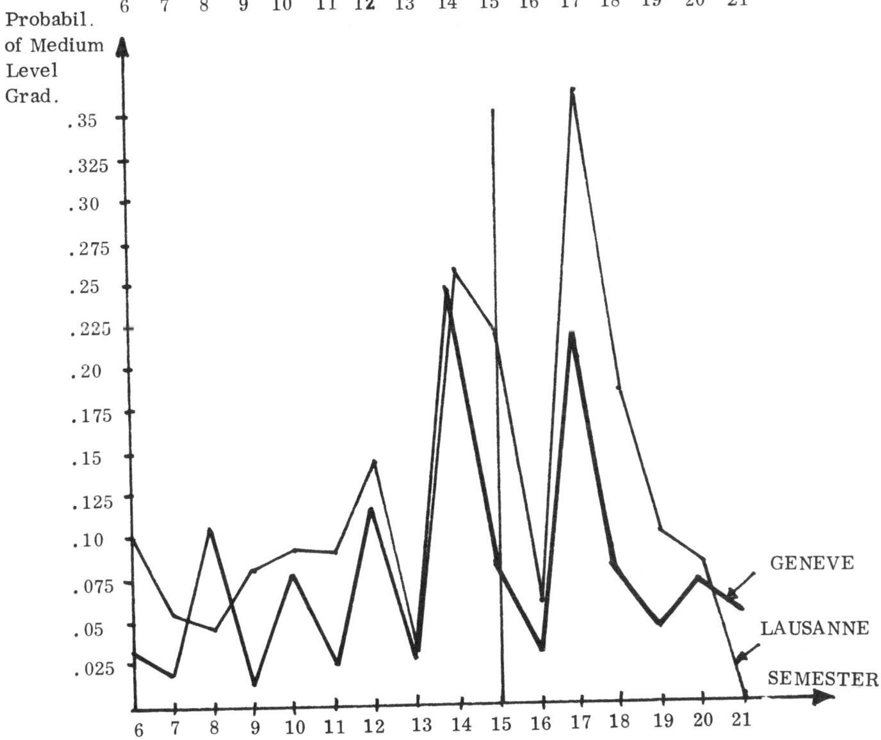

Table 83

Means of Semester Characteristics for the last 6 Semesters

Semester Characteristic	Semester					
	16	17	18	19	20	21
SEMNUM	16	17	18	19	20	21
Log of Semnum	2.773	2.833	2.890	2.944	2.996	3.045
SQ of Semnum	256	289	324	361	400	441
Low	1	1	1	1	1	1
Medium	1	1	1	1	1	1
High	1	1	1	1	1	1
Lin Medium	9	10	11	12	13	14
Lin High	6	7	8	9	10	11
Log Low	2.565	2.639	2.708	2.773	2.833	2.890
Log Medium	2.197	2.303	2.398	2.485	2.565	2.639
Log High	1.609	1.792	1.946	2.079	2.197	2.303
XN Drop	0	0	0	0	0	0
XN Low	0	0	0	0	0	0
XN Medium	.009	.002	0	0	0	0
SQ High	25	36	49	64	81	100

Table 84

Mean Indices of Dissimilarity for the last 6 Semesters between 'Observed' Transition Matrices and Matrices Predicted on the Basis of Independent Determinants, by University

University	Mean Index of Dissimilarity			
	without Modification of Means		with Modifications of Means	
	all cells included	predicted cells only	all cells included	predicted cells only
Basel	55.1	48.2	22.3	13.5
Bern	16.3	16.2	18.5	18.4
ETHL	44.6	32.1	44.6	32.1
ETHZ	23.1	18.4	23.1	18.4
Fribourg	25.1	24.7	25.1	24.7
HSG	31.9	13.8	31.9	13.8
Genève	21.1	15.8	16.1	10.0
Lausanne	55.8	34.1	23.3	10.9
Neuchâtel	18.3	14.7	18.3	14.7
Zürich	27.9	27.8	20.3	20.2

the computations using indices of dissimilarity between the estimated transition matrices and the 'observed' ones (12). The discrepancies are generally enormous, especially when no modifications are made and when all the cells are included in the comparison. Including all the cells instead of only the estimated ones has had virtually no effect on the comparisons for the first 15 semesters (see 6.3.1.). This is no longer the case for the last 6 semesters, because of the general importance of high level graduation during this final period. High level graduation now becomes an issue at those universities for which no pertinent determinants have been estimated, too.

Table 84 shows the positive impact of the modifications: In four out of the five cases where modifications are introduced, the indices of dissimilarity are cut considerably. The only university which shows no sign of improvement is BERN.

In spite of these improvements, the estimates are still rather inaccurate. This has serious consequences for the estimation of medium and high level leaving, as is documented by Tables 85 and 86. These tables correspond to Tables 55/56

Table 85

Pattern of Leaving the System after 21 Semesters on the Basis of Estimated Transition Matrices, by University of Exit

University	Level of Leaving			
	Drop out	Low	Medium	High
Basel	.108	.077	.369	.218
Bern	.087	.169	.596	.278
ETHL	-	-	.596	-
ETHZ	.094	.020	.543	.223
Fribourg	.096	.200	.337	.140
HSG	.075	-	.479	-
Genève	.129	.102	.538	-
Lausanne	.160	.054	.666	-
Neuchâtel	.099	.194	.392	-
Zürich	.118	.112	.499	.225

12) 'Observed' is placed between quotation marks, because these matrices, too, have been inferred on the basis of certain assumptions (see section 2.4.1.).

Table 86

Discrepancy between Estimated and Observed Pattern of Leaving after 21 Semesters, by University of Exit (Estimated-Observed)

University	Level of Leaving			
	Drop out	Low	Medium	High
Basel	-.003	.018	-.011	-.093
Bern	-.006	.003	.058	-.043
ETHL	-	-	-.095	-
ETHZ	-.008	.012	-.071	.081
Fribourg	.002	.001	.011	-.051
HSG	.009	-	-.171	-
Genève	.010	.014	.001	-
Lausanne	.007	.011	-.007	-
Neuchâtel	.009	.031	.022	-
Zürich	.004	.010	.055	-.059

in 6.3.2., which referred to the pattern of leaving after 15 semesters. These previous tables indicated that medium level graduation after 15 semesters is seriously underestimated when using independent determinants at the ETHL and at the HSG. In addition to that deficiency, medium level graduation is now also inaccurately predicted for the ETHZ and for the universities of BERN and ZUERICH.

Both, at the ETHZ and the HSG, moreover, the probability of medium level graduation computed on the basis of independent determinants actually decreases from the 15th to the 21st semester. This ridiculous result is due to violations of the second assumption made above. Thus, at the ETHZ, the impact of the natural sciences changes. While they are negatively related to leaving at the medium level during the first 15 semesters, their impact is non-negative, if not positive, during the final ones. This shift is the more influential, because the relative share of the natural sciences in the final semesters increases considerably at the ETHZ (see Table 76). At the HSG, on the other hand, the negative impact of High and Lin High are altered in the final period.

High level graduation has been estimated correctly at all the five universities for the 15 first semesters (see Table 56). As shown by Table 86, it is now inaccurately estimated for all the five universities. It is overestimated at the ETHZ and underestimated at the other four universities. These inaccuracies were to be expected, since the majority of high level graduations fall in to the period after the 15th semester.

The inaccuracies notwithstanding, I will subsequently rely on these extrapolations, because I have no other choice.

7.1.3.4. A Cross-Sectional Perspective

In the social sciences, researchers often face the problem of having to answer questions that inherently require longitudinal data on the basis of cross-sectional data. The inverse problem is quite uncommon, because cross-sectional data are so much easier to get than longitudinal ones. I have, nevertheless, faced just this inverse problem in trying to estimate the dropouts and graduates in a single semester on the basis of the data for the cohort 1965 (see 3.2.2.). The unwieldy procedure outlined for the solution of this problem in Part I is a result of the difficulty to solve cross-sectional questions on the basis of longitudinal data.

Now, there is an easier solution to that problem which will be discussed below. This solution requires that we take another look at the synthetic population file and at the semester means.

The synthetic population file represents nothing else but a synthetic population of Swiss students in a given semester, constructed on the basis of the numbers of students enrolled in 1965. That is, this file 'cross-sectionalizes' the longitudinal data for the cohort 1965. Now, since the whole synthetic population is subject to the same set of regression equations, it is possible to apply this set not only to the means for given universities in specific semesters, but similarly to the overall means for given universities in the synthetic population. Or, to put it alternatively, we can compute the transition probabilities for the whole synthetic cohort on the basis of the overall means using the same set of regression equations we have applied to the semester means of the different semesters.

These transition probabilities are given by the following equations:

$$p_{ij} = \frac{1}{n_{i\cdot\cdot}} \sum_{t=1}^{21} \sum_{c}^{n_{i\cdot t}} \hat{y}_{ijct} = \hat{a}_{ij} + \sum_{v}^{V} \hat{b}_{ijv} \left(\frac{\sum_{t=1}^{21} n_{i\cdot t} \sum_{c} X'_{vct}/n_{i\cdot t}}{n_{i\cdot\cdot}} \right)$$

$$= \hat{a}_{ij} + \sum_{v}^{V} \hat{b}_{ijv} \left(\frac{\sum_{t=1}^{21} \sum_{c} X'_{vct}}{n_{i\cdot\cdot}} \right),$$

where $\sum_{t}\sum_{c} X'_{vct}/n_{i\cdot\cdot}$ corresponds to the profile of the typical individual in state i in the synthetic population, or, to the overall means of the corresponding independent determinants of flow.

The Tables 87 and 88 give the overall means for the individual characteristics and for the semester characteristics respectively for each university. For the semester characteristics, the modifications introduced in the previous section have been taken into consideration, as is indicated by the 7 underlined cells.

These two tables give the distribution of all the Swiss students at the ten Swiss universities over the relevant independent variables under the assumption of an absence of population shifts since the year 1965. The extent to which the present distribution of the student bodies at the 10 universities differs from the ones given in these tables allows an assessment of the direction and importance of population shifts since 1965.

As indicated above, this cross-sectional perspective is useful in computing drop-outs and graduates in future semesters. Similarly, it also proves to be useful in the context of simulation.

In general, the view introduced in this subsection will prove to be very advantageous in the contexts of both, prediction and simulation.

Table 87

Overall Means (over 21 Semesters) of Individual Characteristics by University

Individual Characteristic	University									
	Basel	Bern	ETHL	ETHZ	Frib	HSG	Genève	Laus	Neuch	Zürich
Sex	.242	.176	.001	.047	.154	.024	.304	.331	.338	.211
Age	6.560	7.079	4.948	6.285	6.921	6.745	6.465	5.559	6.189	6.913
SES	3.422	3.394	3.211	3.246	3.023	3.441	3.471	3.551	3.321	3.409
Origin	.005	.047	.855	.097	.406	.050	.755	.842	.688	.007
Work Load	3.916	4.069	2.907	1.383	3.033	3,240	4.769	3.121	4.995	4.228
Duration Status	7.206	6.606	5.609	6.236	5.509	5.753	6.369	5.919	5.580	6.611
Cantonal A.C.	.043	.033	.010	.005	.054	.022	.020	.018	.092	.038
Committee A.C.	.055	.052	.092	.084	.045	.033	.072	.089	.051	.091
Entrance Exam	.009	.033	.031	.068	.002	.104	.009	.017	.026	.002
Teacher's Cert.	.020	.065	0	.004	.127	0	.007	.004	.018	.104
Business Cert.	.044	.133	0	.003	.129	.535	.080	.088	.121	.072
Medicine	.344	.303	0	.002	.126	0	.252	.292	.103	.246
Law	.109	.146	0	0	.116	.018	.082	.097	.036	.175
Econ/Business	.085	.133	0	.010	.145	.918	.085	.130	.122	.080
Social Sci	.011	.063	0	0	.139	.036	.139	.069	.053	.054
Humanities	.108	.095	0	0	.139	0	.081	.134	.260	.156
Hist Sci	.053	.049	0	0	.096	0	.045	.035	.067	.071
Natural Sci I	.063	.089	.290	.152	.069	0	.084	.047	.122	.049
Natural Sci II	.191	.096	.153	.165	.120	0	.164	.164	.165	.142
Engineering	0	0	.462	.515	0	0	0	0	0	0
Architecture	0	0	.094	.125	0	0	.021	0	0	0

Table 88

Overall Means (over 21 Semesters) of Semester Characteristics by University[1)]

Semester Characteristic	University									
	Basel	Bern	ETHL	ETHZ	Frib	HSG	Genève	Laus	Neuch	Zürich
SEMNUM	8.119	8.137	6.364	6.865	6.483	7.299	8.206	7.163	6.148	8.183
Log of Semnum	1.828	1.865	1.584	1.667	1.623	1.712	1.822	1.856	1.532	1.846
SQ of Semnum	61.701	95.735	58.289	66.026	67.391	74.986	72.725	65.548	56.240	91.783
Low	.921	.928	.898	.903	.898	.905	.918	.910	.883	.924
Medium	.573	.581	.434	.490	.433	.499	.555	.490	.393	.578
High	.442	.459	.266	.331	.303	.334	.433	.350	.265	.448
Lin Medium	2.840	2.854	1.349	1.718	1.802	2.019	2.823	1.964	1.423	2.238
Lin High	1.152	1.216	.485	.567	.737	.847	1.141	.717	.489	1.110
Log Low	1.143	1.187	.801	.923	.861	.975	1.077	.961	.758	1.165
Log Medium	.716	.759	.375	.479	.484	.509	.736	.542	.396	.728
Log High	.308	.324	.127	.155	.198	.229	.337	.196	.139	.300
XN Drop	.069	.067	.089	.080	.084	.082	.068	.080	.086	.067
XN Low	.065	.063	.081	.076	.070	.077	.062	.072	.070	.065
XN Medium	.052	.055	.040	.049	.039	.043	.050	.048	.038	.055
SQ High	2.735	6.185	2.592	2.148	3.499	4.090	7.241	1.970	1.949	1.214

1) The underlined figures are the modified means. The corresponding unmodified means are:
- for SQ Semnum: 91.816, 91.333, 72.290
- for Lin Medium: 2.690
- for SQ High: 5.905, 3.231, 5.386

7.2. Prediction

In order to use the present model for predictive purposes, it had to be assumed in Part I that transition matrices are stable over times. This assumption can now be replaced by a less restrictive one, because the introduction of independent determinants allows to account for population shifts in future cohorts.

By making use of both individual and semester characteristics, it has been possible to consider population shifts as well as structural change in the intracohort analysis. In the context of prediction, however, we turn to intercohort analysis. Structural change can now no longer be taken into account by the semester characteristics. The incorporation of structural change in the intercohort case would require the introduction of additional cohort characteristics into the regression equations; or, alternatively, it would require additional equations that show how the regression coefficients vary as a function of time and of relevant time-dependent characteristics. Important such characteristics may refer to the development of the national economy, changes in the educational policy or changes in the value systems.

Both of these strategies, however, presuppose data on several cohorts. Since the present model is based on the data for a single cohort alone, structural change in the intercohort analysis cannot be taken into account. But, based on the introduction of independent determinants, it is possible to account for population shifts. This is to say that the assumption of stable transition rates is to be replaced by the assumption of stable regression coefficients. Whereas rates may now change due to population shifts, the determinant structure of the cohort 1965 is assumed to persist.

This is one step toward a more realistic model. It is difficult to assess how restrictive the new assumption is. Considering the traditional inability of the Swiss University System to adapt to societal change, it is likely to be a very good approximation to the empirical state of affairs.

7.2.1. Population Shifts

While structural change is rare in the Swiss University System, population shifts have occurred more frequently in the past couple of years. The following tables indicate the direction of these shifts for sex, federal admission certificates and for the most important groups of subjects.

Table 89 shows that with the exception of the university of BASEL, all the Swiss universities have experienced an increase in the share of female students enrolled. In absolute terms, this increase has been largest at the university of GENEVE, followed by the ones at FRIBOURG and the ETHL. In relative terms, it has also been substantial at the ETHZ, the HSG and at BERN and ZUERICH.

Table 89

Female Students at Swiss Universities in the WS73/74: Percentages and Percentage Shifts with respect to the Synthetic Population

University	Female Students	
	Percentage	Percentage Shift
Basel	.239	-.003
Bern	.222	+.046
ETHL	.073	+.072
ETHZ	.080	+.033
Fribourg	.243	+.089
HSG	.043	+.019
Genève	.399	+.095
Lausanne	.320	+.011
Neuchâtel	.314	+.024
Zürich	.264	+.053

Source: ESTA, 1975: Die Studierenden an den Schweizerischen Hochschulen im Wintersemester 1973/74, Table 1.1

Table 90 refers to the population shifts with respect to the federal admission certificates. As will be recalled (see 6.2.1.3.1.), roughly two thirds of the students have entered the university system with a regular federal admission certificate in 1965. Table 90 now shows that within the category of federal admission certificates the minority of students holding a committee type certifi-

Table 90

Population Shifts with respect to Federal Admission Certificates: Total Number of Federal Certificates by Year and Percentage Shares of Committee Type Certificates and of Graduates from the 'Second Course of Education'

Year	Federal Admission Certificates		
	Total Number of Fed. Certificats	Percentage Shares of	
		Committee Type	Grad. of 2nd Course of Ed
1963	3706	9.5	2.4
1964	3974	11.6	2.6
1965	3981	10.1	2.7
1966	4334	12.3	2.7
1967	4328	10.7	4.7
1968	4606	13.4	4.5
1969	4794	11.8	4.8
1970	4900	13.7	5.3

Source: Kurt Sauter, 1972: Der zweite Bildungsweg in der Schweiz, Dissertation, Zürich, p. 128

cate has been growing over the past few years. An important subgroup among the holders of committee type certificates are, in turn, the graduates of the so-called 'second course of education' (2. Bildungsweg). Their share among the federal certificate holders has increased at an even faster pace, although their numbers are still quite small (13).

---

13) There are some indications that this tremendous increase in graduates of the second course of education has reached its peak in the meantime. Thus, the 'Akademikergemeinschaft' (AKAD), the most important institution of the second course of education in Switzerland, reports an all-time high in the number of successful graduates for 1972. For the three following years, graduation of students from this institution has stagnated:

year	65	66	67	68	69	70	71	72	73	74	75
graduates	43	74	102	100	132	141	168	210	190	169	188

I am indebted to Dr. G. Durtschi from the AKAD for providing me with these figures.

This population shift draws its importance from the fact that the graduates of the second course of education are a very select group of students who are likely to experience a rather special flow through the university system. Most of these people have completed an apprenticeship before they began studying for the admission to the university. Thus, their average age lies considerably higher than the age of the average student: The average age of the graduates of the second course of education interviewed by SAUTER (1972) was 24.5, while the average age of the cohort 1965 on entering the university system was 20.5. But age is a very important determinant of flow, probably just because it indicates very special situations, such as the one of the graduates of the second course of education.

Table 91 shows the distribution of Swiss students in the WS73/74 over the 10 groups of subjects considered as independent determinants in this second part. Table 92 indicates how the distribution in Table 91 deviates from the one of the synthetic population (see Table 87). There are again quite substantial shifts, but no uniform pattern emerges. Every university has a unique combination of gaining and losing subjects.

In general, however, architecture, law (except at ZUERICH), medicine (except at ZUERICH and NEUCHATEL) and the humanities (except at LAUSANNE) have been gaining in importance, while the natural sciences I (except at ZUERICH) and II (except at BERN, NEUCHATEL and the ETHZ) have been losing. The most substantial absolute gains are those for law at NEUCHATEL and for the humanities at BERN and FRIBOURG.

These shifts are reflections of the changes in the preferences of students, which in turn are responses, partly, to new fads, partly, to changing demands of the labor market and, partly, to changes in the courses offered at the different universities.

These few tables give only an impressionistic picture of the population shifts

Table 91

Distribution of Swiss Students over 10 Groups of Subjects in the WS73/74 by University

Group of Subjects	University									
	Basel	Bern	ETHL	ETHZ	Frib	HSG	Genève	Laus	Neuch	Zürich
Medicine	.353	.270	-	.033	.122	-	.274	.346	.031	.229
Law	.118	.159	-	-	.127	.040[5]	.109	.152	.141	.151
Econ/Business[1]	.084	.095	-	-	.094	.920[6]	.101	.137	.136	.108
Social Sci	.031	.032	-	-	.170	.040[7]	.134	.060	.045	.108
Humanities[2]	.163	.179	-	-	.248	-	.120	.068	.302	.179
Hist Sci	.053	.026	-	-	.059	-	.049	.051	.040	.067
Natural Sci I[3]	.041	.079	.220	.124	.048	-	.048	.031	.088	.048
Natural Sci II[3]	.135	.131	.098	.193	.083	-	.118	.128	.168	.081
Engineering	-	-	.518[4]	.489[4]	-	-	-	-	-	-
Architecture	-	-	.161	.146	-	-	.017	-	-	-

Source: ESTA, 1975: Die Studierenden an den Schweizerischen Hochschulen im Wintersemester 1973/74, Tables 1.2, 1.5, 1.6, 1.7

1) including sociology
2) including all lower and higher secondary school teachers of the Phil. I Faculty
3) including half of the secondary school teachers of the Phil. II Faculty
4) including Economics/Business
5) 'Staatswissenschaft'
6) including the basic courses
7) 'Wirtschaftspaedagogik'

Table 92

Population Shifts with respect to the Enrolement in the 10 Groups of Subjects by University (WS73/74 - Synthetic Population)

Group of Subjects	University									
	Basel	Bern	ETHL	ETHZ	Frib	HSG	Genève	Laus	Neuch	Zürich
Medicine	+.009	+.038	-	+.031	+.004	-	+.022	+.054	-.072	-.017
Law	+.009	+.013	-	-	+.011	+.022	+.027	+.055	+.105	-.024
Econ/Business	-.001	-.038	-	-	-.051	+.002	+.016	+.007	+.014	+.028
Social Sci	+.020	-.031	-	-	+.031	+.004	-.005	-.009	-.008	+.054
Humanities	+.053	+.084	-	-	+.109	-	+.039	-.066	+.042	+.023
Hist Sci	± 0	-.023	-	-	-.040	-	+.004	+.016	-.017	-.004
Natural Sci I	-.022	-.010	-.070	-.028	-.021	-	-.036	-.016	-.034	+.001
Natural Sci II	-.056	+.035	-.055	+.028	-.037	-	-.046	-.036	+.003	-.061
Engineering	-	-	+.056	-.026	-	-	-	-	-	-
Architecture	-	-	+.067	+.021	-	-	+.004	-	-	-

in the Swiss University System, but they document that such changes do take place with respect to important variables.

In allowing for population shifts, but not for structural change, the new assumption presupposes that population shifts do not affect the determinant structure, as given by the regression coefficients. This implication may well be unrealistic. Consider the case of SES. The impact of this variable has been conspicuously negligible in the context of the present model. This result may be due to the rather homogenous middle/upper class composition of the Swiss student body. If the socio-economic composition of this student body were to change in such a way that the share of the lower classes would increase substantially, SES might well become an issue in the determination of an increasing number of flows.

I have tried to check, whether or not the socio-economic composition of the students enrolled at Swiss universities has changed at all over the past few years. The best figures I could find to answer that question are presented in Table 93. This table shows that the share of lower class students among the graduates of the higher secondary schools of the canton of ZUERICH has not been increasing lately. The corresponding figure for the total cohort 1965 was 9.0 (see RIES/KRIESI 1974d).

Table 93

Population Shifts with respect to SES among the Graduates[1] from Higher Secondary Schools of the Canton of Zürich: Percentage Distribution for the Years 1969 to 1973/74

Socio-Economic Strata	Year				
	1969	1970	1971	1972	1973/74
Upper Class	29.1	26.1	30.0	32.9	33.7
Upper Middle Class	28.0	29.7	32.6	29.9	28.4
Lower Middle Class	32.1	33.3	27.5	26.1	28.7
Lower Class	10.8	10.9	9.9	11.1	9.2

1) Only Schools that lead to an admission certificate to the university are included. The statistics give the figures for the students in the last grade, but since virtually all of those who reach this grade actually graduate, it is safe to speak of 'graduates'.

Source: Erziehungsdirektion des Kt. Zürich, Arbeitsgruppe für Bildungsplanung und Bildungsstatistik: Mittelschulstatistik 1969-1973/74

Since there is no population shift regarding SES in the first place, there is, of course, no inferred change in the determinant structure in the case of SES either. It is, however, still possible that population shifts in some other variables may result in structural change. I will, nevertheless, not consider such an eventuality in the predictions outlined below.

Under the new assumption of stable regression coefficients, the model with independent determinants allows to predict future total numbers of students and graduates on the basis of future stocks of first enrolments, just as did the simple model in Part I.

7.2.2. Prediction of the Total Number of Students

Whether there are population shifts or not, the same general procedure can be used for the prediction of the total number of students in future semesters. Population shifts only affect the specific matrices to be entered into this general procedure.

Without restriction of generality, the procedure outlined here refers to the prediction of students in future winter-semesters. A similar scheme could again be provided for summer-semesters. The present procedure differs from the one discussed in Part I, because the model with independent determinants consists of type 1 matrices only, whereas type 2 matrices have been used for this task in Part I. The present procedure is, however, closely related to the one employed in the prediction of graduates in a single semester by university of exit (see 3.2.2.).

The general procedure consists of three steps:
In a first step, the summer-stocks of first enrolments of the 11 cohorts involved are transformed into winter-stocks and added to the first enrolments in the winter-semesters of the corresponding cohorts. This transformation is performed by the T part of the first transition matrix of type 1 from the simple model of Part I.

In a second step, the resulting yearly stocks in the 11 winter-semesters are projected to the winter-semester we are interested in. This projection follows a successive pattern analogous to the one described in 2.2. for type 1 matrices.

Finally, in a third step, the projected stocks are summed over to yield the required total. For the prediction of the total number of students in WS75, this procedure looks as follows (14):

Step 1:
$$S(65) = S(SS65) \cdot T_{11} + S(WS65)$$
$$S(66) = S(SS66) \cdot T_{11} + S(WS66)$$
$$\vdots$$
$$S(75) = S(SS75) \cdot T_{11} + S(WS75)$$

Step 2:
$$S_{SS66}(65) = S(65) \cdot T_{12}$$
$$S_{WS66}(65) = S_{SS66}(65) \cdot T_{13}$$
$$\vdots$$
$$S_{WS75}(65) = S_{SS75}(65) \cdot T_{121}$$

analogously for all the other cohorts involved:

$$S_{WS75}(66) = S_{SS75}(66) \cdot T_{119}$$
$$S_{WS75}(67) = S_{SS75}(67) \cdot T_{117}$$
$$\vdots$$
$$S_{WS75}(74) = S_{SS75}(74) \cdot T_{13}$$
$$S_{WS75}(75) = S_{WS75}(75) \cdot I$$

Step 3:
$$N(WS75) = \sum_{n=65}^{75} S_{WS75}(n)$$

In the absence of population shifts, the matrices T_{1n} involved are identical with the ones obtained for the cohort 1965 using independent determinants. If there are expected population shifts from one cohort to the other, however, a different set of type 1 matrices has to be used for each cohort involved. To arrive

14) The notation is the same as in 3.2.2.

at these different sets of matrices, the general procedure has to be supplemented by two preliminary steps: First, the successive semester means have to be ascertained for each cohort involved. This is easier said than done, because these future semester means have to be extrapolated from present trends. Secondly, on the basis of these future semester means, which may differ for corresponding semesters from one cohort to another, the different type 1 matrices can be obtained by using program B-9.

In the absence of population shifts, the present procedure should yield the same results as the one outlined in 3.2.1. This is, however, not the case, because the extrapolated type 1 matrices for the last 6 semesters in this second part are, even in their modified form, not of equal quality as the corresponding matrices in Part I (see 7.1.3.3.). To get an idea of the accuracy of the present procedure, I have reestimated the total numbers of students in the WS73/74 using the matrices of the second part.

After having adjusted the matrices for drop-out errors, I have submitted them to the program B-10. This program first corrects for errors in the input data in a manner analogous to the program used in Part I (see 3.2.3.). Then it performs the three steps outlined above to yield the figures corresponding to the final prediction in Table 23.

Table 94 allows to compare the present results with those of Table 23. It shows that, only in three cases, the predictions are seriously aggravated by using matrices estimated on the basis of independent determinants. These cases are the two technical schools and the HSG. For the case of the ETHL, the independent determinants have turned out to be generally inaccurate (see 6.3.1.), whereas for the other two institutions, it is primarily the last 6 matrices, which are incorrect (see 7.1.3.3.).

The impact of the natural sciences leads to an overestimation of the students at the ETHZ, the one of High and Lin High to an underestimation of those at the HSG thus complementing the deviations in 7.1.3.3.

Table 94

Final Prediction of the Total Number of Swiss Students at the Swiss Universities in the WS73/74: Percentages of the Official Figures

University	Total According to ESTA (= 100%)	Final Prediction (Table 23) Part I	Final Prediction Part II
Basel	4244	94.4	90.6
Bern	5790	101.3	101.9
ETHL	994	95.0	109.2
ETHZ	5794	99.4	108.7
Fribourg	2354	109.3	108.2
HSG	1413	95.0	88.2
Genève	4480	102.5	103.3
Lausanne	3057	108.4	107.0
Neuchâtel	1172	88.8	88.0
Zürich	9057	99.2	99.2
Total	38358	100.1	101.2

Apart from these three institutions, the correspondance between the two sets of figures is quite satisfactory. In two instances, the universities of LAUSANNE and FRIBOURG, the present predictions are even a bit more accurate than the previous ones.

This overall correspondance is a result of the fact that the inaccurate matrices for the last 6 semesters are of only limited relevance for the prediction of the total number of students. They apply to the remaining rest of the earliest cohorts which are the least numerous ones in the first place.

Since the present figures are, however, somewhat less accurate than the previous ones, it is advisable to use the procedure of Part I, if one judges population shifts not to be important and/or, if one does not have sufficient information about such population shifts.

The overall accuracy of the prediction of Part I suggests that the population shifts which have taken place in the period covered by that prediction (see 7.2.1.), i.e. between 1965 and 1973, are indeed of negligible importance with regard to its outcome.

7.2.3. Prediction of Dropouts and Graduates

As discussed in Part I, there are two aspects to the prediction of students in different absorbing states:

- prediction of dropouts and graduates in a single cohort
- prediction of dropouts and graduates in a single semester

Since we have only type 1 matrices in this second part, the dropouts and graduates can only be predicted for the university of exit. The procedure employed in the prediction of dropouts and graduates in a single cohort is identical with the one discussed for the same task in Part I. The program in APPENDIX B-7 performs the necessary steps. In the case of population shifts, the matrices involved have again to be reestimated using the new semester means in combination with the given determinant structure.

As indicated earlier (see 7.1.3.4.), in predicting dropouts and graduates in a single semester, we can use the overall means of the synthetic population: To arrive at the transition rates in question, we only have to multiply the pertinent regression coefficients with the corresponding overall means, sum over the resulting products and add the constant term of the regression equation, just as indicated by the formula on p. 248. A simplified version of program B-9 performs the job.

To test the accuracy of the procedure for a single semester, I have stimated graduates and dropouts in the WS73/74. I have proceeded in four different ways:

- simple prediction: The overall means of the cohort 1965 are used and no modifications for semester characteristics in the last 6 semesters are introduced.
- prediction with population shifts: The new distributions for female students and for the ten different groups of subjects as given in Tables 89 and 91 are used instead of the overall means of the cohort 1965, but no modifications for the last 6 semesters are introduced.
- prediction with modifications for semester characteristics: The modifications discussed in 7.1.3.3. are taken into consideration, but no adjustments are made for population shifts.

- prediction based on population shifts and modifications: This prediction combines the two adjustments just discussed.

All of the predicted figures have been adjusted for dropout errors (15). An actual test of the accuracy of the four kinds of predictions can only be made for medium level and high level graduation, because official figures for dropping out and for low level graduation do not exist to my knowledge. For the two higher levels, the official figures have the additional shortcoming that they are not published for a single semester, but only for entire years. Assuming that students are as likely to graduate in the summer-semester as they are in the winter-semester, I have divided the figures by two to arrive at the numbers for the winter-semester 1973/74. Table 95 gives the results for the medium level, Table 96 for the high level.

First, we note that the estimates are rather inaccurate in general. At the medium level, taking both, population shifts and modifications into account, good estimates are obtained only for the three universities of FRIBOURG, GENEVE and NEUCHATEL. For the other universities, the deviations are, although small in absolute terms, enormous in relative terms. Similarly, at the high level, only the results for BERN turn out to be accurate.

Secondly, population shifts with regard to sex and groups of subjects generally are of limited importance. At the medium level, some noticable improvement results from their inclusion for the universities of FRIBOURG, GENEVE and LAUSANNE. At the high level, their impact is virtually non-existant, which reflects the earlier finding that high level graduation is primarily determined by semester characteristics (see 7.1.1.5.4.).

---

15) The unadjusted rates resulting from the above procedure have been multiplied by appropriate adjustment factors. To arrive at these factors, I have gone back to Part I and compared the leaving pattern derived from the unadjusted matrices with the one derived from the adjusted ones given in Table 13. Dividing the adjusted rates by the unadjusted ones has yielded the desired factors.

Table 95

Prediction of Medium Level Graduation in the WS73/74 by University: Percentages of the Total Number of Students Enroled Adjusted for Drop-out Errors (Swiss Students only)

University	Actual Figures According to ESTA 1)	Simple Predict.	Predict. with Pop. Shifts	Predict. with Modification of Semest. Charact.	Predict. with Pop. Shifts & Modifications
Basel	2.3	5.4	5.4	3.3	3.3
Bern	1.9	4.5	4.4	3.5	3.4
ETHL	5.7	3.5	3.5	3.5	3.5
ETHZ	5.9	5.7	5.6	5.7	5.6
Fribourg	5.8	3.6	3.9	3.6	3.9
HSG	8.4	5.0	5.0	5.0	5.0
Genève	4.6	5.2	5.6	4.3	4.8
Lausanne	3.9	7.4	7.1	6.2	5.9
Neuchâtel	5.0	4.7	4.8	4.7	4.8
Zürich	2.5	4.5	4.6	3.0	3.1

1) Source: ESTA, 1974: Statistisches Jahrbuch der Schweiz, p.470-85. The figures for the degrees are only given by years in the ESTA publications. Assuming that students are as likely to graduate in the summer-semester as in the winter-semester, they are divided by 2 to get an estimate of the graduates in the WS73/74.

Table 96

Prediction of High Level Graduation in the WS73/74 by University: Percentages of the Total Number of Students Enroled Adjusted for Drop-out Errors (Swiss Students only)

University	Actual Figures According to ESTA 1)	Simple Predict.	Predict. with Pop. Shifts	Predict. with Modification of Semest. Charact.	Predict. with Pop. Shifts & Modifications
Basel	2.5	4.4	4.4	2.0	2.0
Bern	2.1	2.3	2.2	2.3	2.2
ETHZ	1.5	.8	.8	.8	.8
Fribourg	1.0	1.2	1.2	1.2	1.2
Zürich	2.2	1.7	1.7	1.7	1.7

Source: See Footnote above

As a third result, we note the general improvement of the predictions at the five universities where modifications are introduced at the medium level. At BASEL, this improvement also carries over to the high level.

In spite of theses improvements, the results remain unsatisfactory. There are at least three reasons for this outcome: One of the reasons has to do with the smallness, in absolute terms, of the rates involved. It simply is much more difficult to estimate a small number than to be on target with a large one, such as the total number of students in a single semester. Secondly, the inaccuracy is, of course, a result of the inadequate approximation of the last 6 matrices. While these matrices are of limited relevance for the prediction of the total number of students, they are all the more important for the estimation of medium and high level graduation, as documented in the Tables 85/86.

Finally, graduation rates at the medium and high level are difficult to estimate, because they fluctuate considerably from one semester to the next, as is shown by the following two tables:

Table 97

Actual Figures for Medium Level Graduation by University and Semester: Percentages of Total Number of Students Enrolled (Swiss Students only)

University	Semester				
	WS70	WS71	WS72	WS73	average
Basel	2.5	2.4	2.2	2.3	2.4
Bern	1.9	1.5	1.7	1.9	1.8
ETHL	6.0	5.7	5.7	5.7	5.8
ETHZ	6.3	5.8	5.8	5.9	6.0
Fribourg	5.2	5.8	4.9	5.8	5.4
HSG	5.5	6.4	5.8	8.4	6.5
Genève	4.9	4.6	4.5	4.6	4.7
Lausanne	7.7	6.4	4.9	3.9	5.7
Neuchâtel	6.5	4.8	5.1	5.0	5.4
Zürich	2.9	2.8	3.0	2.5	2.8

Source: ESTA, Statistisches Jahrbuch der Schweiz 1971-1974

Table 98

Actual Figures for High Level Graduation by University and Semester: Percentages of Total Number of Students Enrolled (Swiss Students only)

University	Semester				
	WS70	WS71	WS72	WS73	average
Basel	2.5	2.4	2.5	2.5	2.5
Bern	1.8	1.4	1.4	2.1	1.5
ETHZ	1.4	1.4	1.2	1.5	1.4
Fribourg	1.3	1.4	1.2	1.0	1.2
Zürich	2.2	2.5	2.1	2.2	2.3

Source: See Footnote above

At the two universities with the most erratic pattern on the medium level, the HSG and the university of LAUSANNE, the final estimates of Table 95 lie much closer to the average rate reported in the last column of Table 97 than to the rate for the WS73/74. At least at LAUSANNE, the inaccurate prediction also seems to be a result of a trend toward longer studies for medium level degrees, i.e. a trend toward lower graduation rates per semester. This type of structural change has, of course, not been incorporated in the model.

On the other hand, the only accurate result at the high level, the one for the university of BERN, deviates more heavily from the average rate than from the rate for the WS73/74, indicating that the good fit for the latter was some kind of a chance hit.

The second cause of the inaccuracy at the two higher levels of graduation, which is probably the central one, does, however, not apply to <u>dropping out</u> and to <u>low level graduation</u> (see Tables 85/86). Although the predicted figures for these levels cannot be tested against empirical evidence, they appear to be rather reliable for this fact. <u>Table 99</u> shows the results of the predictions for the two lower levels of leaving. Since the modifications of semester characteristics have virtually no impact at these levels (16), only the first two types of predictions are provided.

---

16) The only exception is low level graduation at LAUSANNE, but there, too, their impact is almost nil.

Table 99

Prediction of Dropping out and Low Level Graduation in the WS73/74 by University: Percentages of the Total Number of Students Enrolled Adjusted for Drop-out Errors (Swiss Students only)

University	Dropping out		Low Level Graduation	
	Simple Prediction	Prediction with Pop. Shifts	Simple Prediction	Prediction with Pop. Shifts
Basel	1.0	1.3	.7	.7
Bern	.9	.9	1.2	1.7
ETHL	-	-	-	-
ETHZ	1.3	1.2	.2	.2
Fribourg	1.3	1.5	2.1	3.0
HSG	1.1	1.1	-	-
Genève	1.4	1.4	.9	1.0
Lausanne	1.8	1.7	.6	.6
Neuchâtel	1.4	1.4	2.3	2.3
Zürich	1.2	1.3	.9	1.0

The results in Table 99 are unique, because they give, for the first time, an idea of the absolute size of the leaving at the two lower levels for a single semester. One final reservation regarding the dropout figures has to be made: Although I have adjusted for dropout errors, recall that these adjustments do not include late dropouts (see 2.4.2.). This means that the actual dropout rates are likely to be somewhat higher than the predicted ones; but the difference will certainly be very small.

7.2.4. Other Predictions for a Single Semester

The same procedure used to predict dropouts and graduates in future semesters can also be used to predict other types of flow for such periods. As in the case of the two lower levels of leaving, the shortcomings of the last 6 matrices are of only minor importance or do not make any difference at all for these other movements. The following three tables give predictions for the WS73/74 for:

- staying at the university of origin (Table 100)
- interruptions and flows abroad (Table 101)
- the most important inter-university flows (Table 102)

Table 100

Prediction of Staying at the University of Origin in WS73/74 by University: Percentages of the Total Number of Students Enroled Adjusted for Drop-out Errors (Swiss Students only)

University	Simple Prediction	Prediction with Pop. Shifts	Prediction with Modifications of Semest. Charact.	Prediction with Pop. Shifts & Modifications
Basel	84.8	84.5	88.1	87.9
Bern	88.3	87.8	88.3	87.8
ETHL[1)]	90.5	82.4	90.5	82.4
ETHZ	85.4	85.4	85.4	85.4
Fribourg	85.5	84.1	85.5	84.1
HSG	86.2	86.2	86.2	86.2
Genève	87.2	85.9	87.2	85.9
Lausanne	83.7	83.7	85.0	85.1
Neuchâtel	85.9	87.4	85.9	87.4
Zürich	86.2	85.6	89.3	88.7

1) unadjusted figures, because no predictions are made for dropping out at the ETHL

Table 101

Predictions of Interruptions and Flows to Foreign Universities in the WS73/74 by University: Percentages of the Total Number of Students Enroled (Swiss Students only)

University	Interruptions		Flows to Foreign U's	
	Simple Prediction	Prediction with Pop. Shifts	Simple Prediction	Prediction with Pop. Shifts
Basel	2.0	2.0	.9	.9
Bern	2.8	2.7	.8	1.0
ETHL	2.4	2.9	-	-
ETHZ	6.8	7.0	.2	.2
Fribourg	2.4	2.4	.7	1.2
HSG	5.3	5.3	-	-
Genève	1.9	1.9	.5	.7
Lausanne	2.8	2.9	.5	.4
Neuchâtel	2.2	2.1	-	-
Zürich	2.5	2.5	.9	1.1

Table 102

Prediction of Inter-University Flows in the WS73/74: Percentages of the Total Number of Students Enrolled at the University of Origin (Swiss Students Only)

Inter-University Flow	Prediction		Inter-University Flow	Prediction	
	simple	with Pop Shift		Simple	with Pop Shift
Basel-Bern	.3	.3	Genève-Basel	.5	.5
Basel-Fribourg	.2	.2	Genève-Zürich	.8	.9
Basel-Zürich	.3	.4	Lausanne-Bern	.6	.7
Bern-Lausanne	.2	.2	Lausanne-Genève	.7	.8
Bern-Zürich	.4	.5	Neuchâtel-Gen.	.7	.4
ETHZ-Zürich	.6	.6	Neuchâtel-Laus.	1.9	.7
Fribourg-Basel	.4	.4	Zürich-Bern	.3	.2
Fribourg-Bern	.8	.7	Zürich-Fribourg	.2	.2
Fribourg-Genève	.3	.3	Zürich-Genève	.2	.2
Fribourg-Laus	.4	.4	Zürich-Lausanne	.2	.3
Fribourg-Zürich	1.1	1.1			

With respect to staying, population shifts are shown to have an impact at all the universities, except at the ETHZ, the HSG and LAUSANNE. The biggest effects result for the ETHL and for NEUCHATEL. While the former, however, cannot be taken seriously, since it is a result of the spurious impact of sex at the ETHL, the latter is a consequence of the lowering of the share of medicine at NEUCHATEL.

Modifications of semester characteristics give higher percentages of stayers at BASEL, ZUERICH and LAUSANNE.

Apart from the ETHL, the final predictions range from 84.1% staying (at FRIBOURG) to 88.7% staying (at ZUERICH). In line with earlier results, the Swiss German universities (BASEL, BERN, and ZUERICH) have a somewhat higher staying rate than the technical schools, the HSG and the French speaking universities. The differences are, however, not sizable.

For the remaining predictions, no adjustments for dropout errors have to be made, because they are unaffected by them. With regard to interruptions, a shift toward architecture results in an increase at the two technical schools. The interruptions at the other institutions are hardly affected by population shifts at all. The figures in Table 101 for a single semester complement the ones for a single cohort given in Table 9.

Regarding the flows to foreign universities, the shift in favor of the humanities makes for a higher rate at BERN, GENEVE and ZUERICH and, above all, at FRIBOURG. On the other hand, the corresponding flows from LAUSANNE decrease in size, because at this university, the humanities have lost in importance. These figures can again be compared with the pertinent ones from Table 9.

As in the case of the single cohort, the inter-university flows are also very limited in a single semester. Only fractions of a percent generally move in a single semester. Similarly, population shifts amount to only a tenth of a percent in general. There is one exception: Due to the smaller share of medical students at NEUCHATEL, fewer students move from this university to the ones of GENEVE and LAUSANNE.

These tables are again unique in that they give for the first time an idea of the size of the respective flows in a single semester.

7.2.5. Conclusion

The prediction procedures developed here are superior to the ones in Part I in that they make a less restrictive assumption regarding the rates of flow: Instead of stable rates of flow, only a stable determinant structure is required. Population shifts can now be incorporated.

In spite of this superiority, the predictions of the total numbers of students using the new methods will in general be less accurate than the ones in Part I, because of the shortcomings of the last 6 matrices involved and because population shifts do not make much difference for this prediction.

Similarly, in predicting graduates at the medium and the high level, one preferably uses the procedures and matrices of Part I, although those procedures are more intricate. For the leaving at the two lower levels as well as for the other types of flows, the methods developed here are, however, sufficiently accurate. They should, therefore, be preferred to the previous ones in these instances, because of the incorporation of population shifts which have some impact on some of these rates.

In general, however, it can be concluded that, although there have been quite substantial population shifts between 1965 and 1973, none of them has reached a size such as to make a decisive impact on the rates of flow. But in the absence of population shifts, it is quite clear that the present model predicts the same rates of flow for any future semester. Thus, all the simple predictions in 7.2.3. and 7.2.4. apply not only to the WS73/74, but to any other semester as well.

7.3. Simulation

In his book on inequality, BOUDON (1974) introduces a methodological innovation: He constructs an ideal-typical society. This allows him not only to incorporate factors that are empirically unmeasurable, above all, this makes it possible for him to work out the implications of a set of assumptions in a systematic fashion.

The present model with independent determinants lacks the syntactic sophistication of BOUDON's ideal-typical society, for reasons explained earlier. But it still represents, however crude, an ideal-typical reconstruction of the system of flows of students in Switzerland. As such it, too, can prove to be a quite versatile tool in the attempt to work out the implications of a set of assumptions.

Assumptions can be introduced into this model by changing the overall means, the semester means, or the determinant structure. Changes in the means, i.e.

population shifts, are preferable, because they are easier to handle. When changing an overall mean or a semester mean and leaving everything else constant, the latter phrase has a precise meaning. First, it implies that all the other semester means will remain unchanged, which is to say that those individuals who replace the original ones have the same combination of characteristics on all the other relevant variables as the ones they are replacing. Secondly, it means that the determinant structure, too, will remain as it is. This in turn says that all the relevant variables will have the same impact as before on the rate of flow in question, and no other variable will become relevant.

The implications of changes in means can be checked in several ways. Changes in semester means and the determinant structure can be checked against shifts in the resulting transition matrices by using the index of dissimilarity. Or, the resulting changes in the pattern of leaving of a single cohort can be used as a suitable criterion. Changes in the overall means, on the other hand, are best checked against shifts in the transition rates predicted for a single semester.

The implications of changes in the overall means can be worked out with paper and pencil - occasionally with the additional help of a pocket calculator. The implications of changes in semester means and in the determinant structure are better calculated by the computer.

The discussion of overall means allows an assessment of the direct impact of a certain characteristic, while the discussion of semester means will show the systematic character of the model, by giving not only the direct impacts, but also the indirect ones.

In the following, I will only deal with population shifts in individual characteristics and neglect shifts in the semester characteristics. Shifts in the latter are more difficult to handle, because, first, a change in the semester mean of a semester characteristic entails changes in a host of other semester characteristics as well. Moreover, since changes in semester characteristics are a result of structural change in the university system, they will most likely go hand in hand with changes in the determinant structure.

7.3.1. Population Shifts with no Direct Impact on the Rates of Flow

First of all, it seems worth asking, what types of population changes can take place without having any direct impact on the rates of flow. Since each transition at every particular university is practically governed by a unique set of determinants, there is also a unique set of irrelevant characteristics in each case. Without going into the details of all these cases, a few general remarks can be made on the basis of overall means.

As has been mentioned before, SES is a rather unimportant determinant of the flow of students. Thus, an increase in educational opportunity for lower classes would not affect the flow of students in general. At the ETHL, FRIBOURG, GENEVE and NEUCHATEL, there are, however, 8 transitions altogether which would be affected. At these universities, a lowering of the average SES to 2.5 (17) would result in the following changes of the corresponding simple predictions:

Table 103

Effects of Increasing Educational Opportunity on the Flow of Students

University	Transition	Simple Prediction	Change due to Shifts in SES	New Prediction
ETHL	Interruption	2.4	+.7	= 3.1
Fribourg	Staying	85.5	-.5	=85.0
Fribourg	Medium Level	5.8	+.3	= 6.1
Genève	Staying	87.2	-.9	=86.3
Genève	Low Level	.9	+.3	= 1.2
Genève	Medium Level	5.2	+.7	= 5.9
Neuchâtel	Interruption	2.2	-.6	= 1.6
Neuchâtel	Dropping out	1.4	+.5	= 1.9

17) This roughly amounts to cutting the share of the upper two classes in half, doubling the one of the two lower classes and increasing the percentage of the medium middle class by as much as to make everything add up to 100 again.

A strong relative increase in interruptions at the ETHL, in low and medium level graduation at GENEVE and in dropouts at NEUCHATEL would take place, while the interruptions at NEUCHATEL would be considerably reduced. The effects at FRIBOURG and for staying at GENEVE would be of only minor importance.

In addition to that reservation, it has to be recalled that a population shift toward the lower classes may well affect the determinant structure. Today, a lower class student entering the university is well adapted to the upper middle class orientation prevalent at this institution. This is a result of anticipatory socialization and social selection on the lower levels of the educational system. A lowering of the selective barriers may bring in students with quite different orientations, and, concomitantly, different types of flows.

With the exception of teacher's certificates, changes in the distribution of the admission certificates would not directly alter the flow of students in general. Thus, the noticable increase in the share of committee type federal certificates (see Table 90) has virtually no impact on the flow. Only three transitions are affected by this type of certificate, and even if its share were to double, a less than 10% change would result in every one of these cases (medium level graduation at the ETHL, dropping out at the ETHZ and interruptions at LAUSANNE) with respect to the figures predicted for a single semester.

Similarly, doubling the share of cantonal certificates or of entrance exams or both has very little effect on the transitions involved (five cases for the cantonal certificates and seven for the entrance exams).

Finally, the business certificate has some noticable impact in only three instances (out of the 11 where it is entered). Doubling the share of business certificates at all the universities just results in a 17% increase of low level graduates in a single semester at BERN, and at LAUSANNE, the dropout rate would increase by 11% and the rate of medium level graduation by 13%.

A change in duration status whatsoever would not affect the graduation pattern in a direct manner at all at 8 out of the 10 universities, the exceptions being FRIBOURG and ZUERICH.

But variables which have been shown to be very important can be disregarded in certain contexts, too, if we restrict discussion to overall means. Thus, sex is no issue at all at the HSG and at NEUCHATEL. Even if the share of female students keeps increasing at the HSG, the model says that the pattern of leaving in a single semester will not be affected at this institution; nor will be any other flow for that matter. Similarly, the dropout rate at BERN, the ETHZ, FRIBOURG and GENEVE, low level graduation at LAUSANNE, medium level graduation at both of the technical schools and at BERN, as well as staying at the ETHZ and at NEUCHATEL would not be directly affected even by very large increases of the share of female students.

Age, too, does not determine everything: Medium level graduation is influenced by age at GENEVE only. Age does not determine dropping out at the ETHZ and at FRIBOURG, it has no direct impact on low level graduation at the ETHZ, at BASEL, GENEVE and LAUSANNE, and the rate of stayers cannot be changed directly by changes in the average age at BASEL, BERN, both of the technical schools, at the HSG and at NEUCHATEL.

Origin, while being important for inter-university flows, has virtually no impact on the leaving of the system, and the composition of students with regard to origin can therefore be altered at will without changing the leaving pattern. The only exception: low level graduation at LAUSANNE.

Even shifts in the distribution of students among the different groups of subjects could take place without directly changing important flows: Except at LAUSANNE, the rate of staying would remain the same, even if there were distributive shifts between law, economics/business, the social sciences and the natural sciences I. Engineering does not have any impact on the leaving pattern at all, and, with respect to low level graduation, shifts between medicine, law and the historical sciences do not have any either.

Finally, as will be recalled, high level graduation is practically unaffected by individual characteristics, and therefore will not be changed by shifts among them.

But not only is there a wide variety of cases, where some variable or another does not have any direct impact at all, even if they do have an impact, changes in these variables often do not make a big difference at all. Changes in sex and the different groups of subjects, important variables indeed, have thus been shown to have a limited import in the previous section. I want to ask now, what would happen, if the three socio-economic characteristics with the greatest effect on the leaving pattern - sex, age and work load - were to undergo more drastic changes.

7.3.2. Drastic Changes in Socio-Economic Characteristics

Drastic changes would be a doubling of the share of female students and of the average work load, as well as an increase of the average age by two years. Although drastic, such changes are not inconceivable. They in part extrapolate existing trends. The share of females students has generally increased since 1965 (see Table 89). A doubling of the percentage of female students of the cohort 1965 would give the female students a majority only at the three French speaking universities. As a consequence of the increasing importance of the second course of education (see Table 90), it is also possible that the average age is to increase. The average age might increase, too, if more graduates of the higher secondary schools postpone their matriculation at the universities than have done so far: Roughly one third of the cohort 1965 (34.5%) have not immediately continued their studies after graduation from the secondary schools. They have either done military service (18.9%), got a job (15%), gone abroad (6.1%), completed some required practical work (5.8%), or acquired another training (4.3%) (18).

---

18) The percentages do not add up to 34.5%, because the same student could have done several things in the meantime.

Table 104

Mean Indices of Dissimilarity between Matrices Based on the Original Socio-Economic Characteristics and Matrices Based on Drastically Changed Socio-Economic Characteristics (Estimated Cells Included only)

University	Index of Dissimilarity [1]			
	double share of females	increase of av. age by 2 years	double av. work load	combination of the 3 changes
Basel	.9	.3	.6	.7
Bern	.3	.9	.6	1.0
ETHL	-	0	.6	.6
ETHZ	.6	.5	.1	.6
Fribourg	.8	1.3	0	2.1
HSG	0	.7	.7	.7
Genève	1.1	2.7	.7	3.3
Lausanne	1.0	1.6	.7	2.6
Neuchâtel	0	1.0	.3	1.3
Zürich	.8	1.4	.4	1.7

1) A 0 signifies that the corresponding item is no issue at the university in question. At the ETHL, the impact of Sex is spurious.

It is quite possible that, depending on the economic development, more students postpone their studies in the future. The economic situation may also require the students to put in more hours into a job to finance their studies.

These changes are introduced into the model by way of changing the semester means. All the means at all the universities in every semester have been changed in the same way to facilitate computations and not to come into conflict with the existing determinant structure. The simulated transition matrices have been computed by using a modified version of program B-9.

Table 104 shows how the transition matrices change, if each one of the modifications is introduced separately, and if all of them are introduced together. It gives the average indices of dissimilarity, averages taken over all 21 semesters. The standard deviations have been omitted, because they are all very small, i.e. smaller than .5, with one exception (19); half of them are either .1

19) This exception is age at the HSG, for which the standard deviation turns out to be .7.

or 0. This indicates that the changes in the transition matrices are similar in all of the semesters, which, of course, is also a result of the way the changes have been introduced.

Moreover, the new transition matrices deviate only slightly from the original ones. Deviations are in general largest for the changes in age and smallest for those in work load, a result that corresponds to earlier findings (see Table 66). The deviations do not add up in the combination of all three changes, because of cancellations.

Although the deviations in the single matrices turn out to be only small, they may cumulate to result in quite different leaving patterns after 21 semesters. The leaving pattern for the simulated matrices can be computed using program B-7. The next four tables show the relative percentage differences in the leaving pattern for each one of the changes (20). Because of the additivity of the model, the shifts in the first three tables add up to the corresponding ones for the combination of all three changes in Table 108, except for rounding errors.

The doubling of the share of female students generally leads to more low level graduates and fewer graduates on the two higher levels. With respect to dropping out, the results are mixed. Only at BASEL (+33%) and at ZUERICH (+19%), the drop out rate would increase substantially. At the rest of the universities, it would either decrease or remain virtually unchanged. Low level graduation would be most affected: At BASEL, the ETHZ and GENEVE, it would be tremendously raised. LAUSANNE, on the other hand, is the only university, where it would be reduced. Dropping out at the HSG, low level graduation at NEUCHATEL and medium level graduation at BASEL and GENEVE would not be affected at all.

---

20) The relative percentage differences can be obtained by dividing the difference between corresponding simulated and original figures through the original ones (given in Table 85).

Table 105

Relative Percentage Difference in the Leaving Pattern when Doubling the Share of Female Students, by University

University	Level of Leaving			
	Drop out	Low	Medium	High
Basel	33	39	0	-12
Bern	-3	10	-7	-8
ETHL	-	-	-2	-
ETHZ	1	45	-2	-4
Fribourg	-7	13	-1	-14
HSG	0	-	1	-
Genève	-6	39	0	-
Lausanne	-7	-9	-1	-
Neuchâtel	1	0	-2	-
Zürich	19	2	-2	-9

Table 106

Relative Percentage Difference in the Leaving Pattern when Increasing the Average Age by 2 Years, by University

University	Level of Leaving			
	Drop out	Low	Medium	High
Basel	44	3	-7	-9
Bern	55	40	-8	-9
ETHL	-	-	-5	-
ETHZ	-4	-5	-8	-14
Fribourg	-9	38	-16	-25
HSG	-1	-	-7	-
Genève	28	-19	-3	-
Lausanne	-16	-22	-30	-
Neuchâtel	61	59	-4	-
Zürich	37	74	-17	-20

Several rates would, however, change with regard to which sex has been shown to have no direct impact in the previous section. Examples are the dropout rates at BERN, the ETHZ, FRIBOURG and GENEVE. This is an indication of the systematic character of the present model.

Table 107

Relative Percentage Difference in the Leaving Pattern when Doubling the Average Work Load, by University

University	Level of Leaving			
	Drop out	Low	Medium	High
Basel	5	-21	-1	-1
Bern	7	-26	12	14
ETHL	-	-	-10	-
ETHZ	0	-10	-3	-2
Fribourg	0	1	1	2
HSG	63	-	-1	-
Genève	4	4	-5	-
Lausanne	0	22	-.2	-
Neuchâtel	0	1	1	-
Zürich	15	-29	1	8

Table 108

Relative Percentage Differences in the Leaving Pattern for the Combination of the 3 Changes, by University

University	Level of Leaving			
	Drop out	Low	Medium	High
Basel	81	16	-7	-20
Bern	59	27	-4	-3
ETHL	-	-	-16	-
ETHZ	-4	10	-13	-15
Fribourg	-14	49	-15	-34
HSG	57	-	-7	-
Genève	25	19	-7	-
Lausanne	-21	-13	-27	-
Neuchâtel	61	39	-4	-
Zürich	69	63	-18	-21

The consequences of an increase in average age are generally more substantial than the ones for sex (Table 106). Graduation at the two higher levels would decrease in all instances for which independent determinants have been estimated. Low level graduation and dropping out would, on the other hand, not generally be raised: At the ETHZ, the HSG, FRIBOURG and LAUSANNE, the dropout rate would also be reduced as would be the low level graduation at the ETHZ, GENEVE and LAUSANNE.

For the other universities, however, drastic increases in the dropout rates would result and the low level graduation would shoot up at FRIBOURG, BERN and, above all, at ZUERICH (+74%). The largest decrease at the higher levels would occur at LAUSANNE (-30% medium level graduation), but higher level graduation would also be considerably cut at FRIBOURG and ZUERICH (decreases ranging from 16% to 25% at both of the higher levels).

The impact of a doubling of the average work load is less pronounced and it does not follow a clear cut pattern (Table 107). It would raise the dropout rate at the HSG by 63% and at the university of ZUERICH by 15%, but leave it pretty much unchanged at the rest of the universities. The low level would again be affected the most with substantial cuts at the three Swiss German universities and a considerable increase at LAUSANNE.

If all the three changes were to occur together, the leaving pattern would be quite drastically changed, as is summarized by Table 108, although the impacts of the three characteristics cancel to some extent. Graduation at the two higher levels would generally be lowered, the one at the lower levels would be raised, except at LAUSANNE, and dropping out would increase at the three Swiss German universities, at the HSG and at GENEVE and NEUCHATEL, while it would decrease at the ETHZ, FRIBOURG and LAUSANNE.

Thus, in spite of the fact that the single transition matrices are only slightly affected even by the induced drastic changes, the impact on the pattern of leaving is quite impressive. Which is to say that population shifts can make a difference, if they are only drastic enough.

7.3.3. Limitations on Inter-University Flows of Medical Students

A final question that shall be asked to illustrate the use of this model for purposes of simulation is concerned with the flow of medical students between the Swiss universities. Together with duration status and origin, medicine has proved to be the most important determinant of inter-university flows. It has also been indicated that the impact of medicine on these flows has undergone change and is likely to change further.

One way to assess the presumable impact of such changes is to ask what would happen, if medical students were no longer allowed to change their university. Now, this question provides an example where the determinant structure has to be manipulated to work out the implications: The coefficient for the characteristic of medicine has to be set to zero, wherever it enters a regression equation for an inter-university transition.

A modified version of program B-9 again allows to make these adjustments and computes new transition matrices on their basis. To check the implications, the number of moves predicted when using the original set of matrices can be compared with the number of moves predicted with the new set. A variant of program B-7 does this job.

Ten of the 23 important transitions between Swiss universities would be strongly affected, if medicine no longer were a factor in the inter-university moving. These ten transitions would be reduced to the following fractions of the original volume:

- NEUCHATEL-LAUSANNE	to 15%
- LAUSANNE-BERN BERN-LAUSANNE	to 20%
- FRIBOURG-BASEL	to 27%
- FRIBOURG-ZUERICH	to 39%
- FRIBOURG-GENEVE	to 40%
- FRIBOURG-LAUSANNE	to 45%
- FRIBOURG-BERN	to 52%
- BASEL-ZUERICH	to 53%
- NEUCHATEL-GENEVE	to 60%

These fractions, however, surely underestimate the amount of the reduction in the ten flows which would occur. Take, for example, the flows from FRIBOURG to BASEL, GENEVE, LAUSANNE and ZUERICH. They are determined by two characteristics - medicine and XN Low (in the case of FRIBOURG-ZUERICH, origin also plays a minor role, see Fig. 27). These two characteristics, however, are not independent from each other, since most of the medical students leave around the 7th semester after having passed their second preliminary exam, a fact which is picked up by the entry of XN Low.

This illustrates how tricky it is in this simple additive model to manipulate the determinant structure in the way indicated. The example nevertheless goes a long way to show the decisive impact the subject of medicine has on the rate of inter-university flows in Switzerland. Any change regarding this impact will deeply affect the volume of those flows.

8. SUMMARY AND DISCUSSION

8.1. The Relationship to Markovian Models

I have started out this second part by motivating the search for independent determinants with a discussion of the relationship of the present model to Markovian type models. The flow of Swiss students through the Swiss University System being a specific kind of intragenerational mobility, it has been tempting to check, whether or not it fulfils the assumptions of the latter models which have been used pervasively for fitting social mobility processes. As has been shown, the flow of Swiss students, however, is neither stationary nor Markovian. Attempts to attribute the failure of Markovian models to the presence of cumulative inertia have generally been unsuccessful. This phenomenon only applied to inter-university flows (and not to the leaving of the system) in the first place, and even regarding these flows, it has proved to be operating to a limited extent only, due to structural constraints. The phenomenon of cumulative inertia implies that the students be free to establish stronger ties to a given state; but the constraints forcing some students to make certain moves are incompatible with that implication.

The relaxation of the third assumption made by the Markovian models - homogeneity - has, eventually, led to the consideration of independent determinants of flow. In resorting to general statistical techniques to introduce them into the model given in Part I, I gave up on the idea to reconstruct the flow of Swiss students on the basis of Markovian models.

8.2. The Techniques

The general technique for the introduction of independent determinants of flow into the model has been provided by SPILERMAN (1972), who used simple additive regression models with dummy variables as indicators of whether a certain transition has been made or not. Direct application of that technique to the present model would, however, have resulted in an enormous amount of calculations: Because of the structural constraints limiting the flow in the Swiss University System, a different set of regression coefficients would hav had to be estimated for every one of the 15 one-step transition matrices involved.

The introduction of a set of independent variables accounting for heterogeneity among the semesters and the construction of a synthetic population file, where every individual is a case for every semester he has been in the university system, have provided an elegant solution to this problem, allowing for just one unique set of regression coefficients for all the semesters.

The pertinent independent determinants have been selected in a two-step inductive procedure. First, on the basis of a preliminary understanding of the determination processes, consisting of preliminary results from this study, of the results reported in the working papers of the team at Aarau, and of personal acquaintance with the Swiss University System, a pool of 36 variables including the most important tentative determinants has been selected. In a second step, these variables have been submitted to a stepwise regression procedure for each one of the 83 transitions in question.

Although the R^2 values resulting for the different regressions have generally been very small, the selected predictors have been shown to be the main determinants of heterogeneity with respect to the flows of students in Switzerland: It was possible to reproduce the rates of flow as well as the patterns of leaving on their basis in a quite satisfactory manner.

Although the selected predictors do not tell much about individual level behavior, they tell a great deal about the one of the student bodies at the different universities.

It has been pointed out several times that, while the stepwise regression procedure employed is well suited for description, prediction and simulation, it can only tentatively be used for theory construction because of its weak syntax. More specifically, the problem with these procedures in the context of theory construction is that they do not allow for assumptions about the causal sequence among the variables to be selected (see BLAU 1973, p. 47). Consequently, as BLAU stresses, "several measures of the same underlying concept may be selected, and factors that exert only indirect effects on the dependent variables are excluded, since they do not improve prediction." But, still according to BLAU, theoretical analysis focuses on the most fundamental conditions affecting a dependent variable and the aim of the analysis is to find intervening variables mediating their influences, i.e. indirect effects are of prime theoretical importance.

BLAU therefore dismisses stepwise regression procedures as theoretically meaningless. In my opinion, this judgement of his is inadequate. First of all, the selection of several measures for the same underlying concept can be avoided, if the stepwise procedure is guided by a careful analysis of preliminary results and if a multicollinearity criterion is specified, as was done in the present study. Secondly, even if one does not get at the most fundamental conditions using stepwise procedures, it is still worthwile using them, because one can detect some of the pertinent intermediary variables. One can also show, in a very general way, where the fundamental conditions ought to be sought.

Moreover, given the present stage of sociology, where, as BLAU also concedes (p. 18), research is largely exploratory, neglecting indirect effects can be advantageous: The consideration of indirect effects leads to increasingly complex influence structures, even if one includes only a rather limited number of independent variables in the analysis: With 3 predictors, there are 4 possible indirect effects, with 4 predictors there are already 11 of them, with 5 there are 26 and so on. In addition, one has to be aware of ever increasing possibilities of spurious effects. Such complex structures easily get out of hand.

In sum: In spite of their syntactical weakness, the stepwise regression procedures employed in this study have resulted in a determinant structure which is at least tentatively useful for the construction of sociological theory.

8.3. Description

8.3.1. The Importance of Structural Characteristics

Much of this second part has been devoted to the description of the selected determinants. Without going into details again, we can but note once more the importance of structural factors of the university system in the determination of the rates of flow: The unique setting of a particular group of subjects at a specific university determines to a large extent, whether or not a student will, on the average, stay at his university of origin and whether or not he will graduate and at what level. Age and sex substantially supplement structural determinants for staying and graduating at the two lower levels. Low level graduation and interruptions are, in part, also determined by structural factors from outside of the university system, such as prior educational institutions attended and exigencies of the military system. Flows to foreign universities primarily depend on whether or not one studies humanities, inter-university flows within Switzerland are above all affected by the subject of medicine, origin and duration status.

The structural influence of the university system is most clearly seen for the graduation at the two higher levels, which is primarily a question of the fulfilment of given requirements which it takes a certain amount of time to accomplish.

The groups of subjects have been used as indicators of different structural conditions. The 60 odd subjects have originally been reclassified into 12 groups to make for more homogenous groupings than the ones provided by the traditional faculties of the Swiss universities (see p. 17). Each one of the 12 groups used is, however, still quite heterogenous. To get at the unique setting which seems to be so important, one would actually have to take up each subject separately.

At Swiss universities, as at most of the European universities, each subject is organized around a so-called chair or a group of such chairs. Each one of these chairs is, as BEN-DAVID (1974, p. 17) notes, a self-contained unit:

> "The university has no authority over this unit, nor in effect has anyone else, because the incumbents of the chairs have tenure, and the resources and facilities for the chairs are rigidly fixed by statute and administrative rules, or, at least, by time-honored usage."

Swiss universities actually are, as BEN-DAVID pointedly puts it for European universities in general, associations of independent practitioners. Each one of these practitioners is endowed with his own resources, facilities and collaborators that form an institute. It is at these institutes, where research and teaching in the particular subject in question is concentrated. At these institutes, actors within the university system can clearly be identified and here the networks of relationships of the university system with other societal systems are discernible (see BINDER et al. 1974, BINDER 1975, WISSENSCHAFTSRAT 1972).

The top leadership at Swiss universities has traditionally been weak. At the top there is no viable organizational unit that could represent the university as a whole. The university as a whole is no 'interlocuteur valable' (DEPPELER 1969, SCHWEIZERISCHE HOCHSCHULKONFERENZ 1972). The academic self-government has traditionally been monopolized by the 'faculty'. This has been meaningful as long as the faculties were able to deal with all substantive matters of the subjects embraced by them. Due to the increasing specialization of the scientific disciplines and the resulting creation of ever new disciplines within the framework of the old faculties, the latter are no longer competent to do so. The result has been an even greater fragmentation of the university structure. A second development has enhanced this trend twoard fragmentation: The traditional organization of the university has been characterized by its autonomy from other societal systems. The increasing relevance of science for those other systems has, however, led to encroachments on this autonomy on their part; as science has become the 'first force of production' (HABERMAS 1970) teaching and research have become increasingly 'vergesellschaftet' (GOLDSCHMIDT 1973). Moreover, the autonomy of the university has also been broken from within the system of science: Social networks that transcend the boundaries of an individual university, so-called invisible colleges (CRANE 1972), have been formed among the scientists. - Increasing differentiation of scientific disciplines and a loss of autonomy combine to break up even more the traditionally lose structure of the university.

Thus, in trying to detect the unique structural factors that are responsible for the different flows at each one of the different universities, one ought to use the institute as the unit of analysis. This means, however, that the complexity of the research problem is greatly increased because of the great number of institutes. BINDER et al. (1974), for example, included 155 institutes - all of them administrative units with an own budget etc., doing at least some research - in their study of the two institutions of higher education at ZUERICH (the University of ZUERICH and the ETHZ). One would have to draw samples.

Once the proper unit of analysis has been selected, the question still remains which particular factors are responsible for the particular setting at an institute and how each one of them does affect the different types of flow. In trying to answer that question it will be most useful to turn to the sociology of organizations. BLAU (1973) in his analysis of the organization of academic work at American universities has shown that the same fundamental social forces appear to govern the development of the formal structure of organizations at universities as well as at government bureaus. While he has analyzed the whole university as an organization, it would be most useful to replicate his analysis in the context of the Swiss University System at the level of the institute.

The analysis of BLAU indicates that the size of an institution has a predominant effect on its character (p. 252):

"The reason is that institutional size comprises several elements, though numerous correlates of it have been controlled. It refers to both economic and manpower resources ..., an impersonal atmosphere, a large administrative work load, and the statistical probability of finding colleagues with common interests."

It is, however, noteworthy that size affects neither of the 2 indicators of educational performance in BLAU's study (college drop out and continuation rates at the graduate level). This leads one to assume that the effect of size on the flow of students is primarily an indirect one.

A second set of important variables is concerned with the quality of the human resources of a university, or of an institute in the Swiss case: Formal qualifications of the faculty, the orientation of the faculty, the institution's reputation and the aptitude of the students. Important are also institutional resources such as the academic tradition, the departmental (or 'faculty') structure, the academic climate and research emphasis. Variables characterizing the economic situation (finances and facilities) as well as the faculty-student relationship (faculty-student ratio, faculty-student contact) are also very influential in this context.

BLAU's study is an abundant source of hypotheses for an analysis of the institutes from an organization sociological point of view. For some of the variables in question data have been obtained by RIES/KRIESI 1974a. Unfortunately, however, those data pertain to the level of the 'faculty' and are, furthermore, subject to some errors. With respect to the faculty-student contacts, it could be shown that a lack of such contacts does contribute to the determination of dropping out at Swiss universities, too (see KRIESI 1975).

The fragmentation within the Swiss universities is complemented by a similar fragmentation between them. It is not astonishing that there are relatively few inter-university moves in Switzerland: As has been pointed out at the outset (see p. 19), the degree structures at the different universities are highly incompatible. It may therefore happen to a student that the degree he has obtained at one university will not be accepted by another one. The situation in Switzerland is a far cry from the standardization of the different degree levels in the United States' system of higher education which makes that system integrated and which allows the student to move with relative ease from level to level and from one part of the system to another (see BEN-DAVID 1974).

One may object that this dominance of structural factors is an inevitable result, since most of the variables in the pool submitted to the regression procedure have been structural in nature in the first place. One may point to a host of studies which have shown that differential ability, differential socio-economic background, differential motivation and other personality factors have a decisive impact on the differential success of students, i.e. on the flows concerning the levels of leaving (1). LAVIN (1965) notes that measures of ability explain between 35% and 45% of the variance of the performance at the university level and he adds that no other single predictor can explain that much.

Partly, this objection is justified, partly it is not: An SES variable has been introduced among the set of variables submitted to the regressions, but it has turned out to be insignificant. Furthermore, with respect to motivational factors, prior analysis has shown that they are insignificant, too, for the determination of leaving the system (see KRIESI/RIES 1974).

Unfortunately, however, no data on other personality characteristics and, above all, no data on intellectual abilities have been collected by the team at Aarau. It is quite possible that at least with respect to leaving at different levels and, complementarily, with respect to staying, the dominance of structural determinants might have been attenuated, had variables measuring individual ability been introduced. Such variables might have raised the R^2, too. But the overall impact of personality factors and background characteristics most likely would

1) For a summary of these results, see LAVIN (1965), SUMMERSKILL (1962) or SAENGER/CEHA (1972). TRENT/MEDSKER (1968) especially stress the importance of motivational factors.

not have changed substantially, because the Swiss student population is rather homogenous with respect to these characteristics except for age and sex. Access to higher education in Switzerland has remained limited to this date. The 42'178 students enrolled at Swiss universities in 1970 make up only 8.2% of the 20-24 age group of the population living in Switzerland (2), as compared with the 43% of the 18-21 age group enrolled at American universities in 1968 (see BEN-DAVID 1974). Moreover, most of these students have been shaped by the common experience of the passage through the higher secondary school system, which two-thirds of them leave with a regular federal admission certificate. The Swiss higher secondary schools are much more homogenous than American high schools. They are generally highly selective and instill a common middle class orientation into their pupils. Only in the last few years the higher secondary school system has been expanding, a development which may, as indicated in this study, result in population shifts that will increase the relevance of personality and background variables. Up to the present, however, the two mechanisms of selectivity and anticipatory socialization have combined to make for a quite homogenous student population in Switzerland, thereby reducing the importance of personality factors and background characteristics.

The importance of structural aspects of a particular subject at a particular university emerging from this study finally has already been shown by a German study: JENNE et al. (1969) have started from the assumption that no general hypothesis applicable to all subjects could be stated due to the vast differences in the contents of the various disciplines of science and the differences in the courses of study deriving therefrom. They have been able to corroborate that basic assumption to a large extent. Numerous American studies have pointed in the same direction, although their focus - due to the divergent situation at American universities - is somewhat different: They have shown that the fit between individual characteristics, on the one hand, and the environment, on the

2) Foreigners are included in both of the figures. Source: Eidgenössisches Statistisches Amt, Statistisches Jahrbuch der Schweiz 1975.

other, is the decisive aspect in determining academic performance (see LAVIN 1965 and NEWCOMB/FELDMAN 1969). Elaborate research programs have been developed in the U.S. to assess the differential environmental impact. The three most prominent measures used are the College Characteristics Index of STERN (1970), the College and University Environment Scale (PACE 1963) and the Inventory of College Activities (ASTIN 1968).

To sum up this section, it may be pointed out that the detection of the prominence of structural conditions for the determination of the flows of Swiss students through the Swiss University System supports the contention of the theoretical usefulness of the techniques employed.

8.3.2. Two Models for the Leaving of the System

In this study, the leaving of the system has been fitted by two different types of models and one may, now, wonder how this has been possible. In the first part, the model of immanent change has applied to most of the leaving patterns at most of the universities; in the second part, the same is true of the simple additive regression model. The two models are the result of two different kinds of investigations of causal relations. To arrive at the model of immanent change, data gathered at different points in time have been used explicitely. To arrive at the regression models, on the other hand, the longitudinal data have, in effect, been cross-sectionalized in the synthetic population file. Moreover, while the former model has been estimated at the aggregate level, the latter is based on individual level data, although it is primarily effective at the aggregate level.

The two models do not contradict each other, but are complementary. The model of immanent change is rather unspecific. The two fundamental parameters of change presuppose no specific substantive interpretation (see COLEMAN 1968, p. 441). ZIEGLER (1972, p. 157) criticizes models of immanent change for this lack of specificity, contending that they amount to a renunciation of any attempt to an exact analysis of the underlying causal mechanisms. He believes these models to be not only theoretically, but also pragmatically worthless.

By introducing exogenous causal determinants, the regression model gives an idea of what the factors implied by the two fundamental parameters of change look like.

8.3.3. Two Additional Descriptive Devices

In addition to regression equations, two other descriptive devices have been discussed: First, it is possible to construct matrices for homogenous subpopulations. For the case of female students, this method has provided replications of earlier results. Then, one can interpret the profiles of the 'typical' individual in SPILERMAN's formula and in the extension thereof as semester means, or as overall means. This alternative perspective not only serves descriptive purposes, it has also some further advantages: It allows to calculate approximations for the last 6 matrices, for which no estimates could have been obtained on the basis of independent determinants. Although these approximations are far from accurate, they still have proved to be useful for the completion of the model with independent determinants. Prediction and simulation are, moreover, greatly facilitated by using this perspective.

8.4. Prediction and Simulation

Although, in allowing for population shifts, they make less restrictive assumptions than the ones in Part I, the prediction procedures developed in this second part have generally been less accurate than the ones from the previous part, due to the shortcomings of the last 6 matrices. The quite substantial population shifts between 1965 and 1973 do, moreover, not make a decisive impact on the estimation of the rates of flow. Except for the two higher levels of graduation, the present methods, however, have given accurate and unique results for flows in a single semester.

To predict future numbers of Swiss students is a tricky business. There have been several attempts to do so, but the models used have generally been inadequate (see MUELLER/JAEGER 1971). In taking the input of students into the university system as given, the present model is much less ambitious than the previous ones which make no such provisions. In this study, I have, moreover, been rather cautious. I have not tried to actually predict future numbers. I have, however, shown that the present model leads to satisfactory results in cases where we already know the actual figures. I have clearly stated the assumptions of the model. Given that the conditions underlying these assumptions do not change substantially, the model should lead to quite accurate predictions.

Should these conditions prove to be too unstable, the model can still be of use, because the implications of changes in these conditions can be worked out by simulation procedures.

In the concluding part on simulation, it has been shown how direct and indirect implications of assumptions about population shifts in individual characteristics which are introduced into the model by changing semester means, overall means or the determinant structure can be worked out. After a discussion of many instances, where changes in independent determinants do not result in any direct changes in the rates of flow, it could be demonstrated that drastic changes in socio-economic characteristics entail drastic changes in the pattern of leaving, although there is little change in the single matrices. Finally, the trickiness of manipulations of the determinant structure has been illustrated by the introduction of limitations of the flow of medical students, which become an issue in the context of the numerus clausus.

8.5. Conclusion

In introducing inependent determinants into the model of flow presented in Part I, I have followed the lead of SPILERMAN (1972). I am not aware of any other author who has explicitly done so. PERRENOUD (1973) discusses the possibility of introducing such determinants and points out some of its pitfalls. The incorporation of both, individual characteristics and characteristics of the time periods covered, never seems to have been done before.

The results presented in this study should have shown the usefulness of this procedure in the fields of description, prediction and simulation. I hope that the final discussion has also given some indication of the theoretical usefulness of this technique.

Astin, A.W. 1968. The College Environment. Washington: American Council of Education.

Bartholomew, D.J. 1967. Stochastic Models for Social Processes. New York: Wiley.

Ben-David, J. 1974. Trends in American Higher Education. Chicago: The University of Chicago Press (1st ed. 1972).

Binder, H. 1975. Forschungsinstitute - interne Struktur und externes Verhalten. Lizentiatsarbeit. Zürich: Soziologisches Institut.

Binder, J.R. Brühlhart, S. Greuter, R. Nef und S. Siebenmann, 1974. Schlussbericht der Forschungsgruppe Hochschule. Bericht an den Schweizerischen Nationalfonds zur Förderung der wissenschaftlichen Forschung (Kredit Nr. 1.473.71 SR). Zürich.

Blalock, H.M. 1961. Causal Inferences in Non-Experimental Research, Chapel Hill: University of North Carolina Press.

Blau, P.M. 1973. The Organization of Academic Work. New York: Wiley.

Blau, P.M. and O.D. Duncan 1967. The American Occupational Structure. New York: Wiley.

Blumen, I., M. Kogan and P.J. McCarthy 1955. The Industrial Mobility of Labor as a Probability Process. Cornell Studies in Industrial and Labor Relations, Vol. 6, Ithaca: New York State School of Industrial and Labor Relations.

Boudon, R. 1973. Mathematical Structures of Social Mobility. Amsterdam: Elsevier.

_______. 1974. Education, Opportunity and Social Inequality. Changing Prospects in Western Society. New York: Wiley.

Brownlee, K.A. 1965. Statistical Theory and Methodology in Science and Engineering, 2nd ed. New York: Wiley.

Campbell, E.Q. and N. Alexander, 1965. "Structural Effects and Interpersonal Relationships". American Journal of Sociology, Vol. 71, p. 284-89.

Coleman, J.S. 1968. The Mathematical Study of Change. In: H.M. Blalock and A.B. Blalock (Eds.): Methodology in Social Research. New York: McGraw-Hill, p. 428-77.

_______. 1973. The Mathematics of Collective Action. Chicago: Aldine.

_______. unpublished. A System of Social Accounts.

Correa, H. 1969. Quantitative Methods of Educational Planning. Scranton, Penn.: International Textbook Comp.

Crane, D. 1972. Invisible Colleges. Diffusion of Knowledge in Scientific Communities. Chicago: The University of Chicago Press.

Cuendet, G. 1975. Modèle TUSS, Version CSS. Genève: Conseil Suisse de la Science, Troisième Rapport sur le Dévelopment des Universités Suisses, Rapport de Travail.

Deppeler, R. 1969. Strukturprobleme der Universität. Bern: Schweizerischer Wissenschaftsrat.

Dickens, D. 1972. Program for the Analysis of Mobility Processes by the Introduction of Independent Variables into a Markov Chain. Ann Arbor: Xerox, University Microfilms.

Draper, N. and H. Smith 1966. Applied Regression Analysis. New York: Wiley.

Duncan, O.D. 1966. Methodological Issues in the Analysis of Social Mobility. In: N.J. Smelser and S.M. Lipset (Eds.): Social Structure and Mobility in Economic Development. Chicago: Aldine.

Duncan, O.D. 1968. "Patterns of Occupational Mobility among Negro Men." Demography, Vol.5, No.1, p.11-21.

Eidgenössisches Statistisches Amt (ESTA). Studienanfängerstatistiken der Semester: WS67/68, W67/68 + S68, W68/69 + S69, W69/70 + S70, W70/71 + S71, W71/72 + S72, W72/73, W73/74. Bern.

_______. Statistisches Jahrbuch der Schweiz, 1971-1974. Bern.

_______. Die Schweizerische Studentendatei, erstes Element des Schweizerischen Universitäts- und Hochschulsystems. Arbeitsbericht zur ersten Ausbauphase 1972-74. Bern 1974.

_______. 1975. Die Studierenden an den Schweizer Hochschulen im Wintersemester 1973/74. Bern.

Erziehungsdirektion des Kantons Zürich, Arbeitsgruppe für Bildungsplanung und Bildungsstatistik. Mittelschulstatistik für 1969 bis 1973/74.

Foerst, R., I. Bark, H.D. Bilsky und G. Viehoff. 1975. Verwaltung, Steuerung und Prognose des Studentendurchlaufs in der Medizin. Studenten-Informationssystem Medizin. Hannover: HIS-Brief 60.

Ginsberg, R.B. 1971. "Semi-Markov Processes and Mobility". Journal of Mathematical Sociology, Vol.1, No.2, p.233-262.

Girod, R. 1971. Mobilité Sociale. Genève: Droz.

Goldschmidt, D. 1973. "Zum Schicksal der deutschen Universität im Ausgang ihrer bürgerlichen Epoche". Neue Sammlung, Vol.13, p.1-25.

Habermas, J. 1970. Technik und Wissenschaft als Ideologie. Frankfurt: Edition Suhrkamp 287.

Henry, N.W. 1971. "The Retention Model: A Markov Chain with Variable Transition Probabilities". Journal of the American Statistical Association, Vol. 66, p.264-67.

Henry, N.W., R. Ginnis and H.W. Tegmeyer. 1971. "A Finite Model of Mobility". Journal of Mathematical Sociology, Vol.1, No.1, p.107-118.

Hitpass, J. 1967. Das Schicksal des Immatrikulationsjahrganges 1958. Gütersloh.

Hodge, R.W. 1966. "Occupational Mobility as a Probability Process." Demography, Vol.3, p.19-34.

Jenne, M., M. Krueger und K. Mueller-Plantenberg. 1969. Student im Studium. Stuttgart: Klett.

Kath, G., C. Oehler und R. Reichwein. 1966. Studienweg und Studienerfolg. Berlin: Institut für Bildungsforschung in der Max-Planck-Gesellschaft.

Kling, A. 1969. Bildungsforschung im Bereich Universität. Meisenheim: Verlag Anton Hain.

Kraehe, H. 1970. Studienabbruch und Studienfachwechsel bei Medizinstudenten. Hannover: HIS.

Kriesi, H. 1975. Integration der Studierenden und ihr Studienerfolg. Hochschule Aargau für Bildungswissenschaften, Vorbereitungsstufe, Arbeitsbericht Nr.10.

Kriesi, H. und H. Ries. 1974. Studienplanung und Studienmotivation. Arbeitsbericht Nr.6 der Hochschule Aargau für Bildungswissenschaften, Vorbereitungsstufe, Aarau.

Kuhn, T.S. 1962. The Structure of Scientific Revolutions. Chicago: The University of Chicago Press.

_______. 1970. Logic of Discovery or Psychology of Research? In: Lakatos, I. and A. Musgrave (Eds.): Criticism and the Growth of Knowledge. Cambridge: Cambridge University Press.

Land, K.C. 1969. "Duration of Residence and Prospective Migration: Further Evidence". Demography, Vol.6, p.133-40.

Lavin, D.E. 1965. The Prediction of Academic Performance: A Theoretical Analysis and Review of Research. New York: Russell Sage Foundation.

Leontief, W.W. 1941. The Structure of American Economy, 1919-1939. New York: Oxford University Press.

Lieberson, S. and G.V. Fuguitt. 1967. "Negro-White Occupational Differences in the Absence of Discrimination". American Journal of Sociology, Vol.73, p.188-200.

Loether, H.J. and D.G. McTavish. 1974. Inferential Statistics for Sociologists. An Introduction. Boston: Allyn & Bacon.

McFarland, D.D. 1970. "Intragenerational Social Mobility as a Markov Process: Including a Time-Stationary Markov Model that Explains Observed Declines in Mobility Rates over Time". American Sociological Review, Vol. 33, p. 712-22.

_______. 1974. "Substantive Contributions of Markov Models to the Study of Social Mobility". Internationale Arbeitstagung zur Analyse sozialer Ungleichheit und sozialer Mobilität, Werner Reimers Stiftung, Bad Homburg.

_______. 1975. "On the Relationship between Mathematical Models and Empirical Data, and the Relationship between Duration of Stay and Subsequent Mobility". American Sociological Review, Vol. 39, p. 883-885.

McGinnis, R. 1968. "A Stochastic Model of Social Mobility". American Sociological Review, Vol. 33, p. 712-22.

Morrison, P.A. 1967. "Duration of Residence and Prospective Migration: The Evaluation of a Stochastic Model". Demography, Vol. 4, p. 553-61.

Müller, B. und C. Jäger. 1971. Schweizerische Studentenprognosen. Bern: Schweizerischer Wissenschaftsrat.

Newcomb, T.M. and K.A. Feldman. 1969. The Impact of College on Students. San Francisco: Jossey-Bass.

Nie, N.H., C.H. Hull, J.G. Jenkins, K. Steinbrenner and D.H. Bent. 1975. SPSS: Statistical Package for the Social Sciences. 2nd edition. New York: New York: McGraw-Hill.

OECD 1969. Modèles Mathématiques pour la Planification de L'Enseignement. Paris.

_______. 1970. M.O.S. Un Modèle Optionnel de Simulation du Système Educatif. Paris.

_______. 1971. L'Utilisation des Modèles de Simulation dans la Planification de L'Enseignement. Une Evaluation Critique de M.O.S. Paris.

Pace, C.R. 1967. Analyses of a National Sample of College Environments. U.S. Department of Health, Education and Welfare, Project 5-0764 (OE-5-1-321). Los Angeles: University of California.

Perrenoud, P. 1973. Modèles de Flux et Theorie du Système d'Enseignement. Réflexions sur les relations entre sociologie et planification de l'éducation. Genève: Service de la Recherche Sociologique.

Perrenoud, P. et W. Hutmacher. 1973. Modèles de Flux et Theorie du Système d'Enseignement. Genève: Service de la Recherche Sociologique.

Prais, S.J. 1955a. "The Formal Theory of Social Mobility". Population Studies, Vol. 9, p. 72-81.

_______. 1955b. "Measuring Social Mobility". Journal of the Royal Statistical Association, A 118, p. 56-66.

Ries, H. 1974. Studienverlauf an Schweizer Hochschulen: Erster Teil A. Theoretischer und technischer Bericht zur empirischen Untersuchung des Studienverlaufs der Studierenden des Immatrikulationsjahrgangs 1965. Arbeitsbericht Nr.4A der Hochschule Aargau für Bildungswissenschaften, Vorbereitungsstufe, Aarau.

_______. 1975. Der Studienerfolg der Studierenden mit einem Maturitätszeugnis der Eidg. Maturitätskommission. Wissenschaftspolitik, Beiheft 6.

Ries, H. und H. Kriesi. 1974a. Die Entwicklung der Schweizer Hochschulen von 1965 bis 1973. Arbeitsbericht Nr.4C der Hochschule Aargau für Bildungswissenschaften, Vorbereitungsstufe, Aarau.

_______. 1974b. Hochschul- und Studienfachwahl der Studierenden des Jahrgangs 1965 an Schweizer Hochschulen. Arbeitsbericht Nr.7 der Hochschule Aargau für Bildungswissenschaften, Vorbereitungsstufe, Aarau.

_______. 1974c. Studiendauer und Studienverlauf. Arbeitsbericht Nr.8 der Hochschule Aargau für Bildungswissenschaften, Vorbereitungsstufe, Aarau.

_______. 1974d. Soziodemographische Merkmale und Mittelschulausbildung der Studierenden des Jahrgangs 1965. Arbeitsbericht Nr.5 der Hochschule Aargau für Bildungswissenschaften, Aarau.

_______. 1974e. Die ökonomische Lage der Studierenden und ihre Wohnsituation. Arbeitsbericht Nr.9 der Hochschule Aargau für Bildungswissenschaften, Vorbereitungsstufe, Aarau.

Saenger-Ceha, M.M.Th. 1972. Psychological and Social Factors in Student Drop-Out. Amsterdam: Swets & Zeitlinger.

Sauter, K. 1972. Der zweite Bildungsweg in der Schweiz. Zürich, Diss.

Schallberger, U. 1971. Studienverlauf und Studienerfolg des Immatrikulationsjahrganges 1961 der Zürcher Studenten. Zürich: Institut für Arbeitspsychologie.

_______. 1975. "Zum Ausmass des Langzeitstudiums und des Studienabbruchs an Schweizer Hochschulen. Neue Zürcher Zeitung, Freitag, 12. Sept.

Schweizerische Hochschulkonferenz, 1972. Ausbildungskapazität, Studienreform, Strukturreform, Ist-Zustand und Planung an den Schweizerischen Universitäten. Bern.

Siegel, S. 1956. Nonparametric Statistics for the Behavioral Sciences. International Student Edition. Tokyo: McGraw-Hill.

SOFI (Soziologisches Forschungsinstitut Göttingen) 1973. Ausbildungsentscheidungen von Abiturienten.

Sørensen, A.B. 1972. The Occupational Mobility Process: An Analysis of Occupational Careers. The Johns Hopkins University, Center for Social Organization of Schools, Report Nr.125, Baltimore.

Sørensen, A.B. 1974. The Organization of Activities in Time. Dept. of Sociology, University of Wisconsin, Madison.

Spilerman, S. 1972. "The Analysis of Mobility Processes by the Introduction of Independent Variables into a Markov Chain". American Sociological Review, Vol. 37, p. 277-94.

Stern, G.G. 1970. People in Context. New York.

Stinchcombe, A.L. 1968. Constructing Social Theories. New York: Harcourt & Brace.

Stone, R. 1966. Mathematics in the Social Sciences and other Essays. London: Chapman and Hall Ltd.

_______. 1971. Demographic Accounting and Model Building. Paris: OECD.

Suits, D.B. 1957. "The Use of Dummy Variables in Regression Equations". Journal of the American Statistical Association, Vol. 52, p. 548.

Summerskill, J. 1962. Dropouts from College. In: N. Sanford (Ed.): The American College: A Psychological and Social Interpretation of Higher Learning. New York: Wiley.

Taeuber, K.E. and A.E. Taeuber. 1965. Negroes in Cities. Chicago: Aldine.

Theil, H. 1971. Econometrics. New York: Wiley.

Trent, J.W. and L.L. Medsker. 1968. Beyond High School. San Francisco: Jossey-Bass.

Weizsäcker, C.C. v. und H.L. Freytag. 1967. "Plan eines Projekts: 'Ein quantitatives Modell des Bildungswesens in der Bundesrepublik Deutschland' ". Statistische Hefte, Vol. 8, p. 32-49.

Wissenschaftsrat. 1968. Empfehlungen des Wissenschaftsrats zur Struktur und Verwaltungsorganisation der Universitäten. Tübingen.

Ziegler, R. 1972. Theorie und Modell. Der Beitrag der Formalisierung zur soziologischen Theoriebildung. München: Oldenbourg.

APPENDIX A: TRANSITION MATRICES

1.1. TRANSITION MATRIX FOR CHANGE OF UNIVERSITY S65/W65

	1	2	3	4	5	6	7	8	9	10	11	12	13	14	15	16
1 BASEL	.940	.0	.0	.0	.0	.0	.0	.0	.0	.0	.0	.056	.003	.0	.0	.0
2 BERN	.0	.996	.0	.0	.0	.0	.0	.0	.0	.0	.0	.0	.0	.0	.0	.0
3 ETHL	.0	.0	.999	.0	.0	.0	.0	.0	.0	.0	.0	.0	.0	.0	.0	.0
4 ETHZ	.0	.0	.0	.999	.0	.0	.0	.0	.0	.0	.0	.0	.0	.0	.0	.0
5 FRIBOURG	.0	.0	.0	.0	.844	.0	.0	.0	.0	.0	.0	.094	.063	.0	.0	.0
6 HSG	.0	.0	.0	.0	.0	.750	.0	.0	.0	.0	.0	.250	.0	.0	.0	.0
7 GENEVE	.0	.0	.0	.0	.0	.0	.941	.0	.0	.0	.0	.0	.059	.0	.0	.0
8 LAUSANNE	.0	.0	.0	.0	.0	.0	.0	****	.0	.0	.0	.0	.0	.0	.0	.0
9 NEUCHATL	.0	.0	.0	.0	.0	.0	.0	.0	.833	.0	.0	.167	.0	.0	.0	.0
10 ZUERICH	.0	.0	.0	.0	.0	.0	.0	.0	.0	.905	.0	.083	.013	.0	.0	.0
11 FOREGN U	.0	.0	.0	.0	.0	.0	.0	.0	.0	.0	.0	.0	.0	.0	.0	.0
12 INTERRUP	.0	.0	.0	.0	.0	.0	.0	.0	.0	.0	.0	.0	.0	.0	.0	.0

1.2. TRANSITION MATRIX FOR CHANGE OF UNIVERSITY W65/S66

	1	2	3	4	5	6	7	8	9	10	11	12	13	14	15	16
1 BASEL	.947	.002	.0	.005	.0	.0	.0	.002	.0	.002	.005	.025	.010	.002	.0	.0
2 BERN	.0	.946	.0	.0	.0	.0	.0	.0	.0	.0	.0	.051	.0	.002	.0	.0
3 ETHL	.0	.0	.887	.0	.0	.0	.0	.031	.0	.0	.0	.072	.010	.0	.0	.0
4 ETHZ	.0	.0	.001	.941	.0	.0	.0	.0	.0	.0	.0	.045	.008	.0	.0	.0
5 FRIBOURG	.0	.0	.0	.0	.976	.0	.0	.0	.0	.0	.003	.015	.006	.0	.0	.0
6 HSG	.0	.007	.0	.0	.0	.827	.0	.0	.0	.014	.007	.137	.007	.0	.0	.0
7 GENEVE	.002	.0	.0	.0	.0	.002	.944	.0	.0	.0	.002	.039	.007	.0	.0	.0
8 LAUSANNE	.0	.0	.003	.0	.006	.0	.0	.932	.0	.0	.0	.041	.013	.003	.0	.0
9 NEUCHATL	.0	.012	.0	.0	.0	.0	.0	.0	.949	.0	.0	.035	.006	.0	.0	.0
10 ZUERICH	.0	.0	.0	.001	.001	.0	.002	.001	.001	.913	.003	.055	.011	.013	.0	.0
11 FOREGN U	.0	.0	.0	.0	.0	.0	.0	.0	.0	.0	.0	.0	.0	.0	.0	.0
12 INTERRUP	.118	.021	.043	.054	.021	.021	.011	.011	.021	.193	.011	.451	.011	.0	.0	.0

1.3.TRANSITION MATRIX FOR CHANGE OF UNIVERSITY S66/W66

	1	2	3	4	5	6	7	8	9	10	11	12	13	14	15	16
1 BASEL	.948	.003	.0	.0	.008	.0	.0	.0	.0	.003	.010	.015	.015	.0	.0	.0
2 BERN	.0	.967	.0	.0	.0	.0	.0	.0	.002	.002	.0	.029	.002	.0	.0	.0
3 ETHL	.0	.0	.946	.022	.0	.0	.0	.022	.0	.0	.0	.011	.0	.0	.0	.0
4 ETHZ	.001	.004	.0	.849	.002	.0	.0	.0	.0	.021	.0	.111	.001	.009	.0	.0
5 FRIBOURG	.003	.013	.0	.0	.939	.0	.006	.0	.0	.006	.009	.016	.006	.003	.0	.0
6 HSG	.0	.0	.0	.0	.0	.805	.008	.0	.0	.017	.017	.136	.034	.0	.0	.0
7 GENEVE	.008	.005	.003	.0	.0	.0	.906	.008	.0	.003	.005	.028	.020	.013	.0	.0
8 LAUSANNE	.003	.007	.0	.0	.003	.0	.0	.910	.0	.020	.007	.026	.020	.003	.0	.0
9 NEUCHATL	.0	.006	.0	.0	.0	.0	.006	.006	.946	.0	.0	.036	.0	.0	.0	.0
10 ZUERICH	.005	.004	.0	.004	.002	.001	.002	.004	.001	.910	.006	.046	.017	.001	.0	.0
11 FOREGN U	.0	.0	.0	.0	.0	.111	.0	.0	.0	.111	.778	.0	.0	.0	.0	.0
12 INTERRUP	.047	.099	.021	.099	.026	.056	.047	.039	.021	.159	.009	.352	.021	.0	.0	.0

1.4.TRANSITION MATRIX FOR CHANGE OF UNIVERSITY W66/S67

	1	2	3	4	5	6	7	8	9	10	11	12	13	14	15	16
1 BASEL	.907	.005	.0	.005	.008	.0	.005	.003	.0	.008	.008	.035	.015	.003	.0	.0
2 BERN	.0	.946	.0	.0	.0	.0	.0	.0	.0	.006	.0	.035	.004	.010	.0	.0
3 ETHL	.0	.0	.924	.021	.011	.011	.0	.032	.0	.0	.0	.0	.0	.0	.0	.0
4 ETHZ	.001	.001	.001	.928	.001	.0	.003	.001	.003	.013	.0	.047	.010	.0	.0	.0
5 FRIBOURG	.0	.0	.0	.003	.929	.0	.0	.006	.0	.010	.010	.019	.010	.013	.0	.0
6 HSG	.0	.0	.0	.0	.0	.864	.0	.0	.009	.018	.009	.082	.027	.0	.0	.0
7 GENEVE	.003	.003	.003	.0	.003	.003	.934	.003	.0	.011	.008	.022	.011	.0	.0	.0
8 LAUSANNE	.0	.014	.0	.0	.0	.0	.003	.893	.003	.010	.003	.048	.024	.0	.003	.0
9 NEUCHATL	.0	.012	.0	.0	.0	.0	.018	.072	.837	.018	.0	.030	.012	.0	.0	.0
10 ZUERICH	.001	.001	.0	.002	.006	.005	.004	.006	.0	.908	.011	.033	.021	.002	.0	.0
11 FOREGN U	.0	.0	.0	.0	.111	.074	.074	.037	.037	.074	.593	.0	.0	.0	.0	.0
12 INTERRUP	.025	.036	.004	.150	.004	.032	.025	.021	.007	.121	.0	.568	.004	.0	.0	.0

1.5. TRANSITION MATRIX FOR CHANGE OF UNIVERSITY S67/W67

	1	2	3	4	5	6	7	8	9	10	11	12	13	14	15	16
1 BASEL	.948	.011	.0	.0	.0	.0	.0	.0	.0	.008	.008	.016	.008	.0	.0	.0
2 BERN	.002	.901	.0	.004	.0	.002	.0	.008	.002	.006	.008	.028	.018	.024	.0	.0
3 ETHL	.0	.0	.989	.0	.0	.0	.0	.0	.0	.0	.0	.011	.0	.0	.0	.0
4 ETHZ	.004	.001	.0	.858	.0	.0	.0	.0	.0	.009	.003	.109	.013	.001	.0	.0
5 FRIBOURG	.007	.007	.0	.0	.887	.0	.0	.0	.0	.003	.003	.029	.020	.046	.003	.0
6 HSG	.0	.0	.0	.0	.0	.929	.0	.0	.0	.0	.009	.063	.0	.0	.0	.0
7 GENEVE	.0	.0	.0	.0	.0	.0	.879	.0	.0	.005	.011	.027	.030	.044	.003	.0
8 LAUSANNE	.0	.0	.0	.0	.003	.0	.003	.908	.0	.020	.007	.020	.031	.003	.003	.0
9 NEUCHATL	.0	.007	.0	.0	.0	.0	.007	.041	.863	.0	.0	.007	.041	.034	.0	.0
10 ZUERICH	.0	.004	.0	.001	.0	.0	.0	.012	.0	.924	.009	.026	.012	.007	.001	.0
11 FOREGN U	.028	.056	.0	.0	.0	.028	.0	.0	.0	.056	.667	.056	.056	.056	.0	.0
12 INTERRUP	.041	.078	.003	.275	.020	.058	.041	.058	.010	.095	.010	.278	.0	.007	.0	.0

1.6. TRANSITION MATRIX FOR CHANGE OF UNIVERSITY W67/S68

	1	2	3	4	5	6	7	8	9	10	11	12	13	14	15	16
1 BASEL	.907	.003	.0	.0	.003	.0	.003	.0	.003	.0	.011	.046	.022	.003	.0	.0
2 BERN	.002	.894	.0	.002	.004	.0	.002	.002	.0	.004	.016	.055	.006	.014	.0	.0
3 ETHL	.0	.0	.910	.022	.0	.0	.022	.011	.0	.0	.0	.022	.011	.0	.0	.0
4 ETHZ	.003	.0	.0	.933	.004	.003	.001	.0	.0	.007	.001	.039	.006	.0	.0	.0
5 FRIBOURG	.007	.007	.0	.0	.874	.0	.004	.0	.0	.014	.014	.047	.004	.025	.004	.0
6 HSG	.0	.0	.0	.0	.0	.967	.0	.0	.0	.0	.008	.024	.0	.0	.0	.0
7 GENEVE	.003	.0	.0	.0	.003	.003	.924	.006	.0	.003	.006	.036	.015	.0	.003	.0
8 LAUSANNE	.0	.010	.0	.0	.0	.0	.0	.898	.003	.023	.0	.033	.020	.010	.003	.0
9 NEUCHATL	.0	.0	.0	.0	.0	.008	.0	.031	.831	.015	.0	.015	.0	.100	.0	.0
10 ZUERICH	.001	.001	.0	.0	.0	.001	.004	.003	.001	.901	.020	.039	.011	.020	.0	.0
11 FOREGN U	.059	.078	.0	.0	.039	.020	.020	.039	.0	.157	.569	.0	.020	.0	.0	.0
12 INTERRUP	.026	.047	.0	.124	.013	.013	.017	.021	.013	.098	.004	.626	.004	.0	.0	.0

1.7. TRANSITION MATRIX FOR CHANGE OF UNIVERSITY S68/W68

	1	2	3	4	5	6	7	8	9	10	11	12	13	14	15	16
1 BASEL	.917	.009	.0	.0	.003	.0	.003	.006	.0	.009	.011	.026	.014	.006	.0	.0
2 BERN	.004	.859	.0	.0	.0	.0	.0	.007	.0	.017	.011	.028	.015	.056	.0	.0
3 ETHL	.0	.0	.939	.0	.0	.0	.0	.012	.0	.0	.0	.049	.0	.0	.0	.0
4 ETHZ	.0	.001	.0	.901	.0	.0	.0	.0	.0	.004	.0	.085	.006	.001	.0	.0
5 FRIBOURG	.016	.020	.0	.0	.798	.0	.008	.008	.0	.035	.008	.016	.023	.059	.016	.0
6 HSG	.0	.0	.0	.0	.0	.968	.0	.0	.0	.0	.0	.023	.008	.0	.0	.0
7 GENEVE	.006	.0	.0	.0	.006	.0	.874	.0	.0	.0	.009	.043	.015	.012	.034	.0
8 LAUSANNE	.0	.010	.0	.0	.0	.0	.003	.804	.003	.010	.0	.045	.017	.007	.100	.0
9 NEUCHATL	.0	.009	.0	.0	.0	.0	.0	.0	.833	.0	.0	.035	.026	.070	.026	.0
10 ZUERICH	.003	.003	.0	.0	.003	.001	.005	.003	.0	.910	.012	.025	.006	.027	.0	.0
11 FOREGN U	.015	.091	.0	.0	.0	.0	.030	.0	.0	.136	.636	.015	.030	.045	.0	.0
12 INTERRUP	.044	.078	.014	.254	.044	.027	.041	.037	.003	.085	.010	.336	.003	.007	.0	.0

1.8. TRANSITION MATRIX FOR CHANGE OF UNIVERSITY W68/S69

	1	2	3	4	5	6	7	8	9	10	11	12	13	14	15	16
1 BASEL	.901	.0	.0	.0	.0	.0	.003	.003	.0	.003	.012	.032	.006	.032	.009	.0
2 BERN	.005	.901	.002	.002	.0	.0	.0	.007	.0	.007	.020	.025	.011	.018	.0	.0
3 ETHL	.0	.0	.962	.0	.0	.0	.0	.012	.0	.0	.0	.012	.0	.0	.012	.0
4 ETHZ	.001	.001	.001	.958	.0	.001	.001	.0	.0	.004	.001	.019	.004	.001	.003	.0
5 FRIBOURG	.014	.009	.0	.0	.819	.0	.014	.009	.0	.027	.009	.027	.005	.014	.054	.0
6 HSG	.0	.0	.0	.0	.0	.744	.0	.0	.0	.0	.008	.030	.008	.0	.218	.0
7 GENEVE	.0	.0	.0	.0	.003	.0	.932	.0	.003	.007	.003	.020	.010	.007	.020	.0
8 LAUSANNE	.0	.004	.0	.004	.0	.0	.004	.879	.0	.008	.012	.012	.020	.004	.055	.0
9 NEUCHATL	.0	.0	.0	.0	.0	.0	.0	.021	.928	.0	.0	.010	.0	.031	.010	.0
10 ZUERICH	.0	.003	.0	.003	.0	.005	.001	.001	.0	.916	.021	.021	.008	.022	.0	.0
11 FOREGN U	.029	.044	.0	.0	.074	.0	.015	.015	.0	.147	.647	.029	.0	.0	.0	.0
12 INTERRUP	.036	.060	.004	.125	.008	.016	.020	.028	.016	.069	.008	.584	.004	.004	.0	.0

1.9. TRANSITION MATRIX FOR CHANGE OF UNIVERSITY S69/W69

	1	2	3	4	5	6	7	8	9	10	11	12	13	14	15	16
1 BASEL	.906	.003	.0	.0	.0	.0	.006	.0	.0	.003	.006	.030	.006	.015	.024	.0
2 BERN	.0	.938	.0	.002	.002	.0	.005	.0	.0	.002	.017	.017	.005	.014	.0	.0
3 ETHL	.0	.0	.887	.0	.0	.0	.0	.0	.0	.0	.0	.062	.0	.0	.050	.0
4 ETHZ	.0	.0	.0	.853	.0	.0	.0	.0	.0	.006	.001	.073	.032	.001	.036	.0
5 FRIBOURG	.0	.005	.0	.0	.763	.0	.0	.0	.0	.021	.0	.032	.021	.021	.126	.011
6 HSG	.009	.009	.0	.0	.0	.778	.0	.0	.0	.009	.009	.019	.0	.0	.167	.0
7 GENEVE	.003	.0	.0	.0	.0	.0	.816	.0	.0	.003	.0	.040	.017	.013	.108	.003
8 LAUSANNE	.008	.004	.0	.0	.0	.0	.029	.838	.004	.008	.0	.046	.008	.008	.046	.0
9 NEUCHATL	.0	.0	.0	.0	.0	.0	.0	.0	.790	.0	.011	.011	.011	.032	.147	.0
10 ZUERICH	.001	.003	.0	.0	.001	.0	.001	.001	.0	.933	.008	.019	.012	.010	.014	.0
11 FOREGN U	.012	.060	.0	.012	.0	.0	.012	.0	.0	.108	.699	.060	.012	.024	.0	.0
12 INTERRUP	.036	.090	.018	.237	.040	.004	.049	.018	.004	.112	.009	.336	.004	.0	.004	.0

1.10. TRANSITION MATRIX FOR CHANGE OF UNIVERSITY W69/S70

	1	2	3	4	5	6	7	8	9	10	11	12	13	14	15	16
1 BASEL	.864	.0	.0	.0	.0	.0	.0	.0	.0	.0	.006	.026	.003	.032	.067	.0
2 BERN	.0	.921	.0	.002	.005	.0	.0	.0	.0	.009	.005	.026	.0	.014	.019	.0
3 ETHL	.0	.0	.480	.0	.0	.0	.0	.0	.0	.0	.0	.013	.0	.0	.507	.0
4 ETHZ	.0	.0	.0	.750	.0	.002	.002	.0	.0	.0	.006	.031	.005	.003	.198	.0
5 FRIBOURG	.0	.013	.0	.0	.833	.0	.006	.006	.0	.013	.013	.019	.019	.019	.051	.006
6 HSG	.012	.012	.0	.0	.0	.706	.0	.0	.0	.0	.0	.047	.024	.0	.200	.0
7 GENEVE	.004	.0	.0	.0	.004	.0	.910	.0	.0	.011	.011	.019	.026	.0	.011	.004
8 LAUSANNE	.0	.010	.0	.0	.0	.0	.019	.845	.0	.0	.010	.019	.010	.005	.082	.0
9 NEUCHATL	.0	.0	.0	.0	.0	.0	.013	.0	.831	.0	.0	.013	.0	.026	.117	.0
10 ZUERICH	.004	.007	.0	.0	.003	.001	.003	.001	.001	.888	.019	.026	.008	.004	.033	.0
11 FOREGN U	.038	.115	.0	.013	.0	.0	.013	.0	.0	.154	.603	.013	.0	.026	.026	.0
12 INTERRUP	.035	.020	.0	.175	.020	.010	.040	.035	.005	.060	.005	.575	.0	.0	.0	.0

1.11. TRANSITION MATRIX FOR CHANGE OF UNIVERSITY S70/W70

	1	2	3	4	5	6	7	8	9	10	11	12	13	14	15	16
1 BASEL	.896	.0	.0	.0	.0	.0	.0	.0	.0	.0	.010	.024	.014	.004	.052	.0
2 BERN	.002	.901	.0	.0	.0	.0	.0	.0	.002	.007	.019	.012	.005	.014	.036	.0
3 ETHL	.0	.0	.972	.0	.0	.0	.0	.0	.0	.0	.0	.028	.0	.0	.0	.0
4 ETHZ	.0	.002	.0	.712	.002	.0	.0	.0	.0	.0	.006	.048	.014	.004	.210	.002
5 FRIBOURG	.0	.029	.0	.0	.755	.0	.0	.014	.0	.014	.0	.029	.022	.0	.129	.007
6 HSG	.016	.0	.0	.0	.0	.781	.0	.0	.0	.0	.016	.016	.0	.016	.141	.016
7 GENEVE	.0	.004	.0	.0	.0	.004	.878	.0	.0	.012	.004	.008	.012	.004	.077	.0
8 LAUSANNE	.0	.005	.0	.0	.0	.0	.0	.847	.0	.0	.005	.043	.005	.0	.092	.0
9 NEUCHATL	.0	.0	.0	.0	.0	.0	.015	.0	.682	.0	.015	.030	.0	.045	.212	.0
10 ZUERICH	.0	.006	.0	.0	.004	.0	.0	.0	.0	.895	.009	.022	.009	.006	.045	.0
11 FOREGN U	.026	.039	.0	.039	.013	.0	.0	.0	.0	.104	.714	.026	.026	.013	.013	.0
12 INTERRUP	.052	.042	.021	.156	.036	.010	.042	.036	.005	.099	.010	.464	.016	.0	.005	.0

1.12. TRANSITION MATRIX FOR CHANGE OF UNIVERSITY W70/S71

	1	2	3	4	5	6	7	8	9	10	11	12	13	14	15	16
1 BASEL	.903	.0	.0	.0	.0	.0	.004	.0	.0	.004	.018	.018	.004	.007	.041	.0
2 BERN	.003	.903	.0	.003	.0	.0	.0	.0	.0	.0	.0	.025	.0	.005	.051	.013
3 ETHL	.0	.0	.462	.0	.0	.0	.026	.0	.0	.0	.0	.0	.0	.0	.513	.0
4 ETHZ	.0	.002	.0	.680	.0	.0	.002	.0	.0	.0	.002	.017	.002	.0	.291	.002
5 FRIBOURG	.0	.0	.009	.009	.864	.0	.009	.0	.0	.0	.0	.026	.0	.043	.034	.009
6 HSG	.0	.0	.0	.0	.0	.887	.0	.019	.0	.0	.0	.038	.0	.0	.057	.0
7 GENEVE	.004	.0	.0	.0	.0	.0	.950	.0	.0	.004	.008	.004	.0	.004	.025	.0
8 LAUSANNE	.0	.006	.0	.0	.0	.0	.0	.854	.0	.006	.024	.006	.012	.0	.091	.0
9 NEUCHATL	.0	.021	.0	.0	.0	.0	.064	.021	.809	.021	.0	.021	.0	.0	.043	.0
10 ZUERICH	.002	.003	.0	.0	.003	.0	.003	.0	.0	.870	.011	.018	.003	.011	.066	.008
11 FOREGN U	.025	.099	.0	.0	.0	.0	.0	.012	.0	.099	.692	.025	.012	.037	.0	.0
12 INTERRUP	.031	.055	.0	.159	.018	.006	.055	.037	.012	.049	.006	.552	.0	.0	.012	.0

1.13. TRANSITION MATRIX FOR CHANGE OF UNIVERSITY S71/W71

	1	2	3	4	5	6	7	8	9	10	11	12	13	14	15	16
1 BASEL	.918	.004	.0	.0	.004	.0	.004	.004	.0	.004	.0	.028	.0	.0	.028	.008
2 BERN	.0	.903	.0	.0	.0	.0	.0	.003	.0	.0	.0	.011	.011	.003	.071	.0
3 ETHL	.0	.0	.947	.0	.0	.0	.0	.0	.0	.0	.0	.0	.0	.0	.053	.0
4 ETHZ	.0	.0	.0	.734	.0	.0	.0	.0	.0	.007	.003	.027	.013	.003	.212	.0
5 FRIBOURG	.0	.0	.0	.0	.792	.0	.009	.0	.0	.0	.0	.038	.0	.038	.104	.019
6 HSG	.0	.0	.0	.0	.0	.917	.0	.0	.0	.0	.0	.0	.0	.0	.083	.0
7 GENEVE	.008	.0	.0	.0	.0	.0	.853	.0	.0	.008	.0	.004	.004	.008	.115	.0
8 LAUSANNE	.0	.007	.0	.007	.007	.0	.007	.794	.0	.0	.0	.020	.007	.0	.147	.007
9 NEUCHATL	.0	.0	.0	.0	.0	.0	.0	.0	.925	.0	.0	.0	.025	.0	.050	.0
10 ZUERICH	.002	.003	.0	.0	.0	.0	.0	.0	.0	.899	.002	.014	.007	.003	.056	.017
11 FOREGN U	.0	.013	.0	.0	.0	.0	.0	.013	.0	.118	.763	.0	.0	.066	.0	.026
12 INTERRUP	.067	.074	.015	.067	.015	.015	.030	.044	.022	.126	.022	.467	.007	.0	.015	.0

1.14. TRANSITION MATRIX FOR CHANGE OF UNIVERSITY W71/S72

	1	2	3	4	5	6	7	8	9	10	11	12	13	14	15	16
1 BASEL	.906	.0	.0	.0	.0	.0	.0	.0	.0	.0	.008	.004	.008	.004	.049	.020
2 BERN	.003	.854	.0	.003	.0	.0	.0	.0	.0	.0	.003	.025	.011	.0	.087	.014
3 ETHL	.0	.0	.650	.0	.0	.0	.0	.0	.0	.0	.0	.0	.0	.050	.300	.0
4 ETHZ	.004	.0	.0	.771	.0	.0	.0	.0	.0	.009	.004	.004	.004	.0	.191	.013
5 FRIBOURG	.0	.0	.0	.0	.818	.0	.0	.011	.0	.011	.011	.023	.0	.023	.045	.068
6 HSG	.0	.022	.0	.0	.0	.848	.0	.0	.0	.0	.0	.043	.0	.0	.065	.022
7 GENEVE	.005	.0	.0	.0	.0	.0	.948	.0	.0	.0	.009	.0	.009	.0	.028	.0
8 LAUSANNE	.0	.0	.0	.0	.0	.0	.008	.937	.0	.008	.0	.0	.016	.0	.031	.0
9 NEUCHATL	.0	.0	.0	.0	.0	.0	.0	.0	.875	.0	.0	.025	.0	.0	.075	.025
10 ZUERICH	.0	.0	.0	.0	.0	.002	.004	.002	.0	.886	.005	.011	.007	.005	.061	.020
11 FOREGN U	.016	.0	.0	.0	.0	.0	.032	.016	.0	.032	.794	.032	.0	.063	.0	.016
12 INTERRUP	.091	.081	.0	.101	.040	.010	.0	.020	.0	.071	.030	.495	.010	.0	.020	.0

1.15. TRANSITION MATRIX FOR CHANGE OF UNIVERSITY S72/W72

	1	2	3	4	5	6	7	8	9	10	11	12	13	14	15	16
1 BASEL	.771	.004	.0	.0	.0	.0	.0	.0	.0	.0	.013	.004	.004	.0	.128	.077
2 BERN	.0	.782	.0	.0	.003	.0	.0	.006	.0	.006	.003	.016	.003	.006	.149	.025
3 ETHL	.0	.0	****	.0	.0	.0	.0	.0	.0	.0	.0	.0	.0	.0	.0	.0
4 ETHZ	.0	.0	.0	.873	.0	.0	.0	.0	.0	.0	.005	.016	.005	.0	.074	.026
5 FRIBOURG	.0	.013	.0	.0	.855	.0	.0	.0	.0	.0	.0	.013	.0	.026	.066	.026
6 HSG	.0	.0	.0	.0	.0	.927	.0	.0	.0	.0	.0	.0	.0	.0	.049	.024
7 GENEVE	.0	.0	.0	.0	.0	.0	.722	.0	.0	.005	.0	.0	.0	.019	.245	.010
8 LAUSANNE	.0	.0	.0	.0	.0	.0	.0	.688	.0	.0	.008	.016	.0	.0	.256	.032
9 NEUCHATL	.0	.0	.0	.0	.0	.0	.0	.0	.800	.0	.0	.0	.029	.029	.143	.0
10 ZUERICH	.0	.0	.0	.0	.0	.0	.0	.0	.0	.804	.006	.006	.004	.004	.119	.057
11 FOREGN U	.0	.016	.0	.0	.0	.0	.0	.016	.0	.032	.730	.016	.016	.143	.016	.016
12 INTERRUP	.054	.108	.0	.081	.068	.014	.014	.027	.014	.081	.027	.473	.014	.0	.014	.0

1.16. TRANSITION MATRIX FOR CHANGE OF UNIVERSITY W72/S73

	1	2	3	4	5	6	7	8	9	10	11	12	13	14	15	16
1 BASEL	.730	.005	.0	.0	.005	.0	.0	.0	.0	.0	.0	.0	.005	.0	.130	.124
2 BERN	.0	.790	.0	.0	.0	.0	.0	.0	.0	.0	.0	.0	.004	.012	.128	.066
3 ETHL	.0	.0	.768	.0	.0	.0	.0	.0	.0	.0	.0	.0	.0	.0	.077	.077
4 ETHZ	.0	.0	.0	.872	.0	.0	.0	.0	.0	.0	.0	.0	.0	.006	.035	.064
5 FRIBOURG	.0	.0	.0	.0	.873	.0	.0	.0	.014	.0	.0	.0	.0	.014	.028	.070
6 HSG	.0	.0	.0	.0	.0	.897	.0	.0	.0	.0	.0	.0	.0	.0	.077	.026
7 GENEVE	.0	.0	.0	.0	.0	.0	.887	.0	.0	.0	.0	.0	.007	.007	.079	.007
8 LAUSANNE	.0	.0	.0	.0	.0	.0	.0	.714	.0	.0	.0	.0	.0	.022	.220	.044
9 NEUCHATL	.0	.0	.0	.0	.0	.0	.0	.0	.828	.0	.0	.0	.034	.0	.103	.034
10 ZUERICH	.0	.0	.0	.0	.002	.0	.0	.0	.0	.780	.0	.0	.007	.005	.139	.060
11 FOREGN U	.0	.0	.0	.0	.0	.0	.0	.0	.0	.0	.789	.0	.0	.088	.0	.0
12 INTERRUP	.0	.0	.0	.0	.0	.0	.0	.0	.0	.0	.0	.961	.0	.0	.0	.0

1.17. TRANSITION MATRIX FOR CHANGE OF UNIVERSITY S73/W73

	1	2	3	4	5	6	7	8	9	10	11	12	13	14	15	16
1 BASEL	.926	.0	.0	.0	.0	.0	.0	.0	.0	.0	.0	.0	.007	.0	.030	.037
2 BERN	.0	.926	.0	.0	.0	.0	.0	.0	.0	.0	.0	.0	.005	.0	.049	.020
3 ETHL	.0	.0	****	.0	.0	.0	.0	.0	.0	.0	.0	.0	.0	.0	.0	.0
4 ETHZ	.0	.0	.0	.980	.0	.0	.0	.0	.0	.0	.0	.0	.0	.007	.0	.013
5 FRIBOURG	.0	.0	.0	.0	.969	.0	.0	.0	.0	.0	.0	.0	.016	.0	.0	.016
6 HSG	.0	.0	.0	.0	.0	.971	.0	.0	.0	.0	.0	.0	.0	.0	.0	.029
7 GENEVE	.0	.0	.0	.0	.0	.0	.963	.0	.0	.0	.0	.0	.0	.0	.030	.007
8 LAUSANNE	.0	.0	.0	.0	.0	.0	.0	.892	.0	.0	.0	.0	.015	.0	.062	.031
9 NEUCHATL	.0	.0	.0	.0	.0	.0	.0	.0	.880	.0	.0	.0	.0	.0	.040	.080
10 ZUERICH	.0	.0	.0	.0	.0	.0	.0	.0	.0	.908	.0	.0	.015	.0	.028	.049
11 FOREGN U	.0	.0	.0	.0	.0	.0	.0	.0	.0	.0	****	.0	.0	.0	.0	.0
12 INTERRUP	.0	.0	.0	.0	.0	.0	.0	.0	.0	.0	.0	.918	.061	.0	.020	.0

1.18. TRANSITION MATRIX FOR CHANGE OF UNIVERSITY W73/S74

	1	2	3	4	5	6	7	8	9	10	11	12	13	14	15	16
1 BASEL	.472	.0	.0	.0	.0	.0	.0	.0	.0	.0	.0	.0	.0	.024	.168	.336
2 BERN	.0	.519	.0	.C	.0	.0	.0	.0	.0	.0	.0	.0	.0	.016	.275	.190
3 ETHL	.0	.0	.0	.500	.0	.0	.0	.0	.0	.0	.0	.0	.0	.100	.100	.300
4 ETHZ	.0	.0	.0	.664	.0	.0	.0	.0	.0	.0	.0	.0	.0	.027	.075	.233
5 FRIBOURG	.0	.0	.0	.0	.661	.0	.0	.0	.0	.0	.0	.0	.0	.0	.129	.210
6 HSG	.0	.0	.0	.0	.0	.471	.0	.0	.0	.0	.0	.0	.0	.0	.147	.382
7 GENEVE	.0	.0	.0	.0	.0	.0	.667	.0	.0	.0	.0	.0	.0	.047	.217	.070
8 LAUSANNE	.0	.0	.0	.0	.0	.0	.0	.466	.0	.0	.0	.0	.0	.0	.362	.172
9 NEUCHATL	.0	.0	.0	.0	.0	.0	.0	.0	.636	.0	.0	.0	.0	.0	.274	.0
10 ZUERICH	.0	.0	.0	.0	.0	.0	.0	.0	.0	.553	.0	.0	.0	.017	.180	.251
11 FOREGN U	.0	.0	.0	.0	.0	.0	.0	.0	.0	.0	.667	.0	.0	.267	.044	.022
12 INTERRUP	.0	.0	.0	.0	.0	.0	.0	.0	.0	.0	.0	.0	.0	.0	.0	.0

1.19. TRANSITION MATRIX FOR CHANGE OF UNIVERSITY S74/W74

	1	2	3	4	5	6	7	8	9	10	11	12	13	14	15	16
1 BASEL	.831	.0	.0	.0	.0	.0	.0	.0	.0	.0	.0	.0	.0	.017	.017	.136
2 BERN	.0	.755	.0	.0	.0	.0	.0	.0	.0	.0	.0	.0	.0	.0	.153	.092
3 ETHL	.0	.0	.800	.0	.0	.0	.0	.0	.0	.0	.0	.0	.0	.0	.200	.0
4 ETHZ	.0	.0	.0	.763	.0	.0	.0	.0	.0	.0	.0	.0	.0	.031	.031	.175
5 FRIBOURG	.0	.0	.0	.0	.561	.0	.0	.0	.0	.0	.0	.0	.0	.0	.098	.341
6 HSG	.0	.0	.0	.0	.0	.688	.0	.0	.0	.0	.0	.0	.0	.0	.0	.313
7 GENEVE	.0	.0	.0	.0	.0	.0	.779	.0	.0	.0	.0	.0	.0	.012	.081	.128
8 LAUSANNE	.0	.0	.0	.0	.0	.0	.0	.741	.0	.0	.0	.0	.0	.0	.185	.074
9 NEUCHATL	.0	.0	.0	.0	.0	.0	.0	.0	.571	.0	.0	.0	.0	.143	.143	.143
10 ZUERICH	.0	.0	.0	.0	.0	.0	.0	.0	.0	.675	.0	.0	.0	.006	.160	.160
11 FOREGN U	.0	.0	.0	.0	.0	.0	.0	.0	.0	.0	.800	.0	.0	.167	.0	.033
12 INTERRUP	.0	.0	.0	.0	.0	.0	.0	.0	.0	.0	.0	.0	.0	.0	.0	.0

1.20. TRANSITION MATRIX FOR CHANGE OF UNIVERSITY W74/S75

	1	2	3	4	5	6	7	8	9	10	11	12	13	14	15	16
1 BASEL	.633	.0	.0	.0	.0	.0	.0	.0	.0	.0	.0	.0	.0	.020	.041	.306
2 BERN	.0	.514	.0	.0	.0	.0	.0	.0	.0	.0	.0	.0	.0	.014	.162	.311
3 ETHL	.0	.0	.0	.0	.0	.0	.0	.0	.0	.0	.0	.0	.0	.0	.500	.500
4 ETHZ	.0	.0	.0	.608	.0	.0	.0	.0	.0	.0	.0	.0	.0	.041	.095	.257
5 FRIBOURG	.0	.0	.0	.0	.565	.0	.0	.0	.0	.0	.0	.0	.0	.0	.043	.391
6 HSG	.0	.0	.0	.0	.0	.727	.0	.0	.0	.0	.0	.0	.0	.0	.0	.273
7 GENEVE	.0	.0	.0	.0	.0	.0	.836	.0	.0	.0	.0	.0	.0	.015	.045	.104
8 LAUSANNE	.0	.0	.0	.0	.0	.0	.0	.600	.0	.0	.0	.0	.0	.0	.100	.300
9 NEUCHATL	.0	.0	.0	.0	.0	.0	.0	.0	.875	.0	.0	.0	.0	.0	.125	.0
10 ZUERICH	.0	.0	.0	.0	.0	.0	.0	.0	.0	.518	.0	.0	.0	.0	.127	.355
11 FOREGN U	.0	.0	.0	.0	.0	.0	.0	.0	.0	.0	.625	.0	.0	.292	.0	.083
12 INTERRUP	.0	.0	.0	.0	.0	.0	.0	.0	.0	.0	.0	.0	.0	.0	.0	.0

1.21. TRANSITION MATRIX FOR CHANGE OF UNIVERSITY S75/W75

	1	2	3	4	5	6	7	8	9	10	11	12	13	14	15	16
1 BASEL	.871	.0	.0	.0	.0	.0	.0	.0	.0	.0	.0	.0	.0	.0	.065	.065
2 BERN	.0	.684	.0	.0	.0	.0	.0	.0	.0	.0	.0	.0	.0	.0	.132	.184
3 ETHL	.0	.0	.0	.0	.0	.0	.0	.0	.0	.0	.0	.0	.0	.0	.0	.0
4 ETHZ	.0	.0	.0	.800	.0	.0	.0	.0	.0	.0	.0	.0	.0	.0	.0	.200
5 FRIBOURG	.0	.0	.0	.0	.615	.0	.0	.0	.0	.0	.0	.0	.0	.0	.077	.308
6 HSG	.0	.0	.0	.0	.0	.750	.0	.0	.0	.0	.0	.0	.0	.0	.125	.125
7 GENEVE	.0	.0	.0	.0	.0	.0	.839	.0	.0	.0	.0	.0	.0	.0	.071	.089
8 LAUSANNE	.0	.0	.0	.0	.0	.0	.0	.750	.0	.0	.0	.0	.0	.0	.083	.167
9 NEUCHATL	.0	.0	.0	.0	.0	.0	.0	.0	.857	.0	.0	.0	.0	.0	.143	.0
10 ZUERICH	.0	.0	.0	.0	.0	.0	.0	.0	.0	.737	.0	.0	.0	.0	.088	.175
11 FOREGN U	.0	.0	.0	.0	.0	.0	.0	.0	.0	.0	.533	.0	.0	.267	.067	.133
12 INTERRUP	.0	.0	.0	.0	.0	.0	.0	.0	.0	.0	.0	.0	.0	.0	.0	.0

1.22. TRANSITION MATRIX FOR CHANGE OF UNIVERSITY W75/S76

	1	2	3	4	5	6	7	8	9	10	11	12	13	14	15	16
1 BASEL	.556	.0	.0	.0	.0	.0	.0	.0	.0	.0	.0	.0	.0	.0	.0	.444
2 BERN	.0	.538	.0	.0	.0	.0	.0	.0	.0	.0	.0	.0	.0	.0	.154	.308
3 ETHL	.0	.0	.0	.0	.0	.0	.0	.0	.0	.0	.0	.0	.0	.0	.0	.0
4 ETHZ	.0	.0	.0	.500	.0	.0	.0	.0	.0	.0	.0	.0	.0	.0	.0	.500
5 FRIBOURG	.0	.0	.0	.0	.625	.0	.0	.0	.0	.0	.0	.0	.0	.0	.0	.375
6 HSG	.0	.0	.0	.0	.0	.333	.0	.0	.0	.0	.0	.0	.0	.0	.167	.500
7 GENEVE	.0	.0	.0	.0	.0	.0	.702	.0	.0	.0	.0	.0	.0	.0	.043	.255
8 LAUSANNE	.0	.0	.0	.0	.0	.0	.0	.778	.0	.0	.0	.0	.0	.0	.0	.222
9 NEUCHATL	.0	.0	.0	.0	.0	.0	.0	.0	.667	.0	.0	.0	.0	.167	.0	.167
10 ZUERICH	.0	.0	.0	.0	.0	.0	.0	.0	.0	.595	.0	.0	.0	.024	.071	.310
11 FOREGN U	.0	.0	.0	.0	.0	.0	.0	.0	.0	.0	.500	.0	.0	.375	.0	.125
12 INTERRUP	.0	.0	.0	.0	.0	.0	.0	.0	.0	.0	.0	.0	.0	.0	.0	.0

2.1. TRANSITION MATRIX FOR CHANGE OF UNIVERSITY W65/S66

	1	2	3	4	5	6	7	8	9	10	11	12	13	14	15	16
1 BASEL	.947	.002	.0	.005	.0	.0	.0	.002	.0	.002	.005	.025	.010	.002	.0	.0
2 BERN	.0	.946	.0	.0	.0	.0	.0	.0	.0	.0	.0	.051	.0	.002	.0	.0
3 ETHL	.0	.0	.887	.0	.0	.0	.0	.031	.0	.0	.0	.072	.010	.0	.0	.0
4 ETHZ	.0	.0	.001	.941	.0	.0	.0	.0	.0	.0	.0	.045	.008	.0	.0	.0
5 FRIBOURG	.0	.0	.0	.0	.976	.0	.0	.0	.0	.0	.003	.015	.006	.0	.0	.0
6 HSG	.0	.007	.0	.0	.0	.827	.0	.0	.0	.014	.007	.137	.007	.0	.0	.0
7 GENEVE	.002	.0	.0	.0	.0	.002	.944	.0	.0	.0	.002	.039	.007	.0	.0	.0
8 LAUSANNE	.0	.0	.003	.0	.006	.0	.0	.932	.0	.0	.0	.041	.013	.003	.0	.0
9 NEUCHATL	.0	.012	.0	.0	.0	.0	.0	.0	.949	.0	.0	.035	.006	.0	.0	.0
10 ZUERICH	.0	.0	.0	.001	.001	.0	.002	.001	.001	.913	.003	.055	.011	.013	.0	.0
11 INTERRUP	.118	.021	.043	.054	.021	.021	.011	.011	.021	.193	.011	.451	.011	.0	.0	.0

2.2. TRANSITION MATRIX FOR CHANGE OF UNIVERSITY W65/W66

	1	2	3	4	5	6	7	8	9	10	11	12	13	14	15	16
1 BASEL	.918	.005	.0	.005	.005	.0	.0	.0	.002	.002	.015	.025	.022	.002	.0	.0
2 BERN	.0	.952	.0	.0	.0	.0	.0	.0	.002	.002	.0	.039	.002	.002	.0	.0
3 ETHL	.0	.0	.918	.0	.0	.0	.0	.052	.0	.0	.0	.021	.010	.0	.0	.0
4 ETHZ	.001	.004	.0	.820	.002	.0	.0	.0	.0	.020	.0	.132	.010	.007	.0	.0
5 FRIBOURG	.0	.012	.0	.0	.930	.0	.003	.0	.0	.006	.015	.015	.015	.003	.0	.0
6 HSG	.0	.007	.0	.0	.0	.762	.007	.0	.0	.022	.014	.173	.029	.0	.0	.0
7 GENEVE	.010	.005	.002	.0	.0	.0	.878	.007	.0	.005	.007	.041	.032	.012	.0	.0
8 LAUSANNE	.0	.006	.003	.0	.009	.0	.0	.879	.0	.016	.006	.041	.031	.006	.0	.0
9 NEUCHATL	.0	.017	.0	.0	.0	.0	.006	.006	.926	.0	.0	.035	.012	.0	.0	.0
10 ZUERICH	.005	.003	.0	.003	.002	.001	.005	.003	.001	.870	.008	.058	.028	.015	.0	.0
11 INTERRUP	.161	.054	.021	.140	.054	.032	.043	.021	.021	.215	.021	.172	.032	.0	.0	.0

2.3. TRANSITION MATRIX FOR CHANGE OF UNIVERSITY W65/S67

	1	2	3	4	5	6	7	8	9	10	11	12	13	14	15	16
1 BASEL	.854	.010	.0	.007	.010	.0	.005	.002	.0	.010	.022	.042	.034	.005	.0	.0
2 BERN	.0	.922	.0	.0	.0	.0	.0	.0	.002	.004	.0	.053	.006	.012	.0	.0
3 ETHL	.0	.010	.866	.021	.0	.010	.0	.062	.0	.0	.0	.021	.021	.0	.0	.0
4 ETHZ	.002	.005	.0C1	.800	.004	.0	.002	.001	.002	.031	.0	.121	.019	.007	.0	.0
5 FRIBOURG	.0	.012	.0	.003	.874	.0	.006	.006	.0	.015	.015	.028	.025	.015	.0	.0
6 HSG	.0	.007	.0	.0	.0	.755	.0	.0	.007	.036	.0	.165	.043	.0	.0	.0
7 GENEVE	.012	.010	.005	.0	.002	.002	.844	.010	.0	.010	.010	.041	.041	.012	.0	.0
8 LAUSANNE	.0	.016	.0	.0	.013	.0	.003	.822	.003	.019	.009	.057	.047	.006	.003	.0
9 NEUCHATL	.0	.023	.0	.0	.0	.0	.023	.075	.788	.012	.0	.058	.023	.0	.0	.0
10 ZUERICH	.005	.002	.0	.006	.009	.005	.006	.007	.002	.829	.014	.055	.047	.017	.0	.0
11 INTERRUP	.129	.043	.021	.161	.054	.011	.043	.0	.021	.247	.032	.161	.075	.0	.0	.0

2.4. TRANSITION MATRIX FOR CHANGE OF UNIVERSITY W65/W67

	1	2	3	4	5	6	7	8	9	10	11	12	13	14	15	16
1 BASEL	.836	.025	.0	.005	.010	.0	.005	.002	.0	.020	.022	.030	.042	.005	.0	.0
2 BERN	.002	.870	.0	.004	.0	.0	.0	.006	.0	.008	.008	.041	.023	.037	.0	.0
3 ETHL	.0	.010	.856	.021	.0	.010	.0	.062	.0	.0	.0	.010	.031	.0	.0	.0
4 ETHZ	.006	.005	.001	.777	.004	.001	.002	.001	.002	.037	.002	.119	.030	.008	.0	.0
5 FRIBOURG	.006	.022	.0	.003	.790	.0	.006	.003	.0	.015	.012	.037	.043	.062	.003	.0
6 HSG	.0	.007	.0	.0	.0	.820	.0	.0	.007	.043	.007	.086	.043	.0	.0	.0
7 GENEVE	.012	.012	.005	.0	.002	.002	.769	.007	.0	.012	.020	.032	.073	.051	.002	.0
8 LAUSANNE	.0	.016	.0	.013	.0	.006	.0	.810	.003	.022	.013	.035	.069	.006	.006	.0
9 NEUCHATL	.0	.029	.0	.0	.0	.029	.0	.121	.707	.012	.006	.023	.058	.017	.0	.0
10 ZUERICH	.005	.003	.0	.006	.007	.005	.005	.011	.001	.805	.018	.047	.059	.030	.001	.0
11 INTERRUP	.140	.043	.032	.204	.064	.021	.043	.011	.021	.236	.021	.075	.075	.0	.0	.0

2.5. TRANSITION MATRIX FOR CHANGE OF UNIVERSITY W65/S68

	1	2	3	4	5	6	7	8	9	10	11	12	13	14	15	16
1 BASEL	.780	.027	.0	.005	.007	.0	.005	.002	.002	.017	.017	.064	.064	.010	.0	.0
2 BERN	.004	.803	.0	.006	.004	.0	.0	.006	.0	.012	.021	.064	.027	.051	.0	.0
3 ETHL	.0	.010	.784	.031	.0	.010	.021	.072	.0	.0	.0	.031	.041	.0	.0	.0
4 ETHZ	.008	.004	.001	.758	.008	.004	.004	.004	.0	.039	.004	.120	.036	.008	.0	.0
5 FRIBOURG	.009	.028	.0	.003	.704	.0	.009	.0	.0	.028	.022	.071	.047	.077	.009	.0
6 HSG	.0	.007	.0	.0	.0	.834	.0	.0	.0	.036	.007	.086	.043	.0	.0	.0
7 GENEVE	.012	.012	.005	.0	.005	.005	.727	.010	.0	.012	.020	.051	.085	.051	.005	.0
8 LAUSANNE	.0	.022	.0	.0	.013	.0	.003	.769	.006	.025	.006	.054	.088	.006	.006	.0
9 NEUCHATL	.0	.029	.0	.0	.0	.0	.029	.138	.621	.012	.006	.017	.058	.092	.0	.0
10 ZUERICH	.006	.002	.0	.006	.007	.005	.007	.003	.001	.763	.026	.055	.071	.051	.001	.0
11 INTERRUP	.140	.043	.021	.215	.054	.021	.043	.011	.021	.247	.032	.054	.075	.011	.0	.0

2.6. TRANSITION MATRIX FOR CHANGE OF UNIVERSITY W65/W68

	1	2	3	4	5	6	7	8	9	10	11	12	13	14	15	16
1 BASEL	.760	.034	.0	.007	.005	.0	.005	.007	.002	.025	.022	.042	.071	.020	.0	.0
2 BERN	.006	.735	.0	.004	.004	.0	.0	.010	.0	.027	.019	.051	.041	.101	.0	.0
3 ETHL	.0	.010	.763	.031	.0	.010	.021	.082	.0	.0	.0	.041	.041	.0	.0	.0
4 ETHZ	.008	.006	.001	.772	.008	.004	.004	.004	.0	.038	.004	.098	.040	.010	.0	.0
5 FRIBOURG	.019	.043	.0	.003	.598	.0	.022	.006	.0	.047	.028	.034	.062	.121	.022	.0
6 HSG	.0	.007	.0	.0	.0	.863	.0	.0	.0	.036	.007	.050	.050	.0	.0	.0
7 GENEVE	.010	.012	.005	.0	.010	.005	.666	.010	.0	.010	.020	.066	.100	.059	.032	.0
8 LAUSANNE	.0	.028	.003	.0	.009	.0	.006	.639	.006	.028	.006	.047	.110	.016	.098	.0
9 NEUCHATL	.0	.029	.0	.0	.0	.0	.029	.127	.529	.012	.006	.040	.075	.138	.017	.0
10 ZUERICH	.007	.005	.0	.005	.008	.006	.008	.003	.0	.736	.026	.041	.079	.079	.001	.0
11 INTERRUP	.129	.054	.021	.193	.043	.021	.043	.011	.021	.247	.021	.075	.086	.021	.0	.0

2.7. TRANSITION MATRIX FOR CHANGE OF UNIVERSITY W65/S69

	1	2	3	4	5	6	7	8	9	10	11	12	13	14	15	16
1 BASEL	.708	.032	.0	.010	.002	.0	.007	.007	.002	.027	.027	.047	.076	.047	.007	.0
2 BERN	.006	.700	.002	.006	.004	.0	.0	.012	.0	.027	.031	.041	.049	.121	.0	.0
3 ETHL	.0	.010	.753	.031	.0	.010	.021	.093	.0	.0	.0	.031	.041	.0	.010	.0
4 ETHZ	.010	.007	.002	.777	.007	.005	.004	.005	.0	.040	.006	.074	.045	.012	.002	.0
5 FRIBOURG	.028	.050	.0	.003	.512	.0	.025	.009	.0	.059	.022	.047	.065	.130	.056	.0
6 HSG	.0	.007	.0	.0	.0	.676	.0	.0	.0	.036	.014	.043	.058	.0	.187	.0
7 GENEVE	.012	.012	.005	.0	.007	.005	.642	.010	.002	.015	.020	.059	.107	.061	.046	.0
8 LAUSANNE	.0	.022	.0	.0	.013	.0	.006	.589	.006	.025	.009	.044	.120	.022	.142	.0
9 NEUCHATL	.0	.029	.0	.0	.0	.0	.035	.132	.512	.012	.006	.012	.086	.155	.023	.0
10 ZUERICH	.007	.006	.0	.006	.006	.007	.007	.001	.0	.701	.031	.046	.087	.095	.005	.0
11 INTERRUP	.118	.032	.021	.193	.043	.011	.043	.011	.021	.258	.032	.097	.086	.032	.011	.0

2.8. TRANSITION MATRIX FOR CHANGE OF UNIVERSITY W65/W69

	1	2	3	4	5	6	7	8	9	10	11	12	13	14	15	16
1 BASEL	.662	.032	.0	.010	.002	.0	.007	.0	.002	.030	.027	.057	.081	.064	.027	.0
2 BERN	.006	.696	.002	.006	.004	.0	.004	.010	.0	.027	.033	.023	.055	.133	.0	.0
3 ETHL	.0	.010	.691	.021	.0	.010	.010	.072	.0	.0	.0	.082	.052	.0	.052	.0
4 ETHZ	.010	.007	.002	.727	.006	.005	.005	.005	.0	.046	.004	.061	.073	.014	.032	.0
5 FRIBOURG	.025	.053	.0	.0	.412	.0	.022	.006	.0	.065	.022	.040	.077	.146	.127	.006
6 HSG	.007	.014	.0	.0	.0	.518	.0	.0	.0	.043	.007	.058	.058	.0	.316	.0
7 GENEVE	.012	.012	.005	.0	.007	.005	.551	.010	.002	.015	.015	.056	.120	.071	.122	.002
8 LAUSANNE	.0	.025	.003	.0	.009	.0	.016	.523	.006	.019	.009	.060	.123	.025	.180	.0
9 NEUCHATL	.0	.029	.0	.0	.0	.0	.058	.104	.408	.012	.012	.012	.092	.173	.104	.0
10 ZUERICH	.008	.007	.0	.002	.007	.006	.005	.0	.0	.692	.028	.035	.093	.101	.017	.0
11 INTERRUP	.129	.054	.021	.183	.032	.011	.043	.011	.021	.226	.032	.064	.118	.032	.021	.0

2.9. TRANSITION MATRIX FOR CHANGE OF UNIVERSITY W65/S70

	1	2	3	4	5	6	7	8	9	10	11	12	13	14	15	16
1 BASEL	.595	.027	.0	.005	.002	.0	.005	.0	.002	.022	.020	.059	.084	.091	.089	.0
2 BERN	.006	.677	.002	.008	.008	.0	.0	.004	.0	.033	.019	.035	.055	.140	.014	.0
3 ETHL	.0	.0	.299	.0	.0	.010	.010	.062	.0	.0	.0	.093	.052	.0	.474	.0
4 ETHZ	.010	.007	.002	.587	.005	.006	.006	.005	.0	.046	.007	.042	.076	.018	.180	.0
5 FRIBOURG	.022	.059	.0	.0	.360	.0	.022	.006	.0	.071	.022	.040	.087	.152	.152	.009
6 HSG	.014	.022	.0	.0	.0	.352	.0	.0	.0	.043	.014	.058	.072	.0	.439	.0
7 GENEVE	.012	.010	.005	.0	.007	.005	.534	.007	.002	.012	.017	.044	.134	.076	.129	.005
8 LAUSANNE	.0	.022	.0	.0	.009	.0	.019	.466	.006	.022	.009	.050	.129	.028	.236	.0
9 NEUCHATL	.0	.029	.0	.0	.0	.0	.081	.086	.334	.012	.006	.017	.092	.190	.155	.0
10 ZUERICH	.009	.009	.0	.002	.006	.007	.003	.002	.001	.636	.034	.044	.100	.105	.042	.0
11 INTERRUP	.118	.054	.021	.183	.032	.011	.032	.021	.032	.215	.032	.064	.129	.032	.021	.0

2.10. TRANSITION MATRIX FOR CHANGE OF UNIVERSITY W65/W70

	1	2	3	4	5	6	7	8	9	10	11	12	13	14	15	16
1 BASEL	.558	.032	.0	.005	.002	.0	.005	.0	.002	.025	.022	.042	.093	.093	.123	.0
2 BERN	.008	.642	.002	.008	.004	.0	.0	.002	.002	.029	.027	.029	.057	.148	.041	.0
3 ETHL	.0	.0	.330	.010	.0	.010	.021	.041	.0	.0	.0	.041	.052	.0	.495	.0
4 ETHZ	.010	.007	.002	.439	.005	.006	.006	.005	.0	.043	.007	.045	.086	.023	.312	.001
5 FRIBOURG	.019	.062	.0	.003	.298	.0	.022	.012	.0	.071	.019	.028	.096	.155	.205	.012
6 HSG	.022	.014	.0	.0	.007	.280	.0	.0	.0	.043	.022	.043	.072	.007	.496	.007
7 GENEVE	.012	.010	.005	.002	.007	.005	.483	.007	.0	.012	.017	.029	.139	.081	.185	.005
8 LAUSANNE	.0	.022	.0	.0	.009	.0	.019	.416	.003	.019	.009	.041	.139	.028	.293	.0
9 NEUCHATL	.0	.035	.0	.0	.0	.0	.081	.081	.230	.012	.012	.029	.092	.201	.230	.0
10 ZUERICH	.008	.006	.0	.007	.006	.007	.002	.001	.001	.601	.033	.036	.107	.108	.074	.005
11 INTERRUP	.107	.043	.021	.193	.021	.0	.011	.021	.032	.193	.021	.075	.172	.043	.043	.0

2.11. TRANSITION MATRIX FOR CHANGE OF UNIVERSITY W65/S71

	1	2	3	4	5	6	7	8	9	10	11	12	13	14	15	16
1 BASEL	.529	.027	.C	.005	.002	.0	.005	.0	.002	.022	.027	.039	.096	.101	.145	.002
2 BERN	.008	.614	.002	.010	.004	.0	.002	.002	.0	.023	.014	.031	.057	.152	.078	.002
3 ETHL	.0	.0	.144	.010	.0	.010	.031	.041	.0	.0	.0	.041	.052	.0	.670	.0
4 ETHZ	.008	.008	.002	.318	.005	.006	.007	.005	.0	.036	.007	.027	.087	.027	.450	.002
5 FRIBOURG	.022	.053	.003	.006	.263	.0	.019	.012	.0	.068	.012	.028	.096	.174	.223	.022
6 HSG	.014	.014	.0	.0	.007	.244	.0	.007	.0	.036	.022	.043	.072	.007	.532	.014
7 GENEVE	.012	.007	.0	.002	.007	.005	.481	.007	.0	.012	.022	.017	.139	.081	.203	.005
8 LAUSANNE	.0	.013	.0	.0	.006	.0	.022	.365	.003	.019	.019	.028	.145	.028	.350	.0
9 NEUCHATL	.0	.040	.0	.0	.0	.0	.086	.086	.207	.012	.012	.017	.092	.201	.247	.0
10 ZUERICH	.008	.010	.0	.007	.008	.007	.005	.001	.0	.546	.030	.038	.111	.113	.112	.008
11 INTERRUP	.075	.043	.011	.183	.011	.0	.021	.011	.021	.161	.021	.043	.172	.054	.150	.021

2.12. TRANSITION MATRIX FOR CHANGE OF UNIVERSITY W65/W71

	1	2	3	4	5	6	7	8	9	10	11	12	13	14	15	16
1 BASEL	.514	.030	.0	.002	.005	.0	.007	.002	.002	.020	.025	.027	.096	.101	.167	.005
2 BERN	.006	.569	.002	.006	.004	.0	.002	.002	.0	.025	.010	.023	.064	.154	.127	.004
3 ETHL	.0	.0	.155	.021	.0	.0	.021	.021	.0	.0	.0	.010	.052	.0	.722	.0
4 ETHZ	.010	.010	.002	.242	.005	.004	.006	.002	.0	.037	.010	.023	.090	.030	.525	.002
5 FRIBOURG	.019	.040	.003	.006	.211	.0	.015	.012	.0	.059	.009	.034	.096	.189	.273	.034
6 HSG	.007	.014	.0	.0	.007	.252	.0	.007	.0	.029	.022	.022	.072	.007	.554	.022
7 GENEVE	.015	.010	.0	.002	.005	.005	.427	.007	.0	.012	.015	.010	.144	.093	.251	.005
8 LAUSANNE	.0	.016	.0	.0	.009	.0	.016	.309	.003	.016	.016	.013	.148	.028	.422	.003
9 NEUCHATL	.0	.035	.0	.0	.0	.0	.081	.081	.201	.012	.012	.006	.098	.201	.276	.0
10 ZUERICH	.007	.013	.0	.007	.007	.007	.002	.001	.0	.521	.022	.027	.115	.115	.142	.018
11 INTERRUP	.064	.043	.011	.140	.0	.0	.021	.011	.032	.140	.021	.054	.183	.054	.204	.021

2.13. TRANSITION MATRIX FOR CHANGE OF UNIVERSITY W65/S72

	1	2	3	4	5	6	7	8	9	10	11	12	13	14	15	16
1 BASEL	.487	.027	.0	.0	.005	.0	.005	.002	.002	.015	.022	.017	.101	.108	.194	.017
2 BERN	.010	.509	.002	.006	.0	.0	.002	.002	.0	.023	.012	.016	.068	.158	.179	.014
3 ETHL	.0	.0	.082	.021	.0	.0	.010	.021	.0	.0	.0	.010	.052	.010	.784	.0
4 ETHZ	.010	.008	.002	.203	.005	.005	.006	.002	.0	.037	.011	.011	.092	.030	.568	.007
5 FRIBOURG	.019	.028	.003	.003	.186	.0	.019	.012	.0	.053	.015	.028	.099	.189	.294	.053
6 HSG	.007	.022	.0	.007	.0	.209	.007	.0	.029	.014	.036	.0	.072	.007	.575	.029
7 GENEVE	.012	.010	.0	.002	.002	.005	.415	.005	.0	.012	.015	.010	.149	.093	.266	.005
8 LAUSANNE	.0	.013	.0	.0	.009	.0	.013	.299	.0	.016	.016	.006	.151	.031	.441	.003
9 NEUCHATL	.0	.023	.0	.0	.0	.0	.075	.081	.178	.012	.012	.012	.098	.201	.305	.006
10 ZUERICH	.007	.008	.0	.002	.006	.007	.005	.002	.0	.469	.018	.026	.122	.120	.182	.031
11 INTERRUP	.064	.043	.011	.097	.0	.0	.021	.011	.032	.129	.032	.021	.183	.054	.279	.021

2.14. TRANSITION MATRIX FOR CHANGE OF UNIVERSITY W65/W72

	1	2	3	4	5	6	7	8	9	10	11	12	13	14	15	16
1 BASEL	.381	.027	.0	.002	.005	.0	.005	.002	.0	.010	.025	.010	.108	.111	.256	.062
2 BERN	.008	.402	.002	.006	.002	.0	.002	.002	.0	.019	.012	.014	.070	.162	.269	.031
3 ETHL	.0	.0	.082	.021	.0	.0	.010	.021	.0	.0	.0	.010	.052	.010	.784	.0
4 ETHZ	.008	.007	.002	.186	.002	.004	.004	.004	.0	.032	.010	.006	.094	.033	.590	.014
5 FRIBOURG	.012	.025	.003	.003	.177	.0	.006	.012	.0	.037	.009	.012	.099	.198	.335	.071
6 HSG	.007	.022	.0	.0	.007	.209	.0	.0	.0	.029	.022	.022	.072	.007	.575	.043
7 GENEVE	.007	.007	.0	.002	.002	.002	.310	.005	.0	.012	.012	.005	.149	.105	.371	.010
8 LAUSANNE	.0	.013	.0	.0	.009	.0	.009	.211	.0	.016	.013	.009	.151	.031	.526	.009
9 NEUCHATL	.0	.017	.0	.0	.0	.0	.040	.052	.150	.006	.0	.0	.098	.219	.408	.012
10 ZUERICH	.006	.010	.0	.001	.005	.007	.003	.001	.0	.384	.019	.021	.124	.122	.244	.057
11 INTERRUP	.064	.043	.011	.064	.0	.0	.021	.011	.032	.129	.011	.021	.183	.064	.312	.032

2.15. TRANSITION MATRIX FOR CHANGE OF UNIVERSITY W65/S73

	1	2	3	4	5	6	7	8	9	10	11	12	13	14	15	16
1 BASEL	.276	.025	.0	.002	.005	.0	.005	.0	.002	.005	.020	.010	.111	.113	.308	.118
2 BERN	.006	.308	.002	.006	.0	.0	.002	.002	.0	.016	.010	.012	.070	.170	.335	.060
3 ETHL	.0	.0	.052	.010	.0	.0	.010	.010	.0	.0	.0	.010	.052	.010	.814	.010
4 ETHZ	.007	.005	.002	.162	.002	.004	.004	.002	.0	.030	.005	.006	.095	.038	.603	.029
5 FRIBOURG	.006	.025	.003	.003	.155	.0	.006	.006	.0	.022	.006	.012	.099	.201	.363	.090
6 HSG	.007	.022	.0	.0	.007	.201	.0	.0	.0	.029	.022	.022	.072	.007	.575	.050
7 GENEVE	.007	.005	.0	.0	.002	.002	.271	.002	.0	.007	.010	.005	.151	.110	.407	.017
8 LAUSANNE	.0	.013	.0	.0	.009	.0	.009	.157	.0	.009	.013	.009	.151	.035	.570	.022
9 NEUCHATL	.0	.017	.0	.0	.0	.0	.040	.035	.132	.006	.0	.0	.098	.219	.431	.023
10 ZUERICH	.003	.010	.0	.001	.005	.003	.002	.001	.0	.299	.016	.019	.125	.124	.304	.081
11 INTERRUP	.054	.032	.011	.064	.011	.0	.021	.011	.011	.107	.011	.021	.215	.064	.322	.043

2.16. TRANSITION MATRIX FOR CHANGE OF UNIVERSITY W65/W73

	1	2	3	4	5	6	7	8	9	10	11	12	13	14	15	16
1 BASEL	.256	.022	.0	.002	.005	.0	.005	.0	.002	.005	.017	.010	.113	.113	.317	.130
2 BERN	.006	.283	.002	.006	.0	.0	.002	.002	.0	.012	.008	.010	.072	.170	.355	.068
3 ETHL	.0	.0	.052	.010	.0	.0	.010	.010	.0	.0	.0	.010	.052	.022	.814	.010
4 ETHZ	.007	.005	.002	.159	.002	.004	.004	.002	.0	.030	.005	.006	.095	.038	.600	.031
5 FRIBOURG	.006	.025	.003	.003	.149	.0	.006	.006	.0	.022	.006	.012	.102	.201	.363	.093
6 HSG	.007	.022	.0	.0	.007	.194	.0	.0	.0	.029	.022	.022	.072	.007	.575	.058
7 GENEVE	.007	.005	.0	.0	.002	.002	.261	.002	.0	.007	.010	.005	.151	.110	.417	.020
8 LAUSANNE	.0	.009	.0	.0	.009	.0	.009	.142	.0	.009	.009	.009	.154	.035	.583	.028
9 NEUCHATL	.0	.017	.0	.0	.0	.0	.035	.035	.121	.006	.0	.0	.098	.219	.437	.035
10 ZUERICH	.003	.009	.0	.001	.005	.003	.002	.001	.0	.270	.015	.017	.131	.124	.315	.099
11 INTERRUP	.0	.0	.0	.0	.0	.0	.0	.0	.0	.0	.0	.0	.0	.0	.0	.0

2.17 TRANSITION MATRIX FOR CHANGE OF UNIVERSITY W65/S74

	1	2	3	4	5	6	7	8	9	10	11	12	13	14	15	16
1 BASEL	.138	.012	.0	.002	.002	.0	.002	.0	.0	.002	.010	.005	.113	.121	.364	.226
2 BERN	.004	.146	.0	.004	.0	.0	.002	.002	.0	.006	.004	.006	.072	.174	.447	.133
3 ETHL	.0	.0	.031	.0	.0	.0	.010	.010	.0	.0	.0	.010	.052	.032	.825	.021
4 ETHZ	.005	.002	.001	.094	.001	.002	.002	.001	.0	.017	.002	.004	.095	.050	.632	.080
5 FRIBOURG	.003	.019	.003	.003	.105	.0	.0	.0	.0	.015	.009	.009	.102	.201	.400	.127
6 HSG	.007	.014	.0	.0	.007	.122	.0	.0	.0	.014	.014	.007	.072	.014	.597	.144
7 GENEVE	.005	.002	.0	.0	.002	.002	.171	.0	.0	.005	.007	.002	.151	.124	.478	.046
8 LAUSANNE	.0	.003	.0	.0	.003	.0	.003	.060	.0	.003	.006	.006	.154	.038	.643	.079
9 NEUCHATL	.0	.012	.0	.0	.0	.0	.023	.023	.081	.006	.0	.0	.098	.219	.500	.040
10 ZUERICH	.002	.006	.0	.002	.002	.001	.0	.0	.0	.154	.008	.009	.131	.136	.365	.179
11 INTERRUP	.0	.0	.0	.0	.0	.0	.0	.0	.0	.0	.0	.0	.0	.0	.0	.0

2.18 TRANSITION MATRIX FOR CHANGE OF UNIVERSITY W65/W74

	1	2	3	4	5	6	7	8	9	10	11	12	13	14	15	16
1 BASEL	.128	.010	.0	.002	.002	.0	.002	.0	.0	.002	.010	.005	.113	.121	.364	.239
2 BERN	.002	.107	.0	.004	.0	.0	.002	.002	.0	.004	.002	.006	.072	.174	.470	.152
3 ETHL	.0	.0	.010	.0	.0	.0	.0	.010	.0	.0	.0	.0	.052	.032	.856	.021
4 ETHZ	.004	.001	.001	.071	.001	.001	.002	.001	.0	.013	.002	.002	.095	.052	.639	.101
5 FRIBOURG	.0	.012	.003	.003	.065	.0	.0	.0	.0	.009	.006	.006	.102	.205	.419	.167
6 HSG	.007	.007	.0	.0	.007	.093	.0	.0	.0	.007	.014	.007	.072	.014	.604	.180
7 GENEVE	.002	.002	.0	.0	.002	.002	.132	.0	.0	.002	.005	.002	.151	.132	.495	.068
8 LAUSANNE	.0	.003	.0	.0	.003	.0	.003	.054	.0	.003	.006	.006	.154	.038	.646	.082
9 NEUCHATL	.0	.006	.0	.0	.0	.0	.012	.012	.052	.006	.0	.0	.098	.230	.518	.069
10 ZUERICH	.001	.003	.0	.001	.001	.001	.0	.0	.0	.100	.006	.006	.131	.140	.392	.212
11 INTERRUP	.0	.0	.0	.0	.0	.0	.0	.0	.0	.0	.0	.0	.0	.0	.0	.0

2.19.TRANSITION MATRIX FOR CHANGE OF UNIVERSITY W65/S75

	1	2	3	4	5	6	7	8	9	10	11	12	13	14	15	16
1 BASEL	.091	.007	.0	.0	.002	.0	.002	.0	.0	.002	.007	.002	.113	.121	.369	.280
2 BERN	.0	.064	.0	.002	.0	.0	.002	.0	.0	.002	.002	.004	.072	.179	.489	.179
3 ETHL	.0	.0	.0	.0	.0	.0	.0	.0	.0	.0	.0	.0	.052	.032	.856	.041
4 ETHZ	.002	.001	.0	.039	.0	.001	.001	.001	.0	.007	.001	.001	.095	.057	.649	.132
5 FRIBOURG	.0	.006	.003	.003	.037	.0	.0	.0	.0	.006	.003	.003	.102	.205	.425	.205
6 HSG	.007	.007	.0	.0	.0	.072	.0	.0	.0	.007	.007	.0	.072	.029	.611	.201
7 GENEVE	.002	.0	.0	.0	.002	.002	.105	.0	.0	.002	.002	.002	.151	.137	.503	.088
8 LAUSANNE	.0	.003	.0	.0	.0	.0	.003	.031	.0	.0	.003	.003	.154	.038	.655	.107
9 NEUCHATL	.0	.006	.0	.0	.0	.0	.006	.012	.046	.006	.0	.0	.098	.230	.523	.075
10 ZUERICH	.001	.002	.0	.0	.0	.0	.0	.0	.0	.052	.003	.003	.131	.142	.408	.251
11 INTERRUP	.0	.0	.0	.0	.0	.0	.0	.0	.0	.0	.0	.0	.0	.0	.0	.0

2.20.TRANSITION MATRIX FOR CHANGE OF UNIVERSITY W65/W75

	1	2	3	4	5	6	7	8	9	10	11	12	13	14	15	16
1 BASEL	.086	.005	.0	.0	.002	.0	.002	.0	.0	.002	.005	.002	.113	.121	.371	.288
2 BERN	.0	.037	.0	.0	.0	.0	.002	.0	.0	.002	.002	.002	.072	.179	.505	.195
3 ETHL	.0	.0	.0	.0	.0	.0	.0	.0	.0	.0	.0	.0	.052	.032	.856	.041
4 ETHZ	.001	.001	.0	.030	.0	.001	.001	.001	.0	.006	.001	.0	.095	.057	.651	.143
5 FRIBOURG	.0	.003	.003	.003	.022	.0	.0	.0	.0	.003	.0	.003	.102	.208	.434	.217
6 HSG	.007	.007	.0	.0	.0	.065	.0	.0	.0	.007	.007	.0	.072	.029	.611	.201
7 GENEVE	.002	.0	.0	.0	.002	.0	.090	.0	.0	.002	.002	.002	.151	.139	.508	.098
8 LAUSANNE	.0	.003	.0	.0	.0	.0	.003	.019	.0	.0	.003	.0	.154	.041	.658	.117
9 NEUCHATL	.0	.006	.0	.0	.0	.0	.006	.012	.040	.006	.0	.0	.098	.230	.529	.075
10 ZUERICH	.001	.001	.0	.0	.0	.0	.0	.0	.0	.039	.002	.003	.131	.144	.412	.262
11 INTERRUP	.0	.0	.0	.0	.0	.0	.0	.0	.0	.0	.0	.0	.0	.0	.0	.0

2.21. TRANSITION MATRIX FOR CHANGE OF UNIVERSITY W65/S76

		1	2	3	4	5	6	7	8	9	10	11	12	13	14	15	16
1	BASEL	.057	.002	.0	.0	.002	.0	.002	.0	.0	.002	.002	.0	.113	.125	.374	.317
2	BERN	.0	.023	.0	.0	.0	.0	.002	.0	.0	.0	.002	.0	.072	.179	.511	.207
3	ETHL	.0	.0	.0	.0	.0	.0	.0	.0	.0	.0	.0	.0	.052	.032	.856	.041
4	ETHZ	.001	.001	.0	.013	.0	.0	.0	.001	.0	.002	.0	.0	.095	.058	.652	.164
5	FRIBOURG	.0	.003	.0	.0	.015	.0	.0	.0	.0	.003	.0	.003	.102	.208	.434	.229
6	HSG	.007	.0	.0	.0	.0	.043	.0	.0	.0	.0	.007	.0	.072	.029	.611	.230
7	GENEVE	.002	.0	.0	.0	.002	.0	.066	.0	.0	.002	.002	.0	.151	.139	.510	.122
8	LAUSANNE	.0	.003	.0	.0	.0	.0	.003	.016	.0	.0	.003	.0	.154	.041	.658	.120
9	NEUCHATL	.0	.006	.0	.0	.0	.0	.006	.006	.029	.006	.0	.0	.098	.236	.529	.086
10	ZUERICH	.001	.0	.0	.0	.0	.0	.0	.0	.0	.019	.001	.002	.131	.145	.417	.278
11	INTERRUP	.0	.0	.0	.0	.0	.0	.0	.0	.0	.0	.0	.0	.0	.0	.0	.0

3.1. TRANSITION MATRIX FOR CHANGE OF SUBJECT S65/W65

	1	2	3	4	5	6	7	8	9	10	11	12	13	14	15	16	17
1 THEOLOGY	.778	.0	.0	.0	.0	.0	.0	.0	.0	.0	.0	.0	.111	.111	.0	.0	.0
2 MEDICINE	.0	.956	.0	.0	.0	.0	.0	.0	.0	.0	.0	.0	.044	.0	.0	.0	.0
3 LAW	.0	.0	.898	.0	.0	.014	.0	.0	.0	.0	.0	.0	.087	.0	.0	.0	.0
4 ECON/BUS	.0	.0	.0	.884	.0	.0	.0	.0	.0	.0	.0	.0	.117	.0	.0	.0	.0
5 SOC SCI	.0	.0	.017	.0	.932	.0	.0	.0	.0	.0	.0	.0	.034	.017	.0	.0	.0
6 HUMANIT	.0	.0	.0	.0	.0	.933	.0	.0	.0	.0	.0	.0	.046	.020	.0	.0	.0
7 HSIT SCI	.0	.0	.0	.0	.0	.0	.943	.0	.0	.0	.0	.0	.057	.0	.0	.0	.0
8 NAT S I	.0	.0	.0	.0	.0	.0	.0	.961	.0	.0	.0	.0	.039	.0	.0	.0	.0
9 NAT S II	.0	.0	.0	.0	.0	.0	.0	.0	.924	.0	.0	.0	.067	.010	.0	.0	.0
10 ENGINEER	.0	.0	.0	.0	.0	.0	.0	.0	.0	.999	.0	.0	.0	.0	.0	.0	.0
11 ARCHITEC	.0	.0	.0	.0	.0	.0	.0	.0	.0	.0	.0	.0	.0	.0	.0	.0	.0
12 OTHERS	.0	.0	.0	.0	.0	.0	.0	.0	.0	.0	.0	.929	.071	.0	.0	.0	.0
13 INTERRUP	.0	.0	.0	.0	.0	.0	.0	.0	.0	.0	.0	.0	.0	.0	.0	.0	.0

3.2. TRANSITION MATRIX FOR CHANGE OF SUBJECT W65/S66

	1	2	3	4	5	6	7	8	9	10	11	12	13	14	15	16	17
1 THEOLOGY	.942	.0	.0	.0	.0	.0	.014	.0	.0	.0	.0	.0	.029	.014	.0	.0	.0
2 MEDICINE	.001	.967	.0	.0	.0	.001	.0	.0	.004	.0	.0	.0	.025	.003	.0	.0	.0
3 LAW	.0	.0	.927	.0	.006	.0	.0	.003	.0	.0	.0	.0	.064	.0	.0	.0	.0
4 ECON/BUS	.0	.0	.002	.911	.0	.0	.0	.0	.0	.002	.0	.0	.082	.002	.0	.0	.0
5 SOC SCI	.0	.004	.0	.0	.880	.007	.0	.0	.0	.004	.0	.0	.054	.007	.040	.0	.0
6 HUMANIT	.0	.0	.002	.0	.002	.925	.004	.0	.006	.0	.002	.0	.038	.018	.004	.0	.0
7 HSIT SCI	.0	.006	.0	.006	.0	.0	.919	.0	.0	.0	.0	.0	.043	.025	.0	.0	.0
8 NAT S I	.0	.003	.0	.0	.0	.0	.003	.939	.008	.003	.0	.0	.040	.008	.0	.0	.0
9 NAT S II	.0	.006	.0	.0	.002	.0	.0	.0	.956	.0	.002	.0	.029	.006	.0	.0	.0
10 ENGINEER	.0	.0	.0	.0	.0	.0	.0	.004	.0	.949	.0	.0	.043	.006	.0	.0	.0
11 ARCHITEC	.0	.008	.0	.0	.0	.0	.0	.008	.0	.0	.858	.0	.102	.024	.0	.0	.0
12 OTHERS	.0	.018	.0	.018	.0	.0	.0	.0	.0	.0	.0	.859	.053	.035	.018	.0	.0
13 INTERRUP	.011	.054	.054	.054	.021	.086	.011	.054	.129	.021	.011	.021	.451	.011	.0	.0	.0

3.3. TRANSITION MATRIX FOR CHANGE OF SUBJECT S66/W66

	1	2	3	4	5	6	7	8	9	10	11	12	13	14	15	16	17
1 THEOLOGY	.956	.0	.0	.0	.0	.0	.0	.0	.0	.0	.0	.0	.030	.015	.0	.0	.0
2 MEDICINE	.003	.946	.003	.001	.006	.003	.001	.003	.001	.0	.0	.0	.024	.007	.0	.0	.0
3 LAW	.0	.003	.932	.006	.0	.003	.003	.0	.0	.0	.0	.0	.047	.006	.0	.0	.0
4 ECON/BUS	.0	.0	.007	.915	.002	.002	.002	.0	.0	.0	.0	.0	.057	.019	.0	.0	.0
5 SOC SCI	.008	.004	.008	.004	.900	.008	.0	.0	.0	.0	.0	.0	.020	.020	.028	.0	.0
6 HUMANIT	.002	.002	.002	.002	.008	.916	.015	.002	.0	.0	.0	.002	.034	.013	.0	.0	.0
7 HSIT SCI	.0	.0	.0	.0	.0	.020	.935	.0	.013	.0	.0	.0	.013	.020	.0	.0	.0
8 NAT S I	.0	.0	.008	.008	.005	.005	.005	.868	.013	.005	.0	.005	.058	.016	.003	.0	.0
9 NAT S II	.0	.006	.006	.004	.002	.002	.002	.010	.907	.002	.0	.0	.052	.010	.002	.0	.0
10 ENGINEER	.0	.006	.002	.006	.0	.002	.0	.002	.006	.888	.0	.0	.087	.0	.0	.0	.0
11 ARCHITEC	.0	.0	.009	.0	.0	.009	.0	.009	.0	.0	.813	.0	.188	.0	.0	.0	.0
12 OTHERS	.0	.0	.0	.0	.0	.0	.0	.0	.0	.0	.0	.765	.039	.020	.118	.0	.0
13 INTERRUP	.009	.060	.090	.129	.039	.069	.026	.064	.039	.060	.030	.009	.352	.021	.0	.0	.0

3.4. TRANSITION MATRIX FOR CHANGE OF SUBJECT W66/S67

	1	2	3	4	5	6	7	8	9	10	11	12	13	14	15	16	17
1 THEOLOGY	.915	.0	.014	.0	.0	.014	.0	.0	.0	.0	.0	.0	.028	.028	.0	.0	.0
2 MEDICINE	.0	.948	.005	.0	.0	.002	.0	.0	.008	.0	.002	.0	.024	.012	.0	.0	.0
3 LAW	.0	.0	.940	.003	.003	.003	.0	.003	.0	.0	.0	.003	.045	.006	.0	.0	.0
4 ECON/BUS	.0	.002	.005	.935	.0	.0	.0	.0	.0	.0	.002	.0	.042	.014	.0	.0	.0
5 SOC SCI	.0	.024	.004	.008	.881	.004	.008	.004	.008	.004	.0	.0	.029	.016	.008	.0	.0
6 HUMANIT	.0	.004	.0	.006	.0	.907	.006	.002	.0	.0	.0	.0	.039	.019	.011	.002	.0
7 HSIT SCI	.0	.0	.0	.0	.019	.006	.932	.0	.006	.0	.0	.0	.025	.006	.006	.0	.0
8 NAT S I	.003	.003	.003	.003	.003	.0	.003	.897	.017	.003	.003	.0	.037	.023	.006	.0	.0
9 NAT S II	.0	.004	.002	.0	.002	.0	.0	.006	.942	.002	.0	.0	.031	.010	.002	.0	.0
10 ENGINEER	.0	.007	.0	.005	.0	.002	.0	.002	.0	.938	.002	.0	.034	.007	.0	.0	.0
11 ARCHITEC	.0	.010	.0	.010	.010	.0	.010	.0	.0	.020	.836	.0	.082	.020	.0	.0	.0
12 OTHERS	.0	.0	.0	.0	.044	.0	.022	.0	.0	.0	.0	.819	.044	.022	.022	.0	.0
13 INTERRUP	.011	.050	.050	.050	.014	.032	.011	.039	.046	.061	.054	.007	.568	.004	.0	.0	.0

3.5. TRANSITION MATRIX FOR CHANGE OF SUBJECT S67/W67

	1	2	3	4	5	6	7	8	9	10	11	12	13	14	15	16	17
1 THEOLOGY	.898	.0	.0	.0	.0	.0	.0	.0	.0	.0	.0	.0	.072	.014	.014	.0	.0
2 MEDICINE	.0	.955	.003	.002	.002	.003	.0	.0	.003	.0	.0	.0	.023	.011	.0	.0	.0
3 LAW	.003	.0	.962	.0	.0	.0	.0	.0	.003	.0	.0	.0	.025	.006	.0	.0	.0
4 ECON/BUS	.0	.0	.0	.929	.0	.0	.0	.0	.0	.0	.0	.0	.030	.016	.0	.0	.0
5 SOC SCI	.0	.0	.0	.0	.815	.0	.0	.0	.0	.0	.0	.004	.043	.034	.086	.009	.0
6 HUMANIT	.0	.005	.0	.002	.005	.869	.009	.0	.005	.0	.002	.0	.032	.028	.039	.002	.0
7 HSIT SCI	.0	.0	.0	.0	.0	.006	.938	.0	.018	.0	.0	.0	.018	.012	.006	.0	.0
8 NAT S I	.0	.0	.003	.003	.003	.003	.0	.879	.012	.015	.0	.0	.027	.039	.018	.0	.0
9 NAT S II	.002	.002	.002	.0	.004	.002	.0	.004	.916	.002	.0	.0	.031	.016	.014	.002	.0
10 ENGINEER	.0	.002	.0	.007	.0	.0	.0	.0	.002	.895	.0	.0	.089	.007	.0	.0	.0
11 ARCHITEC	.0	.0	.0	.0	.0	.0	.0	.0	.0	.0	.774	.0	.206	.020	.0	.0	.0
12 OTHERS	.0	.0	.0	.0	.0	.0	.0	.0	.0	.0	.0	.846	.026	.0	.128	.0	.0
13 INTERRUP	.0	.054	.054	.108	.020	.064	.017	.081	.088	.136	.054	.010	.278	.0	.007	.0	.0

3.6. TRANSITION MATRIX FOR CHANGE OF SUBJECT W67/S68

	1	2	3	4	5	6	7	8	9	10	11	12	13	14	15	16	17
1 THEOLOGY	.953	.0	.0	.0	.016	.0	.0	.0	.0	.0	.0	.0	.031	.0	.0	.0	.0
2 MEDICINE	.0	.927	.0	.002	.002	.002	.002	.0	.005	.002	.0	.0	.050	.003	.006	.0	.0
3 LAW	.0	.003	.945	.0	.003	.0	.0	.0	.0	.0	.0	.0	.047	.003	.0	.0	.0
4 ECON/BUS	.0	.0	.002	.924	.007	.0	.0	.002	.002	.0	.0	.0	.041	.020	.0	.0	.0
5 SOC SCI	.0	.005	.0	.015	.891	.015	.005	.0	.0	.0	.0	.0	.020	.040	.010	.0	.0
6 HUMANIT	.0	.007	.0	.0	.002	.884	.007	.002	.0	.0	.0	.0	.035	.007	.050	.005	.0
7 HSIT SCI	.0	.0	.0	.0	.0	.0	.919	.0	.006	.0	.0	.0	.043	.031	.0	.0	.0
8 NAT S I	.0	.0	.0	.009	.003	.0	.0	.895	.006	.009	.0	.0	.050	.006	.022	.0	.0
9 NAT S II	.0	.004	.0	.0	.002	.002	.0	.0	.939	.0	.0	.002	.020	.004	.025	.002	.0
10 ENGINEER	.0	.0	.0	.002	.005	.0	.0	.007	.002	.934	.0	.002	.037	.012	.0	.0	.0
11 ARCHITEC	.0	.0	.010	.010	.0	.0	.0	.021	.0	.021	.886	.0	.052	.0	.0	.0	.0
12 OTHERS	.0	.0	.0	.0	.0	.027	.0	.0	.027	.0	.0	.811	.081	.027	.027	.0	.0
13 INTERRUP	.026	.064	.038	.026	.013	.060	.017	.009	.043	.060	.021	.0	.626	.004	.0	.0	.0

3.7. TRANSITION MATRIX FOR CHANGE OF SUBJECT S68/W68

	1	2	3	4	5	6	7	8	9	10	11	12	13	14	15	16	17
1 THEOLOGY	.985	.0	.0	.0	.0	.0	.0	.0	.0	.0	.0	.0	.0	.0	.0	.015	.0
2 MEDICINE	.0	.969	.003	.0	.0	.0	.0	.0	.002	.0	.0	.0	.019	.002	.003	.0	.0
3 LAW	.0	.003	.937	.0	.0	.003	.0	.0	.0	.0	.0	.0	.040	.009	.0	.011	.0
4 ECON/BUS	.0	.0	.0	.882	.0	.002	.0	.002	.002	.0	.0	.0	.031	.017	.0	.064	.0
5 SOC SCI	.0	.005	.0	.0	.813	.0	.0	.0	.0	.0	.0	.005	.057	.031	.041	.047	.0
6 HUMANIT	.0	.0	.0	.0	.003	.804	.0	.0	.0	.0	.0	.0	.040	.027	.125	.003	.0
7 HSIT SCI	.0	.0	.0	.0	.006	.006	.918	.0	.0	.0	.0	.0	.025	.019	.025	.0	.0
8 NAT S I	.0	.0	.0	.003	.003	.0	.0	.898	.003	.0	.0	.0	.037	.014	.031	.010	.0
9 NAT S II	.0	.004	.0	.004	.0	.0	.0	.0	.943	.0	.0	.0	.019	.008	.023	.0	.0
10 ENGINEER	.0	.0	.0	.0	.002	.0	.0	.002	.0	.916	.002	.0	.076	.002	.0	.0	.0
11 ARCHITEC	.0	.0	.0	.0	.0	.011	.011	.0	.0	.0	.755	.0	.200	.022	.0	.0	.0
12 OTHERS	.0	.0	.0	.0	.0	.031	.0	.0	.0	.0	.0	.813	.063	.063	.031	.031	.0
13 INTERRUP	.014	.088	.058	.081	.031	.031	.020	.047	.044	.139	.078	.007	.336	.003	.007	.0	.0

3.8. TRANSITION MATRIX FOR CHANGE OF SUBJECT W68/S69

	1	2	3	4	5	6	7	8	9	10	11	12	13	14	15	16	17
1 THEOLOGY	.943	.0	.0	.0	.0	.0	.0	.0	.0	.0	.0	.0	.029	.014	.0	.014	.0
2 MEDICINE	.002	.973	.002	.0	.0	.0	.002	.0	.0	.0	.0	.0	.013	.005	.002	.002	.0
3 LAW	.0	.0	.924	.0	.0	.0	.0	.0	.0	.0	.0	.0	.022	.006	.0	.028	.0
4 ECON/BUS	.0	.0	.0	.856	.003	.0	.0	.0	.0	.0	.003	.0	.025	.015	.0	.103	.0
5 SOC SCI	.0	.0	.006	.0	.900	.012	.0	.0	.0	.006	.0	.0	.012	.012	.006	.047	.0
6 HUMANIT	.0	.0	.003	.003	.0	.862	.003	.0	.0	.0	.0	.0	.032	.010	.082	.003	.0
7 HSIT SCI	.0	.0	.007	.0	.0	.007	.961	.0	.0	.0	.0	.0	.007	.013	.007	.0	.0
8 NAT S I	.0	.0	.0	.004	.004	.0	.004	.948	.004	.0	.0	.0	.014	.004	.014	.007	.0
9 NAT S II	.0	.002	.0	.002	.0	.0	.002	.0	.920	.002	.0	.0	.039	.006	.021	.004	.0
10 ENGINEER	.0	.0	.0	.007	.0	.0	.0	.002	.0	.968	.002	.0	.009	.007	.0	.005	.0
11 ARCHITEC	.0	.011	.0	.0	.011	.0	.011	.0	.0	.0	.913	.0	.054	.0	.0	.0	.0
12 OTHERS	.0	.034	.0	.0	.0	.0	.0	.0	.0	.0	.0	.793	.069	.0	.103	.0	.0
13 INTERRUP	.0	.073	.044	.040	.028	.040	.004	.032	.040	.060	.020	.008	.584	.004	.004	.0	.0

3.9. TRANSITION MATRIX FOR CHANGE OF SUBJECT S69/W69

	1	2	3	4	5	6	7	8	9	10	11	12	13	14	15	16	17
1 THEOLOGY	.821	.0	.0	.0	.0	.0	.0	.0	.0	.0	.0	.0	.030	.0	.015	.119	.015
2 MEDICINE	.0	.980	.0	.0	.0	.0	.002	.002	.002	.0	.0	.0	.008	.005	.002	.002	.0
3 LAW	.0	.0	.911	.003	.0	.003	.0	.0	.0	.0	.0	.0	.023	.0	.003	.058	.0
4 ECON/BUS	.0	.0	.003	.826	.003	.0	.0	.0	.0	.0	.0	.003	.014	.022	.0	.126	.003
5 SOC SCI	.0	.0	.0	.0	.815	.0	.0	.0	.0	.0	.0	.0	.018	.012	.031	.116	.006
6 HUMANIT	.0	.004	.0	.0	.0	.842	.004	.0	.0	.0	.0	.0	.074	.021	.049	.007	.0
7 HSIT SCI	.0	.0	.0	.007	.0	.0	.869	.0	.0	.0	.0	.0	.059	.013	.0	.053	.0
8 NAT S I	.0	.0	.0	.004	.0	.0	.0	.912	.0	.0	.0	.0	.043	.018	.018	.007	.0
9 NAT S II	.0	.0	.0	.0	.002	.0	.0	.0	.928	.0	.0	.0	.021	.002	.011	.036	.0
10 ENGINEER	.0	.0	.0	.002	.0	.0	.0	.0	.0	.827	.0	.0	.079	.049	.002	.040	.0
11 ARCHITEC	.0	.0	.0	.0	.011	.0	.0	.0	.0	.0	.824	.0	.154	.0	.0	.011	.0
12 OTHERS	.0	.0	.0	.040	.0	.040	.0	.0	.0	.0	.0	.600	.040	.0	.040	.240	.0
13 INTERRUP	.013	.040	.045	.063	.022	.067	.018	.022	.094	.130	.090	.013	.336	.0	.0	.004	.0

3.10. TRANSITION MATRIX FOR CHANGE OF SUBJECT W69/S70

	1	2	3	4	5	6	7	8	9	10	11	12	13	14	15	16	17
1 THEOLOGY	.845	.0	.0	.0	.034	.0	.0	.0	.0	.0	.0	.0	.017	.0	.0	.103	.0
2 MEDICINE	.0	.976	.0	.002	.0	.0	.0	.0	.0	.0	.002	.0	.014	.005	.0	.003	.0
3 LAW	.0	.0	.856	.006	.0	.0	.0	.0	.0	.0	.0	.0	.043	.003	.0	.090	.003
4 ECON/BUS	.0	.0	.0	.824	.003	.0	.0	.0	.0	.0	.0	.003	.035	.016	.0	.118	.0
5 SOC SCI	.0	.0	.007	.0	.900	.0	.0	.0	.0	.0	.0	.0	.021	.007	.014	.050	.0
6 HUMANIT	.0	.0	.0	.008	.008	.844	.012	.0	.0	.0	.0	.0	.031	.016	.047	.035	.0
7 HSIT SCI	.0	.0	.0	.0	.0	.0	.921	.0	.0	.0	.0	.029	.022	.007	.014	.007	.0
8 NAT S I	.0	.0	.0	.004	.0	.0	.0	.830	.0	.0	.0	.0	.012	.004	.019	.132	.0
9 NAT S II	.0	.007	.0	.002	.0	.0	.0	.0	.909	.0	.0	.002	.023	.005	.012	.040	.0
10 ENGINEER	.0	.0	.0	.0	.0	.0	.0	.0	.003	.647	.0	.0	.021	.008	.003	.317	.0
11 ARCHITEC	.0	.0	.0	.0	.0	.0	.0	.0	.0	.0	.874	.0	.042	.0	.0	.084	.0
12 OTHERS	.0	.0	.0	.0	.0	.0	.0	.0	.0	.0	.0	.737	.053	.053	.158	.053	.0
13 INTERRUP	.0	.015	.020	.035	.010	.045	.020	.045	.055	.115	.035	.005	.575	.0	.0	.0	.0

3.11. TRANSITION MATRIX FOR CHANGE OF SUBJECT S70/W70

	1	2	3	4	5	6	7	8	9	10	11	12	13	14	15	16	17
1 THEOLOGY	.857	.0	.0	.0	.0	.0	.0	.0	.0	.0	.0	.0	.0	.0	.0	.143	.0
2 MEDICINE	.0	.966	.0	.0	.0	.002	.0	.0	.0	.0	.0	.0	.013	.003	.0	.014	.0
3 LAW	.0	.0	.831	.0	.004	.0	.0	.0	.0	.0	.0	.0	.035	.007	.0	.121	.004
4 ECON/BUS	.0	.0	.004	.823	.0	.0	.0	.0	.0	.0	.0	.0	.015	.018	.004	.132	.004
5 SOC SCI	.0	.0	.0	.007	.880	.0	.0	.0	.0	.0	.0	.0	.007	.015	.030	.052	.007
6 HUMANIT	.0	.0	.0	.0	.004	.840	.0	.0	.0	.0	.0	.0	.044	.013	.027	.062	.013
7 HSIT SCI	.0	.0	.0	.0	.0	.0	.879	.0	.0	.0	.0	.0	.030	.015	.007	.067	.0
8 NAT S I	.0	.0	.0	.0	.004	.0	.0	.842	.004	.0	.0	.0	.018	.018	.009	.103	.0
9 NAT S II	.0	.0	.0	.0	.0	.0	.0	.0	.926	.0	.0	.0	.020	.007	.002	.045	.0
10 ENGINEER	.0	.0	.0	.0	.0	.0	.0	.0	.004	.644	.0	.0	.044	.018	.0	.287	.004
11 ARCHITEC	.0	.0	.0	.0	.0	.0	.0	.0	.0	.0	.769	.0	.110	.0	.0	.121	.0
12 OTHERS	.0	.0	.0	.0	.0	.0	.0	.0	.0	.0	.0	.588	.059	.0	.235	.118	.0
13 INTERRUP	.0	.042	.057	.026	.026	.083	.036	.047	.042	.094	.052	.0	.464	.016	.0	.005	.0

3.12. TRANSITION MATRIX FOR CHANGE OF SUBJECT W70/S71

	1	2	3	4	5	6	7	8	9	10	11	12	13	14	15	16	17
1 THEOLOGY	.857	.0	.0	.0	.0	.0	.0	.0	.0	.0	.0	.0	.0	.0	.024	.119	.0
2 MEDICINE	.0	.937	.0	.0	.003	.002	.0	.0	.002	.0	.0	.0	.016	.0	.002	.034	.002
3 LAW	.0	.004	.879	.012	.0	.0	.0	.004	.0	.0	.0	.0	.016	.0	.0	.069	.016
4 ECON/BUS	.0	.0	.0	.857	.004	.0	.0	.0	.0	.0	.0	.0	.035	.004	.004	.095	.0
5 SOC SCI	.0	.0	.0	.008	.850	.0	.008	.0	.0	.0	.0	.0	.016	.0	.032	.071	.016
6 HUMANIT	.0	.0	.0	.0	.0	.874	.005	.0	.0	.0	.0	.0	.010	.010	.034	.063	.005
7 HSIT SCI	.0	.0	.0	.008	.008	.0	.920	.0	.0	.0	.0	.008	.016	.008	.0	.016	.016
8 NAT S I	.0	.0	.0	.0	.0	.0	.0	.853	.005	.005	.0	.0	.005	.0	.015	.117	.0
9 NAT S II	.0	.0	.0	.0	.0	.0	.0	.0	.925	.0	.0	.0	.021	.008	.005	.039	.003
10 ENGINEER	.0	.0	.0	.010	.005	.005	.0	.0	.0	.534	.0	.0	.010	.0	.0	.430	.005
11 ARCHITEC	.0	.0	.0	.0	.0	.0	.0	.0	.0	.012	.562	.0	.037	.0	.0	.387	.0
12 OTHERS	.0	.0	.0	.0	.100	.0	.0	.0	.0	.0	.0	.700	.100	.0	.100	.0	.0
13 INTERRUP	.012	.074	.025	.031	.0	.049	.043	.025	.061	.049	.049	.012	.552	.0	.0	.012	.0

3.13. TRANSITION MATRIX FOR CHANGE OF SUBJECT S71/W71

	1	2	3	4	5	6	7	8	9	10	11	12	13	14	15	16	17
1 THEOLOGY	.895	.0	.0	.0	.0	.0	.0	.0	.0	.0	.0	.0	.0	.0	.026	.079	.0
2 MEDICINE	.0	.903	.0	.002	.0	.0	.0	.0	.0	.0	.0	.0	.017	.002	.0	.075	.002
3 LAW	.0	.0	.864	.0	.0	.0	.0	.0	.0	.0	.0	.0	.018	.0	.023	.077	.018
4 ECON/BUS	.0	.0	.005	.847	.0	.0	.0	.0	.0	.0	.0	.0	.0	.010	.010	.114	.014
5 SOC SCI	.009	.0	.009	.0	.867	.0	.0	.0	.0	.0	.009	.0	.018	.009	.035	.044	.009
6 HUMANIT	.0	.0	.0	.0	.010	.870	.005	.005	.0	.0	.0	.0	.010	.010	.0	.073	.021
7 HSIT SCI	.0	.0	.0	.0	.0	.008	.862	.0	.0	.0	.0	.0	.032	.008	.0	.089	.0
8 NAT S I	.0	.0	.0	.0	.0	.0	.0	.838	.0	.0	.0	.0	.029	.012	.006	.116	.0
9 NAT S II	.0	.0	.0	.0	.0	.0	.0	.003	.923	.0	.0	.0	.014	.005	.003	.041	.011
10 ENGINEER	.0	.0	.0	.009	.0	.0	.0	.0	.0	.673	.0	.0	.0	.018	.009	.292	.0
11 ARCHITEC	.0	.0	.0	.0	.0	.0	.0	.0	.0	.0	.679	.0	.057	.019	.0	.245	.0
12 OTHERS	.0	.0	.0	.0	.0	.0	.0	.0	.0	.0	.0	.999	.0	.0	.0	.0	.0
13 INTERRUP	.0	.074	.089	.067	.030	.052	.015	.022	.067	.037	.030	.007	.467	.007	.0	.015	.0

3.14. TRANSITION MATRIX FOR CHANGE OF SUBJECT W71/S72

	1	2	3	4	5	6	7	8	9	10	11	12	13	14	15	16	17
1 THEOLOGY	.743	.0	.0	.0	.0	.0	.0	.0	.0	.0	.0	.0	.029	.0	.0	.229	.0
2 MEDICINE	.0	.934	.002	.0	.002	.0	.0	.0	.0	.0	.0	.0	.006	.002	.0	.052	.004
3 LAW	.0	.0	.843	.0	.0	.0	.0	.005	.005	.0	.0	.0	.020	.0	.0	.083	.044
4 ECON/BUS	.0	.0	.0	.825	.011	.0	.0	.005	.0	.0	.0	.0	.016	.021	.0	.090	.032
5 SOC SCI	.0	.0	.0	.010	.885	.0	.0	.0	.0	.0	.0	.0	.038	.019	.048	.010	.0
6 HUMANIT	.0	.0	.006	.0	.006	.886	.0	.006	.0	.0	.0	.0	.012	.006	.017	.040	.023
7 HSIT SCI	.009	.0	.0	.009	.0	.0	.864	.0	.009	.0	.0	.0	.027	.027	.009	.036	.009
8 NAT S I	.0	.007	.0	.0	.0	.0	.0	.867	.007	.0	.0	.0	.0	.007	.007	.087	.020
9 NAT S II	.0	.0	.0	.003	.0	.0	.0	.0	.931	.0	.0	.0	.009	.006	.0	.029	.023
10 ENGINEER	.0	.0	.0	.012	.0	.0	.0	.0	.0	.667	.0	.0	.0	.012	.012	.296	.0
11 ARCHITEC	.0	.0	.0	.0	.0	.0	.0	.0	.0	.0	.561	.0	.0	.0	.0	.439	.0
12 OTHERS	.0	.0	.0	.0	.0	.091	.0	.0	.0	.0	.0	.909	.0	.0	.0	.0	.0
13 INTERRUP	.010	.111	.071	.020	.020	.020	.040	.020	.061	.030	.040	.0	.495	.010	.0	.020	.0

3.15. TRANSITION MATRIX FOR CHANGE OF SUBJECT S72/W72

	1	2	3	4	5	6	7	8	9	10	11	12	13	14	15	16	17
1 THEOLOGY	.678	.0	.0	.0	.0	.0	.0	.0	.0	.0	.0	.0	.036	.0	.071	.214	.0
2 MEDICINE	.0	.616	.0	.0	.0	.0	.0	.0	.0	.0	.0	.0	.006	.002	.0	.303	.071
3 LAW	.0	.0	.861	.0	.0	.0	.0	.0	.0	.0	.0	.0	.022	.0	.017	.066	.033
4 ECON/BUS	.0	.0	.0	.839	.0	.0	.0	.0	.0	.006	.0	.0	.006	.006	.025	.062	.056
5 SOC SCI	.0	.0	.0	.0	.847	.0	.0	.0	.0	.0	.0	.0	.010	.0	.051	.092	.0
6 HUMANIT	.0	.0	.0	.0	.0	.898	.0	.0	.0	.0	.0	.0	.0	.006	.013	.070	.013
7 HSIT SCI	.0	.0	.010	.0	.0	.0	.919	.0	.0	.0	.0	.0	.010	.010	.0	.040	.010
8 NAT S I	.0	.0	.0	.0	.0	.0	.0	.845	.007	.0	.0	.0	.007	.015	.015	.096	.022
9 NAT S II	.0	.0	.0	.0	.0	.0	.0	.0	.912	.0	.0	.0	.006	.003	.0	.042	.036
10 ENGINEER	.0	.0	.0	.0	.0	.0	.0	.0	.0	.947	.0	.0	.0	.0	.0	.053	.0
11 ARCHITEC	.0	.0	.0	.0	.0	.0	.0	.0	.0	.0	.667	.0	.074	.0	.0	.259	.0
12 OTHERS	.0	.0	.0	.0	.0	.0	.0	.0	.0	.0	.0	.700	.0	.0	.200	.100	.0
13 INTERRUP	.014	.054	.068	.068	.054	.054	.054	.014	.081	.0	.014	.0	.473	.014	.0	.014	.0

3.16. TRANSITION MATRIX FOR CHANGE OF SUBJECT W72/S73

	1	2	3	4	5	6	7	8	9	10	11	12	13	14	15	16	17
1 THEOLOGY	.900	.0	.0	.0	.0	.0	.0	.0	.0	.0	.0	.0	.0	.0	.0	.050	.0
2 MEDICINE	.0	.586	.0	.0	.0	.0	.0	.0	.0	.0	.0	.0	.0	.0	.0	.313	.102
3 LAW	.0	.0	.858	.006	.0	.0	.0	.0	.0	.0	.0	.0	.0	.006	.025	.056	.049
4 ECON/BUS	.0	.0	.007	.879	.0	.0	.0	.0	.0	.0	.0	.0	.0	.007	.007	.071	.028
5 SOC SCI	.0	.0	.0	.0	.896	.011	.0	.0	.0	.0	.0	.0	.0	.0	.011	.046	.034
6 HUMANIT	.0	.007	.0	.0	.007	.821	.0	.0	.0	.0	.0	.0	.0	.028	.014	.076	.048
7 HSIT SCI	.0	.011	.0	.0	.0	.0	.863	.0	.0	.0	.0	.0	.0	.011	.011	.063	.042
8 NAT S I	.0	.0	.0	.0	.026	.0	.0	.879	.0	.0	.0	.0	.0	.0	.009	.052	.035
9 NAT S II	.0	.0	.0	.003	.0	.0	.0	.0	.891	.003	.0	.0	.0	.0	.003	.026	.075
10 ENGINEER	.0	.0	.0	.0	.018	.0	.0	.0	.0	.927	.0	.0	.0	.0	.018	.018	.018
11 ARCHITEC	.0	.0	.0	.0	.0	.0	.0	.0	.0	.0	.737	.0	.0	.0	.053	.211	.0
12 OTHERS	.0	.0	.0	.0	.0	.0	.0	.0	.0	.0	.0	.571	.0	.0	.143	.143	.143
13 INTERRUP	.0	.0	.0	.0	.0	.0	.0	.0	.0	.0	.0	.0	.961	.0	.0	.0	.0

3.17. TRANSITION MATRIX FOR CHANGE OF SUBJECT S73/W73

	1	2	3	4	5	6	7	8	9	10	11	12	13	14	15	16	17
1 THEOLOGY	.889	.0	.0	.0	.0	.0	.0	.0	.0	.0	.0	.0	.0	.0	.0	.111	.0
2 MEDICINE	.0	.874	.0	.0	.0	.0	.0	.0	.0	.0	.0	.0	.0	.0	.0	.094	.031
3 LAW	.0	.0	.957	.0	.0	.0	.0	.0	.0	.0	.0	.0	.0	.0	.0	.007	.035
4 ECON/BUS	.0	.0	.0	.960	.0	.0	.0	.0	.0	.0	.0	.0	.0	.0	.0	.008	.032
5 SOC SCI	.0	.0	.0	.0	.952	.0	.0	.0	.0	.0	.0	.0	.0	.012	.0	.012	.024
6 HUMANIT	.0	.0	.0	.0	.0	.917	.0	.0	.0	.0	.0	.0	.0	.017	.0	.042	.025
7 HSIT SCI	.0	.0	.0	.0	.0	.0	.951	.0	.0	.0	.0	.0	.0	.012	.0	.0	.037
8 NAT S I	.0	.0	.0	.0	.0	.0	.0	.911	.0	.0	.0	.0	.0	.030	.0	.010	.050
9 NAT S II	.0	.0	.0	.0	.0	.0	.0	.0	.960	.0	.0	.0	.0	.007	.0	.011	.022
10 ENGINEER	.0	.0	.0	.0	.0	.0	.0	.0	.0	.999	.0	.0	.0	.0	.0	.0	.0
11 ARCHITEC	.0	.0	.0	.0	.0	.0	.0	.0	.0	.0	.999	.0	.0	.0	.0	.0	.0
12 OTHERS	.0	.0	.0	.0	.0	.0	.0	.0	.0	.0	.0	.500	.0	.250	.250	.0	.0
13 INTERRUP	.0	.0	.0	.0	.0	.0	.0	.0	.0	.0	.0	.0	.918	.061	.0	.020	.0

3.18. TRANSITION MATRIX FOR CHANGE OF SUBJECT W73/S74

	1	2	3	4	5	6	7	8	9	10	11	12	13	14	15	16	17
1 THEOLOGY	.500	.0	.0	.0	.0	.0	.0	.0	.0	.0	.0	.0	.0	.0	.0	.438	.063
2 MEDICINE	.0	.383	.0	.0	.0	.0	.0	.0	.0	.0	.0	.0	.0	.0	.006	.455	.156
3 LAW	.0	.0	.570	.0	.0	.0	.0	.0	.0	.0	.0	.0	.0	.0	.015	.074	.341
4 ECON/BUS	.0	.0	.0	.554	.0	.0	.0	.0	.0	.0	.0	.0	.0	.0	.017	.165	.264
5 SOC SCI	.0	.0	.0	.0	.633	.0	.0	.0	.0	.0	.0	.0	.0	.0	.038	.241	.089
6 HUMANIT	.0	.0	.0	.0	.0	.609	.0	.0	.0	.0	.0	.0	.0	.0	.027	.218	.145
7 HSIT SCI	.0	.0	.0	.0	.0	.0	.667	.0	.0	.0	.0	.0	.0	.0	.038	.141	.154
8 NAT S I	.0	.0	.0	.0	.0	.0	.0	.717	.0	.0	.0	.0	.0	.0	.022	.120	.141
9 NAT S II	.0	.0	.0	.0	.0	.0	.0	.0	.614	.0	.0	.0	.0	.0	.027	.076	.284
10 ENGINEER	.0	.0	.0	.0	.0	.0	.0	.0	.0	.731	.0	.0	.0	.0	.058	.077	.135
11 ARCHITEC	.0	.0	.0	.0	.0	.0	.0	.0	.0	.0	.429	.0	.0	.0	.143	.429	.0
12 OTHERS	.0	.0	.0	.0	.0	.0	.0	.0	.0	.0	.0	.0	.0	.0	.999	.0	.0
13 INTERRUP	.0	.0	.0	.0	.0	.0	.0	.0	.0	.0	.0	.0	.0	.0	.0	.0	.0

3.19. TRANSITION MATRIX FOR CHANGE OF SUBJECT S74/W74

	1	2	3	4	5	6	7	8	9	10	11	12	13	14	15	16	17
1 THEOLOGY	.750	.0	.0	.0	.0	.0	.0	.0	.0	.0	.0	.0	.0	.0	.0	.250	.0
2 MEDICINE	.0	.656	.0	.0	.0	.0	.0	.0	.0	.0	.0	.0	.0	.0	.0	.250	.094
3 LAW	.0	.0	.688	.0	.0	.0	.0	.0	.0	.0	.0	.0	.0	.0	.013	.091	.208
4 ECON/BUS	.0	.0	.0	.672	.0	.0	.0	.0	.0	.0	.0	.0	.0	.0	.0	.119	.149
5 SOC SCI	.0	.0	.0	.0	.820	.0	.0	.0	.0	.0	.0	.0	.0	.0	.040	.060	.080
6 HUMANIT	.0	.0	.0	.0	.0	.716	.0	.0	.0	.0	.0	.0	.0	.0	.045	.149	.090
7 HSIT SCI	.0	.0	.0	.0	.0	.0	.750	.0	.0	.0	.0	.0	.0	.0	.019	.115	.115
8 NAT S I	.0	.0	.0	.0	.0	.0	.0	.848	.0	.0	.0	.0	.0	.0	.0	.061	.091
9 NAT S II	.0	.0	.0	.0	.0	.0	.0	.0	.722	.0	.0	.0	.0	.0	.019	.025	.235
10 ENGINEER	.0	.0	.0	.0	.0	.0	.0	.0	.0	.763	.0	.0	.0	.0	.079	.079	.079
11 ARCHITEC	.0	.0	.0	.0	.0	.0	.0	.0	.0	.0	.999	.0	.0	.0	.0	.0	.0
12 OTHERS	.0	.0	.0	.0	.0	.0	.0	.0	.0	.0	.0	.0	.0	.0	.0	.0	.0
13 INTERRUP	.0	.0	.0	.0	.0	.0	.0	.0	.0	.0	.0	.0	.0	.0	.0	.0	.0

3.20. TRANSITION MATRIX FOR CHANGE OF SUBJECT W74/S75

	1	2	3	4	5	6	7	8	9	10	11	12	13	14	15	16	17
1 THEOLOGY	.999	.0	.0	.0	.0	.0	.0	.0	.0	.0	.0	.0	.0	.0	.0	.0	.0
2 MEDICINE	.0	.762	.0	.0	.0	.0	.0	.0	.0	.0	.0	.0	.0	.0	.0	.143	.095
3 LAW	.0	.0	.472	.0	.0	.0	.0	.0	.0	.0	.0	.0	.0	.0	.019	.019	.491
4 ECON/BUS	.0	.0	.0	.511	.0	.0	.0	.0	.0	.0	.0	.0	.0	.0	.0	.111	.378
5 SOC SCI	.0	.0	.0	.0	.683	.0	.0	.0	.0	.0	.0	.0	.0	.0	.049	.098	.171
6 HUMANIT	.0	.0	.0	.0	.0	.646	.0	.0	.0	.0	.0	.0	.0	.0	.063	.146	.146
7 HSIT SCI	.0	.0	.0	.0	.0	.0	.410	.0	.0	.0	.0	.0	.0	.0	.026	.154	.410
8 NAT S I	.0	.0	.0	.0	.0	.0	.0	.696	.0	.0	.0	.0	.0	.0	.0	.054	.250
9 NAT S II	.0	.0	.0	.0	.0	.0	.0	.0	.709	.0	.0	.0	.0	.0	.026	.043	.222
10 ENGINEER	.0	.0	.0	.0	.0	.0	.0	.0	.0	.483	.0	.0	.0	.0	.103	.172	.241
11 ARCHITEC	.0	.0	.0	.0	.0	.0	.0	.0	.0	.0	.500	.0	.0	.0	.0	.333	.167
12 OTHERS	.0	.0	.0	.0	.0	.0	.0	.0	.0	.0	.0	.0	.0	.0	.0	.0	.0
13 INTERRUP	.0	.0	.0	.0	.0	.0	.0	.0	.0	.0	.0	.0	.0	.0	.0	.0	.0

3.21. TRANSITION MATRIX FOR CHANGE OF SUBJECT S75/W75

	1	2	3	4	5	6	7	8	9	10	11	12	13	14	15	16	17
1 THEOLOGY	.333	.0	.0	.0	.0	.0	.0	.0	.0	.0	.0	.0	.0	.0	.167	.167	.333
2 MEDICINE	.0	.688	.0	.0	.0	.0	.0	.0	.0	.0	.0	.0	.0	.0	.031	.219	.063
3 LAW	.0	.0	.640	.0	.0	.0	.0	.0	.0	.0	.0	.0	.0	.0	.040	.080	.240
4 ECON/BUS	.0	.0	.0	.739	.0	.0	.0	.0	.0	.0	.0	.0	.0	.0	.0	.087	.174
5 SOC SCI	.0	.0	.0	.0	.857	.0	.0	.0	.0	.0	.0	.0	.0	.0	.0	.036	.107
6 HUMANIT	.0	.0	.0	.0	.0	.903	.0	.0	.0	.0	.0	.0	.0	.0	.032	.032	.032
7 HSIT SCI	.0	.0	.0	.0	.0	.0	.750	.0	.0	.0	.0	.0	.0	.0	.0	.188	.063
8 NAT S I	.0	.0	.0	.0	.0	.0	.0	.846	.0	.0	.0	.0	.0	.0	.0	.051	.103
9 NAT S II	.0	.0	.0	.0	.0	.0	.0	.0	.771	.0	.0	.0	.0	.0	.0	.012	.217
10 ENGINEER	.0	.0	.0	.0	.0	.0	.0	.0	.0	.929	.0	.0	.0	.0	.0	.071	.0
11 ARCHITEC	.0	.0	.0	.0	.0	.0	.0	.0	.0	.0	.667	.0	.0	.0	.0	.0	.333
12 OTHERS	.0	.0	.0	.0	.0	.0	.0	.0	.0	.0	.0	.0	.0	.0	.0	.0	.0
13 INTERRUP	.0	.0	.0	.0	.0	.0	.0	.0	.0	.0	.0	.0	.0	.0	.0	.0	.0

3.22. TRANSITION MATRIX FOR CHANGE OF SUBJECT W75/S76

	1	2	3	4	5	6	7	8	9	10	11	12	13	14	15	16	17
1 THEOLOGY	.999	.0	.0	.0	.0	.0	.0	.0	.0	.0	.0	.0	.0	.0	.0	.0	.0
2 MEDICINE	.0	.682	.0	.0	.0	.0	.0	.0	.0	.0	.0	.0	.0	.0	.0	.091	.227
3 LAW	.0	.0	.375	.0	.0	.0	.0	.0	.0	.0	.0	.0	.0	.0	.0	.063	.563
4 ECON/BUS	.0	.0	.0	.529	.0	.0	.0	.0	.0	.0	.0	.0	.0	.0	.0	.059	.412
5 SOC SCI	.0	.0	.0	.0	.792	.0	.0	.0	.0	.0	.0	.0	.0	.0	.083	.083	.042
6 HUMANIT	.0	.0	.0	.0	.0	.643	.0	.0	.0	.0	.0	.0	.0	.0	.036	.0	.321
7 HSIT SCI	.0	.0	.0	.0	.0	.0	.750	.0	.0	.0	.0	.0	.0	.0	.0	.083	.167
8 NAT S I	.0	.0	.0	.0	.0	.0	.0	.697	.0	.0	.0	.0	.0	.0	.0	.030	.273
9 NAT S II	.0	.0	.0	.0	.0	.0	.0	.0	.578	.0	.0	.0	.0	.0	.031	.047	.344
10 ENGINEER	.0	.0	.0	.0	.0	.0	.0	.0	.0	.385	.0	.0	.0	.0	.0	.0	.615
11 ARCHITEC	.0	.0	.0	.0	.0	.0	.0	.0	.0	.0	.999	.0	.0	.0	.0	.0	.0
12 OTHERS	.0	.0	.0	.0	.0	.0	.0	.0	.0	.0	.0	.0	.0	.0	.0	.0	.0
13 INTERRUP	.0	.0	.0	.0	.0	.0	.0	.0	.0	.0	.0	.0	.0	.0	.0	.0	.0

4.1. TRANSITION MATRIX FOR CHANGE OF SUBJECT W65/S66

	1	2	3	4	5	6	7	8	9	10	11	12	13	14	15	16	17
1 THEOLOGY	.942	.0	.0	.0	.0	.0	.014	.0	.0	.0	.0	.0	.029	.014	.0	.0	.0
2 MEDICINE	.001	.967	.0	.0	.0	.001	.0	.0	.004	.0	.0	.0	.025	.003	.0	.0	.0
3 LAW	.0	.0	.927	.0	.006	.0	.0	.003	.0	.0	.0	.0	.064	.0	.0	.0	.0
4 ECON/BUS	.0	.0	.002	.911	.0	.0	.0	.0	.0	.002	.0	.0	.082	.002	.0	.0	.0
5 SOC SCI	.0	.004	.0	.0	.880	.007	.0	.0	.0	.004	.0	.0	.054	.007	.040	.0	.0
6 HUMANIT	.0	.0	.002	.0	.002	.925	.004	.0	.006	.0	.002	.0	.038	.018	.004	.0	.0
7 HSIT SCI	.0	.006	.0	.006	.0	.0	.919	.0	.0	.0	.0	.0	.043	.025	.0	.0	.0
8 NAT S I	.0	.003	.0	.0	.0	.0	.003	.939	.008	.003	.0	.0	.040	.008	.0	.0	.0
9 NAT S II	.0	.006	.0	.0	.002	.0	.0	.0	.956	.0	.002	.0	.029	.006	.0	.0	.0
10 ENGINEER	.0	.0	.0	.0	.0	.0	.0	.004	.0	.949	.0	.0	.043	.006	.0	.0	.0
11 ARCHITEC	.0	.008	.0	.0	.0	.0	.0	.008	.0	.0	.858	.0	.102	.024	.0	.0	.0
12 OTHERS	.0	.018	.0	.018	.0	.0	.0	.0	.0	.0	.0	.859	.053	.035	.018	.0	.0
13 INTERRUP	.011	.054	.054	.054	.021	.086	.011	.054	.129	.021	.011	.021	.451	.011	.0	.0	.0

4.2. TRANSITION MATRIX FOR CHANGE OF SUBJECT W65/W66

	1	2	3	4	5	6	7	8	9	10	11	12	13	14	15	16	17
1 THEOLOGY	.927	.0	.0	.0	.0	.0	.014	.0	.0	.0	.0	.0	.043	.014	.0	.0	.0
2 MEDICINE	.003	.936	.004	.0	.006	.003	.0	.003	.006	.0	.0	.0	.029	.012	.0	.0	.0
3 LAW	.0	.003	.910	.006	.006	.003	.003	.003	.0	.0	.0	.0	.062	.006	.0	.0	.0
4 ECON/BUS	.0	.0	.0	.888	.002	.002	.002	.0	.0	.002	.0	.0	.080	.018	.0	.0	.0
5 SOC SCI	.007	.004	.007	.004	.814	.014	.0	.0	.004	.004	.0	.0	.040	.036	.065	.0	.0
6 HUMANIT	.002	.002	.002	.002	.012	.878	.018	.002	.004	.0	.002	.002	.038	.030	.006	.0	.0
7 HSIT SCI	.0	.0	.0	.006	.0	.019	.882	.0	.012	.0	.0	.0	.037	.043	.0	.0	.0
8 NAT S I	.0	.003	.008	.010	.005	.005	.008	.822	.020	.008	.0	.005	.058	.023	.003	.0	.0
9 NAT S II	.0	.012	.006	.002	.002	.002	.002	.010	.883	.0	.002	.0	.065	.017	.0	.0	.0
10 ENGINEER	.0	.006	.002	.006	.0	.002	.0	.006	.006	.857	.0	.0	.110	.006	.0	.0	.0
11 ARCHITEC	.0	.008	.0	.0	.0	.0	.0	.008	.0	.0	.740	.0	.220	.024	.0	.0	.0
12 OTHERS	.0	.018	.0	.018	.0	.018	.0	.018	.0	.0	.0	.702	.053	.053	.123	.0	.0
13 INTERRUP	.021	.064	.107	.107	.043	.097	.043	.064	.107	.097	.021	.011	.172	.032	.0	.0	.0

4.3. TRANSITION MATRIX FOR CHANGE OF SUBJECT W65/S67

	1	2	3	4	5	6	7	8	9	10	11	12	13	14	15	16	17
1 THEOLOGY	.884	.0	.014	.0	.0	.014	.014	.0	.0	.0	.0	.0	.043	.029	.0	.0	.0
2 MEDICINE	.003	.908	.007	.0	.007	.004	.0	.003	.013	.0	.001	.0	.032	.022	.0	.0	.0
3 LAW	.0	.0	.904	.008	.011	.003	.003	.0	.0	.0	.0	.0	.062	.008	.0	.0	.0
4 ECON/BUS	.0	.002	.013	.850	.002	.002	.002	.0	.0	.002	.002	.0	.088	.029	.0	.0	.0
5 SOC SCI	.007	.022	.011	.011	.728	.018	.007	.004	.011	.007	.0	.0	.054	.047	.072	.0	.0
6 HUMANIT	.002	.006	.002	.006	.010	.810	.022	.004	.004	.0	.002	.0	.060	.052	.018	.002	.0
7 HSIT SCI	.0	.0	.0	.006	.019	.025	.845	.0	.019	.0	.0	.0	.037	.043	.006	.0	.0
8 NAT S I	.003	.008	.010	.015	.003	.003	.010	.787	.030	.010	.003	.005	.066	.043	.008	.0	.0
9 NAT S II	.002	.013	.006	.002	.004	.002	.004	.013	.860	.002	.002	.0	.061	.029	.002	.0	.0
10 ENGINEER	.0	.012	.002	.010	.0	.004	.0	.008	.006	.836	.002	.0	.108	.012	.0	.0	.0
11 ARCHITEC	.0	.016	.0	.008	.008	.008	.008	.008	.0	.016	.724	.008	.157	.039	.0	.0	.0
12 OTHERS	.0	.018	.0	.018	.035	.018	.018	.018	.0	.0	.0	.614	.070	.070	.123	.0	.0
13 INTERRUP	.011	.075	.075	.097	.043	.107	.032	.064	.097	.107	.043	.011	.161	.075	.0	.0	.0

4.4. TRANSITION MATRIX FOR CHANGE OF SUBJECT W65/W67

	1	2	3	4	5	6	7	8	9	10	11	12	13	14	15	16	17
1 THEOLOGY	.782	.0	.014	.0	.0	.014	.014	.0	.0	.0	.0	.0	.116	.043	.014	.0	.0
2 MEDICINE	.003	.885	.010	.001	.009	.007	.0	.003	.013	.0	.001	.0	.037	.032	.0	.0	.0
3 LAW	.003	.0	.907	.008	.011	.006	.006	.0	.003	.0	.0	.0	.042	.014	.0	.0	.0
4 ECON/BUS	.0	.002	.015	.836	.002	.002	.002	.0	.0	.002	.0	.0	.046	.044	.0	.0	.0
5 SOC SCI	.018	.025	.011	.011	.608	.014	.007	.004	.011	.007	.0	.0	.065	.080	.145	.007	.0
6 HUMANIT	.002	.010	.002	.006	.014	.748	.026	.004	.010	.0	.002	.0	.042	.074	.056	.004	.0
7 HSIT SCI	.0	.0	.0	.012	.019	.012	.814	.0	.031	.0	.0	.0	.043	.056	.012	.0	.0
8 NAT S I	.003	.008	.013	.018	.005	.005	.010	.744	.040	.020	.003	.005	.028	.076	.025	.0	.0
9 NAT S II	.004	.013	.008	.002	.008	.004	.004	.017	.835	.004	.0	.0	.042	.044	.013	.002	.0
10 ENGINEER	.0	.014	.002	.014	.0	.004	.0	.008	.008	.820	.004	.0	.108	.018	.0	.0	.0
11 ARCHITEC	.0	.008	.0	.008	.008	.0	.008	.0	.0	.024	.685	.008	.189	.055	.0	.0	.0
12 OTHERS	.0	.018	.0	.018	.018	.018	.018	.018	.018	.0	.0	.561	.035	.070	.210	.0	.0
13 INTERRUP	.011	.097	.086	.107	.043	.086	.043	.075	.097	.129	.043	.021	.075	.075	.0	.0	.0

4.5. TRANSITION MATRIX FOR CHANGE OF SUBJECT W65/S68

	1	2	3	4	5	6	7	8	9	10	11	12	13	14	15	16	17
1 THEOLOGY	.811	.0	.014	.0	.014	.014	.014	.0	.0	.0	.0	.0	.072	.043	.014	.0	.0
2 MEDICINE	.003	.841	.010	.003	.009	.009	.001	.003	.016	.001	.001	.0	.062	.037	.006	.0	.0
3 LAW	.003	.003	.876	.008	.011	.006	.008	.0	.003	.0	.0	.0	.067	.017	.0	.0	.0
4 ECON/BUS	.0	.002	.018	.831	.007	.002	.002	.002	.002	.002	.0	.0	.073	.062	.0	.0	.0
5 SOC SCI	.007	.025	.011	.022	.568	.022	.007	.004	.011	.007	.0	.0	.062	.101	.148	.007	.0
6 HUMANIT	.002	.016	.002	.004	.012	.685	.032	.006	.008	.0	.002	.0	.046	.082	.096	.008	.0
7 HSIT SCI	.0	.0	.0	.012	.019	.012	.764	.0	.031	.0	.0	.0	.062	.087	.012	.0	.0
8 NAT S I	.005	.008	.013	.023	.005	.005	.010	.665	.043	.025	.003	.008	.061	.086	.043	.0	.0
9 NAT S II	.004	.017	.008	.002	.008	.004	.004	.017	.801	.004	.0	.0	.044	.048	.036	.004	.0
10 ENGINEER	.0	.014	.002	.018	.004	.002	.0	.014	.008	.792	.004	.002	.114	.027	.0	.0	.0
11 ARCHITEC	.0	.016	.008	.016	.0	.0	.008	.016	.0	.039	.637	.008	.197	.063	.0	.0	.0
12 OTHERS	.0	.018	.0	.018	.018	.035	.018	.018	.018	.0	.0	.456	.088	.088	.228	.0	.0
13 INTERRUP	.011	.097	.086	.097	.043	.097	.032	.064	.140	.129	.043	.011	.054	.075	.011	.0	.0

4.6. TRANSITION MATRIX FOR CHANGE OF SUBJECT W65/W68

	1	2	3	4	5	6	7	8	9	10	11	12	13	14	15	16	17
1 THEOLOGY	.855	.0	.014	.0	.014	.0	.014	.0	.0	.0	.0	.0	.014	.058	.014	.014	.0
2 MEDICINE	.003	.857	.012	.003	.007	.010	.001	.003	.016	.001	.001	.0	.041	.037	.009	.0	.0
3 LAW	.003	.003	.862	.008	.014	.008	.006	.0	.003	.0	.0	.0	.056	.028	.0	.011	.0
4 ECON/BUS	.0	.002	.018	.780	.007	.002	.002	.002	.002	.002	.004	.0	.044	.077	.0	.060	.0
5 SOC SCI	.007	.029	.011	.022	.478	.018	.007	.004	.011	.007	.0	.004	.069	.119	.174	.040	.0
6 HUMANIT	.002	.016	.002	.004	.014	.569	.030	.004	.008	.0	.002	.0	.048	.103	.191	.010	.0
7 HSIT SCI	.0	.0	.0	.006	.019	.012	.739	.0	.037	.0	.0	.0	.043	.106	.037	.0	.0
8 NAT S I	.005	.008	.013	.020	.008	.008	.010	.627	.046	.025	.003	.003	.056	.096	.068	.008	.0
9 NAT S II	.004	.019	.008	.006	.006	.004	.002	.017	.772	.004	.002	.0	.038	.056	.060	.004	.0
10 ENGINEER	.0	.014	.002	.018	.006	.002	.0	.016	.010	.804	.006	.002	.092	.029	.0	.0	.0
11 ARCHITEC	.0	.016	.008	.016	.0	.008	.016	.016	.0	.039	.630	.008	.165	.079	.0	.0	.0
12 OTHERS	.0	.018	.0	.018	.018	.018	.018	.018	.018	.0	.0	.421	.088	.123	.246	.018	.0
13 INTERRUP	.011	.086	.097	.097	.043	.064	.032	.075	.140	.118	.032	.011	.075	.086	.021	.0	.0

4.7. TRANSITION MATRIX FOR CHANCE CF SUBJECT W65/S69

	1	2	3	4	5	6	7	8	9	10	11	12	13	14	15	16	17
1 THEOLOGY	.811	.0	.014	.0	.014	.0	.014	.0	.0	.0	.0	.0	.029	.072	.014	.029	.0
2 MEDICINE	.004	.861	.012	.003	.007	.010	.003	.003	.015	.001	.001	.0	.026	.043	.010	.001	.0
3 LAW	.003	.0	.843	.003	.008	.003	.006	.0	.003	.0	.0	.0	.056	.039	.0	.042	.0
4 ECON/BUS	.0	.002	.018	.676	.009	.002	.002	.002	.002	.002	.004	.0	.049	.088	.0	.146	.0
5 SOC SCI	.007	.029	.014	.025	.445	.022	.004	.004	.011	.007	.0	.0	.054	.123	.185	.069	.0
6 HUMANIT	.002	.016	.002	.008	.014	.507	.032	.002	.006	.0	.002	.0	.044	.109	.245	.012	.0
7 HSIT SCI	.0	.0	.006	.006	.019	.012	.714	.0	.037	.0	.0	.0	.043	.118	.043	.0	.0
8 NAT S I	.003	.008	.013	.023	.013	.005	.013	.620	.046	.028	.003	.005	.040	.096	.078	.013	.0
9 NAT S II	.004	.021	.008	.008	.006	.004	.004	.015	.728	.006	.002	.0	.050	.063	.075	.008	.0
10 ENGINEER	.0	.014	.002	.024	.004	.004	.0	.018	.012	.806	.008	.002	.067	.035	.0	.004	.0
11 ARCHITEC	.0	.024	.008	.016	.008	.008	.024	.016	.0	.039	.614	.008	.157	.079	.0	.0	.0
12 OTHERS	.0	.035	.0	.018	.018	.035	.018	.018	.018	.0	.0	.351	.070	.123	.281	.018	.0
13 INTERRUP	.011	.086	.097	.086	.054	.075	.032	.064	.118	.118	.032	.011	.097	.086	.032	.011	.0

4.8. TRANSITION MATRIX FOR CHANGE OF SUBJECT W65/W69

	1	2	3	4	5	6	7	8	9	10	11	12	13	14	15	16	17
1 THEOLOGY	.667	.0	.014	.0	.014	.0	.014	.0	.0	.0	.0	.0	.029	.072	.029	.145	.014
2 MEDICINE	.004	.858	.010	.033	.007	.009	.004	.004	.019	.001	.0	.0	.019	.047	.013	.003	.0
3 LAW	.003	.0	.792	.036	.011	.008	.003	.0	.003	.0	.0	.0	.034	.042	.003	.095	.0
4 ECON/BUS	.0	.002	.020	.575	.009	.002	.002	.002	.002	.002	.007	.0	.029	.104	.0	.245	.002
5 SOC SCI	.007	.025	.014	.022	.366	.014	.004	.004	.011	.004	.0	.0	.054	.130	.203	.138	.004
6 HUMANIT	.002	.018	.0	.008	.012	.450	.028	.002	.004	.0	.002	.002	.064	.117	.269	.022	.0
7 HSIT SCI	.0	.0	.0	.012	.019	.019	.640	.0	.031	.0	.0	.0	.062	.124	.043	.050	.0
8 NAT S I	.005	.008	.013	.023	.013	.010	.013	.579	.046	.023	.0	.003	.056	.109	.091	.018	.0
9 NAT S II	.004	.021	.008	.008	.006	.002	.004	.012	.705	.004	.002	.0	.040	.065	.086	.035	.0
10 ENGINEER	.0	.014	.002	.027	.006	.004	.0	.016	.012	.726	.008	.002	.067	.078	.0	.039	.0
11 ARCHITEC	.0	.024	.008	.016	.008	.008	.024	.008	.0	.024	.661	.008	.110	.087	.008	.008	.0
12 OTHERS	.0	.035	.0	.035	.018	.035	.018	.018	.018	.0	.0	.246	.035	.123	.298	.123	.0
13 INTERRUP	.011	.075	.097	.086	.043	.064	.032	.075	.129	.118	.021	.011	.064	.118	.032	.021	.0

4.9. TRANSITION MATRIX FOR CHANGE OF SUBJECT W65/S70

	1	2	3	4	5	6	7	8	9	10	11	12	13	14	15	16	17
1 THEOLOGY	.536	.0	.014	.0	.0	.0	.014	.0	.0	.0	.0	.0	.043	.072	.029	.232	.014
2 MEDICINE	.004	.842	.010	.003	.007	.006	.004	.004	.016	.001	.001	.0	.028	.054	.015	.004	.0
3 LAW	.003	.0	.672	.008	.008	.008	.003	.0	.003	.0	.0	.0	.070	.045	.003	.174	.003
4 ECON/BUS	.0	.002	.018	.482	.011	.002	.002	.002	.002	.0	.004	.002	.035	.113	.0	.325	.002
5 SOC SCI	.007	.022	.018	.022	.340	.007	.004	.004	.011	.004	.0	.0	.054	.130	.210	.163	.004
6 HUMANIT	.002	.018	.002	.006	.014	.398	.034	.002	.004	.0	.0	.002	.060	.123	.291	.044	.0
7 HSIT SCI	.0	.0	.0	.012	.019	.019	.602	.0	.031	.0	.0	.0	.056	.143	.050	.062	.006
8 NAT S I	.005	.005	.013	.028	.013	.008	.013	.493	.038	.018	.0	.0	.043	.111	.106	.111	.0
9 NAT S II	.004	.027	.008	.010	.004	.002	.004	.012	.662	.004	.002	.002	.035	.071	.092	.063	.0
10 ENGINEER	.0	.014	.002	.022	.002	.004	.0	.012	.014	.506	.006	.002	.037	.084	.004	.292	.0
11 ARCHITEC	.0	.024	.008	.016	.008	.008	.024	.008	.0	.016	.645	.008	.087	.087	.008	.055	.0
12 OTHERS	.0	.035	.0	.035	.018	.035	.018	.018	.0	.0	.0	.210	.018	.123	.351	.140	.0
13 INTERRUP	.011	.075	.097	.086	.043	.064	.021	.086	.129	.118	.021	.0	.064	.129	.032	.021	.0

4.10. TRANSITION MATRIX FOR CHANGE OF SUBJECT W65/W70

	1	2	3	4	5	6	7	8	9	10	11	12	13	14	15	16	17
1 THEOLOGY	.435	.0	.014	.0	.043	.0	.0	.0	.0	.0	.0	.0	.058	.072	.029	.333	.014
2 MEDICINE	.004	.826	.009	.003	.007	.009	.004	.004	.013	.001	.003	.0	.024	.060	.015	.019	.0
3 LAW	.003	.0	.588	.008	.011	.006	.003	.0	.003	.0	.0	.0	.064	.050	.003	.255	.006
4 ECON/BUS	.0	.002	.018	.400	.011	.002	.002	.002	.002	.0	.004	.002	.027	.122	.002	.402	.004
5 SOC SCI	.007	.022	.018	.022	.308	.011	.0	.004	.011	.007	.0	.0	.040	.138	.221	.188	.004
6 HUMANIT	.002	.018	.0	.006	.016	.364	.032	.002	.002	.0	.0	.0	.048	.129	.301	.076	.006
7 HSIT SCI	.0	.0	.0	.012	.019	.019	.571	.0	.031	.0	.0	.0	.031	.149	.056	.106	.006
8 NAT S I	.005	.005	.008	.025	.010	.005	.013	.428	.033	.020	.003	.0	.028	.121	.114	.182	.003
9 NAT S II	.004	.027	.006	.010	.004	.002	.004	.015	.636	.006	.002	.0	.029	.071	.094	.092	.0
10 ENGINEER	.0	.014	.002	.018	.002	.004	.0	.008	.016	.335	.004	.0	.037	.098	.008	.453	.002
11 ARCHITEC	.0	.024	.008	.016	.008	.008	.024	.008	.0	.024	.551	.0	.087	.087	.016	.142	.0
12 OTHERS	.0	.035	.0	.035	.018	.018	.018	.018	.0	.0	.0	.158	.018	.123	.368	.193	.0
13 INTERRUP	.011	.064	.086	.064	.043	.043	.011	.086	.107	.129	.021	.0	.075	.172	.043	.043	.0

4.11. TRANSITION MATRIX FOR CHANGE OF SUBJECT W65/S71

	1	2	3	4	5	6	7	8	9	10	11	12	13	14	15	16	17
1 THEOLOGY	.377	.0	.014	.0	.043	.0	.014	.0	.0	.0	.0	.0	.014	.072	.043	.406	.014
2 MEDICINE	.004	.789	.009	.003	.009	.009	.004	.004	.012	.001	.003	.0	.025	.060	.015	.053	.001
3 LAW	.003	.003	.524	.020	.011	.006	.003	.003	.003	.0	.0	.0	.062	.050	.003	.300	.011
4 ECON/BUS	.0	.002	.015	.345	.009	.002	.002	.002	.002	.0	.004	.002	.035	.124	.004	.446	.007
5 SOC SCI	.007	.025	.018	.022	.261	.007	.004	.004	.007	.004	.0	.0	.043	.138	.235	.214	.011
6 HUMANIT	.002	.018	.0	.004	.014	.336	.034	.002	.002	.0	.0	.002	.030	.135	.314	.105	.006
7 HSIT SCI	.0	.0	.0	.019	.025	.019	.553	.0	.019	.0	.0	.006	.019	.149	.062	.112	.019
8 NAT S I	.005	.005	.008	.025	.008	.005	.010	.377	.033	.018	.003	.0	.020	.121	.116	.243	.005
9 NAT S II	.004	.027	.006	.010	.004	.002	.004	.013	.612	.004	.0	.0	.019	.077	.096	.123	.0
10 ENGINEER	.0	.012	.002	.024	.004	.006	.0	.006	.014	.184	.004	.0	.024	.098	.010	.608	.004
11 ARCHITEC	.0	.024	.008	.016	.008	.008	.024	.008	.0	.031	.331	.0	.063	.087	.016	.378	.0
12 OTHERS	.0	.035	.0	.035	.018	.0	.018	.0	.0	.0	.0	.123	.018	.123	.421	.210	.0
13 INTERRUP	.011	.064	.064	.032	.043	.032	.011	.064	.107	.086	.043	.0	.043	.172	.054	.150	.021

TRANSITION MATRIX FOR CHANGE OF SUBJECT W65/W71

	1	2	3	4	5	6	7	8	9	10	11	12	13	14	15	16	17
1 THEOLOGY	.333	.0	.014	.0	.043	.0	.014	.0	.0	.0	.0	.0	.014	.072	.058	.435	.014
2 MEDICINE	.004	.726	.009	.004	.007	.007	.003	.004	.015	.001	.003	.0	.018	.062	.015	.121	.003
3 LAW	.003	.003	.487	.017	.008	.008	.003	.003	.003	.0	.0	.0	.039	.050	.017	.336	.022
4 ECON/BUS	.0	.002	.015	.312	.007	.002	.0	.0	.002	.0	.004	.002	.022	.124	.007	.486	.013
5 SOC SCI	.007	.022	.018	.022	.239	.011	.0	.004	.004	.004	.0	.0	.036	.141	.246	.235	.014
6 HUMANIT	.002	.018	.002	.004	.016	.299	.032	.004	.002	.0	.0	.002	.022	.141	.316	.131	.016
7 HSIT SCI	.0	.0	.0	.019	.025	.019	.484	.0	.019	.0	.0	.006	.025	.155	.062	.168	.019
8 NAT S I	.005	.003	.005	.018	.008	.005	.010	.321	.030	.018	.003	.0	.020	.126	.119	.306	.005
9 NAT S II	.004	.027	.002	.010	.002	.002	.006	.012	.580	.004	.002	.0	.013	.081	.098	.152	.008
10 ENGINEER	.0	.014	.002	.022	.004	.006	.0	.006	.010	.124	.006	.0	.012	.100	.014	.675	.004
11 ARCHITEC	.0	.024	.0	.0	.008	.016	.024	.008	.0	.024	.236	.0	.047	.094	.016	.504	.0
12 OTHERS	.0	.018	.0	.035	.018	.0	.018	.0	.0	.0	.0	.140	.018	.123	.421	.210	.0
13 INTERRUP	.011	.043	.064	.032	.043	.021	.011	.064	.107	.064	.021	.0	.054	.183	.054	.204	.021

4.13. TRANSITION MATRIX FOR CHANGE OF SUBJECT W65/S72

	1	2	3	4	5	6	7	8	9	10	11	12	13	14	15	16	17
1 THEOLOGY	.246	.0	.014	.0	.029	.0	.014	.0	.0	.0	.0	.0	.0	.072	.072	.536	.014
2 MEDICINE	.003	.691	.009	.004	.009	.007	.003	.003	.018	.0	.001	.0	.009	.063	.016	.160	.006
3 LAW	.003	.003	.423	.014	.008	.008	.003	.006	.003	.0	.0	.0	.031	.050	.020	.384	.048
4 ECON/BUS	.0	.002	.011	.263	.009	.002	.0	.0	.002	.0	.002	.002	.024	.133	.007	.517	.027
5 SOC SCI	.007	.025	.018	.022	.224	.011	.0	.004	.004	.0	.0	.0	.022	.148	.253	.246	.014
6 HUMANIT	.0	.018	.004	.004	.012	.267	.034	.004	.002	.0	.0	.002	.018	.145	.324	.149	.022
7 HSIT SCI	.006	.0	.0	.019	.012	.019	.410	.0	.025	.0	.0	.006	.043	.174	.068	.193	.025
8 NAT S I	.005	.005	.008	.013	.010	.005	.010	.283	.035	.010	.0	.0	.008	.132	.119	.347	.013
9 NAT S II	.004	.021	.002	.013	.002	.002	.004	.010	.543	.002	.002	.0	.019	.083	.098	.175	.023
10 ENGINEER	.0	.012	.002	.016	.004	.004	.0	.008	.008	.094	.004	.0	.008	.102	.016	.716	.006
11 ARCHITEC	.0	.024	.0	.0	.008	.008	.024	.016	.0	.024	.165	.0	.016	.094	.016	.606	.0
12 OTHERS	.0	.035	.0	.035	.018	.018	.018	.0	.0	.0	.0	.123	.0	.123	.421	.210	.0
13 INTERRUP	.011	.054	.064	.021	.043	.021	.021	.054	.107	.032	.011	.0	.021	.183	.054	.279	.021

4.14. TRANSITION MATRIX FOR CHANGE OF SUBJECT W65/W72

	1	2	3	4	5	6	7	8	9	10	11	12	13	14	15	16	17
1 THEOLOGY	.174	.0	.014	.0	.014	.0	.0	.0	.0	.0	.0	.0	.0	.072	.101	.594	.014
2 MEDICINE	.004	.412	.009	.004	.007	.007	.003	.003	.016	.0	.001	.0	.006	.066	.018	.388	.057
3 LAW	.003	.003	.370	.008	.006	.008	.003	.006	.003	.0	.0	.0	.028	.050	.031	.417	.064
4 ECON/BUS	.0	.002	.013	.236	.009	.002	.0	.0	.002	.0	.002	.002	.015	.133	.011	.530	.042
5 SOC SCI	.0	.025	.014	.018	.199	.011	.0	.0	.004	.0	.0	.0	.014	.148	.264	.282	.018
6 HUMANIT	.0	.016	.004	.004	.012	.245	.030	.004	.002	.0	.0	.002	.012	.147	.326	.175	.026
7 HSIT SCI	.006	.0	.006	.019	.019	.019	.404	.0	.025	.0	.0	.0	.012	.180	.075	.211	.025
8 NAT S I	.003	.005	.005	.008	.010	.003	.010	.240	.030	.010	.0	.0	.010	.137	.124	.387	.023
9 NAT S II	.002	.017	.002	.012	.002	.002	.004	.012	.505	.002	.002	.0	.012	.083	.098	.200	.050
10 ENGINEER	.0	.012	.002	.010	.002	.004	.0	.004	.012	.090	.004	.0	.004	.106	.018	.726	.006
11 ARCHITEC	.0	.024	.0	.0	.008	.008	.024	.008	.0	.024	.110	.0	.016	.094	.024	.661	.0
12 OTHERS	.0	.035	.0	.035	.018	.018	.018	.0	.0	.0	.0	.088	.0	.123	.438	.228	.0
13 INTERRUP	.011	.043	.064	.021	.032	.021	.011	.054	.097	.032	.0	.0	.021	.183	.064	.312	.032

4.15. TRANSITION MATRIX FOR CHANGE OF SUBJECT W65/S73

	1	2	3	4	5	6	7	8	9	10	11	12	13	14	15	16	17
1 THEOLOGY	.159	.0	.014	.0	.014	.0	.014	.0	.0	.0	.0	.0	.0	.072	.101	.609	.014
2 MEDICINE	.003	.222	.007	.004	.006	.006	.003	.003	.013	.0	.001	.0	.006	.066	.018	.538	.106
3 LAW	.003	.003	.316	.011	.006	.008	.003	.006	.003	.0	.0	.0	.028	.050	.042	.437	.084
4 ECON/BUS	.0	.002	.013	.208	.009	.002	.0	.0	.002	.0	.002	.002	.013	.133	.013	.550	.049
5 SOC SCI	.0	.025	.014	.018	.170	.014	.0	.0	.004	.0	.0	.0	.014	.148	.268	.293	.029
6 HUMANIT	.0	.016	.006	.002	.014	.205	.026	.004	.002	.0	.0	.0	.012	.151	.328	.197	.042
7 HSIT SCI	.006	.006	.006	.019	.019	.019	.335	.0	.025	.0	.0	.0	.012	.186	.081	.242	.043
8 NAT S I	.003	.005	.005	.008	.018	.0	.010	.207	.025	.010	.0	.0	.008	.137	.129	.400	.038
9 NAT S II	.002	.013	.002	.010	.002	.0	.004	.012	.453	.004	.002	.0	.012	.083	.098	.217	.090
10 ENGINEER	.0	.008	.002	.010	.004	.002	.0	.002	.008	.082	.002	.0	.004	.106	.022	.738	.010
11 ARCHITEC	.0	.024	.0	.0	.008	.0	.024	.008	.0	.024	.079	.0	.016	.102	.031	.685	.0
12 OTHERS	.0	.035	.0	.035	.018	.018	.018	.0	.0	.0	.0	.053	.0	.123	.456	.228	.018
13 INTERRUP	.011	.043	.043	.011	.032	.011	.011	.054	.086	.032	.0	.0	.021	.215	.064	.322	.043

4.16. TRANSITION MATRIX FOR CHANGE OF SUBJECT W65/W73

	1	2	3	4	5	6	7	8	9	10	11	12	13	14	15	16	17
1 THEOLOGY	.130	.0	.014	.0	.014	.0	.014	.0	.0	.0	.0	.0	.0	.072	.101	.638	.014
2 MEDICINE	.003	.194	.006	.004	.004	.006	.003	.003	.012	.0	.001	.0	.004	.066	.018	.564	.115
3 LAW	.003	.003	.302	.011	.006	.008	.003	.006	.003	.0	.0	.0	.025	.050	.042	.440	.098
4 ECON/BUS	.0	.002	.013	.197	.009	.002	.0	.0	.002	.0	.002	.002	.013	.133	.013	.553	.057
5 SOC SCI	.0	.025	.014	.018	.159	.014	.0	.0	.004	.0	.0	.0	.011	.152	.268	.297	.036
6 HUMANIT	.0	.014	.006	.002	.014	.191	.024	.004	.002	.0	.0	.0	.010	.155	.328	.207	.048
7 HSIT SCI	.006	.006	.006	.019	.019	.012	.317	.0	.025	.0	.0	.0	.012	.193	.081	.242	.062
8 NAT S I	.003	.005	.005	.008	.015	.0	.010	.190	.023	.010	.0	.0	.008	.144	.129	.402	.051
9 NAT S II	.002	.013	.002	.010	.002	.0	.004	.012	.434	.004	.002	.0	.010	.086	.098	.223	.102
10 ENGINEER	.0	.008	.002	.010	.004	.002	.0	.002	.008	.082	.002	.0	.004	.106	.022	.738	.010
11 ARCHITEC	.0	.024	.0	.0	.008	.0	.024	.008	.0	.024	.079	.0	.016	.102	.031	.685	.0
12 OTHERS	.0	.035	.0	.018	.018	.018	.018	.0	.0	.0	.0	.035	.0	.140	.456	.246	.018
13 INTERRUP	.0	.0	.0	.0	.0	.0	.0	.0	.0	.0	.0	.0	.0	.0	.0	.0	.0

4.17. TRANSITION MATRIX FOR CHANGE OF SUBJECT W65/S74

	1	2	3	4	5	6	7	8	9	10	11	12	13	14	15	16	17
1 THEOLOGY	.058	.0	.014	.0	.014	.0	.0	.0	.0	.0	.0	.0	.0	.072	.101	.725	.014
2 MEDICINE	.001	.081	.003	.001	.001	.003	.001	.001	.004	.0	.0	.0	.001	.066	.021	.663	.213
3 LAW	.003	.0	.174	.006	.003	.006	.003	.003	.003	.0	.0	.0	.014	.050	.053	.465	.218
4 ECON/BUS	.0	.002	.009	.122	.004	.002	.0	.0	.002	.0	.0	.002	.007	.133	.018	.579	.119
5 SOC SCI	.0	.014	.007	.011	.087	.007	.0	.0	.004	.0	.0	.0	.004	.152	.282	.369	.069
6 HUMANIT	.0	.010	.004	.002	.010	.127	.016	.002	.002	.0	.0	.0	.006	.155	.336	.253	.082
7 HSIT SCI	.006	.006	.006	.012	.012	.006	.205	.0	.012	.0	.0	.0	.006	.193	.093	.323	.118
8 NAT S I	.003	.003	.003	.005	.010	.0	.008	.129	.015	.008	.0	.0	.005	.144	.137	.445	.089
9 NAT S II	.002	.008	.002	.006	.0	.0	.002	.008	.259	.002	.002	.0	.006	.086	.111	.265	.244
10 ENGINEER	.0	.006	.002	.006	.002	.002	.0	.0	.006	.053	.002	.0	.002	.106	.031	.755	.029
11 ARCHITEC	.0	.016	.0	.0	.0	.0	.016	.0	.0	.016	.039	.0	.008	.102	.055	.748	.0
12 OTHERS	.0	.018	.0	.0	.018	.0	.018	.0	.0	.0	.0	.018	.0	.140	.474	.281	.035
13 INTERRUP	.0	.0	.0	.0	.0	.0	.0	.0	.0	.0	.0	.0	.0	.0	.0	.0	.0

4.18. TRANSITION MATRIX FOR CHANGE OF SUBJECT W65/W74

	1	2	3	4	5	6	7	8	9	10	11	12	13	14	15	16	17
1 THEOLOGY	.043	.0	.014	.0	.014	.0	.0	.0	.0	.0	.0	.0	.0	.072	.101	.739	.014
2 MEDICINE	.001	.060	.001	.001	.001	.003	.001	.001	.003	.0	.0	.0	.0	.066	.024	.678	.220
3 LAW	.003	.0	.120	.003	.003	.003	.003	.003	.003	.0	.0	.0	.008	.050	.053	.484	.263
4 ECON/BUS	.0	.002	.007	.086	.004	.002	.0	.0	.0	.0	.0	.002	.004	.133	.020	.597	.141
5 SOC SCI	.0	.011	.004	.007	.065	.004	.0	.0	.004	.0	.0	.0	.004	.152	.290	.380	.087
6 HUMANIT	.0	.008	.002	.002	.008	.096	.012	.002	.002	.0	.0	.0	.004	.155	.338	.277	.098
7 HSIT SCI	.006	.006	.006	.006	.006	.006	.161	.0	.012	.0	.0	.0	.006	.193	.099	.348	.143
8 NAT S I	.003	.003	.003	.005	.008	.0	.005	.101	.013	.005	.0	.0	.003	.144	.142	.460	.109
9 NAT S II	.002	.006	.002	.004	.0	.0	.0	.004	.171	.002	.002	.0	.004	.086	.115	.278	.307
10 ENGINEER	.0	.004	.002	.004	.002	.002	.0	.0	.004	.041	.0	.0	.002	.106	.035	.761	.039
11 ARCHITEC	.0	.016	.0	.0	.0	.0	.016	.0	.0	.008	.031	.0	.008	.102	.055	.756	.008
12 OTHERS	.0	.0	.0	.0	.018	.0	.018	.0	.0	.0	.0	.018	.0	.140	.474	.298	.035
13 INTERRUP	.0	.0	.0	.0	.0	.0	.0	.0	.0	.0	.0	.0	.0	.0	.0	.0	.0

4.19. TRANSITION MATRIX FOR CHANGE OF SUBJECT W65/S75

	1	2	3	4	5	6	7	8	9	10	11	12	13	14	15	16	17
1 THEOLOGY	.029	.0	.014	.0	.014	.0	.0	.0	.0	.0	.0	.0	.0	.072	.101	.739	.029
2 MEDICINE	.001	.051	.001	.001	.0	.001	.001	.001	.003	.0	.0	.0	.0	.066	.024	.684	.226
3 LAW	.003	.0	.062	.0	.003	.003	.0	.003	.0	.0	.0	.0	.003	.050	.053	.501	.330
4 ECON/BUS	.0	.002	.004	.046	.002	.002	.0	.0	.0	.0	.0	.0	.002	.133	.024	.606	.177
5 SOC SCI	.0	.007	.0	.004	.040	.0	.0	.0	.0	.0	.0	.0	.007	.152	.297	.387	.112
6 HUMANIT	.0	.004	.002	.002	.004	.064	.008	.002	.002	.0	.0	.0	.004	.155	.340	.299	.119
7 HSIT SCI	.006	.006	.0	.006	.0	.006	.081	.0	.006	.0	.0	.0	.0	.193	.112	.385	.199
8 NAT S I	.003	.003	.003	.003	.005	.0	.003	.061	.008	.003	.0	.0	.0	.144	.147	.473	.149
9 NAT S II	.0	.004	.002	.002	.0	.0	.0	.004	.119	.002	.002	.0	.002	.086	.117	.284	.359
10 ENGINEER	.0	.002	.002	.002	.0	.002	.0	.0	.002	.018	.0	.0	.0	.106	.041	.769	.057
11 ARCHITEC	.0	.008	.0	.0	.0	.0	.008	.0	.0	.008	.016	.0	.008	.102	.055	.779	.016
12 OTHERS	.0	.0	.0	.0	.0	.0	.018	.0	.0	.0	.0	.018	.0	.140	.474	.298	.053
13 INTERRUP	.0	.0	.0	.0	.0	.0	.0	.0	.0	.0	.0	.0	.0	.0	.0	.0	.0

4.20. TRANSITION MATRIX FOR CHANGE OF SUBJECT W65/W75

	1	2	3	4	5	6	7	8	9	10	11	12	13	14	15	16	17
1 THEOLOGY	.014	.0	.0	.0	.014	.0	.0	.0	.0	.0	.0	.0	.0	.072	.116	.739	.043
2 MEDICINE	.001	.037	.001	.001	.0	.001	.0	.001	.001	.0	.0	.0	.0	.066	.025	.692	.234
3 LAW	.0	.0	.045	.0	.003	.003	.0	.003	.0	.0	.0	.0	.003	.050	.056	.501	.347
4 ECON/BUS	.0	.002	.002	.038	.002	.002	.0	.0	.0	.0	.0	.0	.002	.133	.027	.606	.190
5 SOC SCI	.0	.004	.0	.004	.029	.0	.0	.0	.004	.0	.0	.0	.004	.152	.297	.391	.123
6 HUMANIT	.0	.002	.002	.002	.004	.058	.006	.002	.002	.0	.0	.0	.004	.155	.342	.306	.121
7 HSIT SCI	.006	.006	.0	.0	.0	.006	.056	.0	.006	.0	.0	.0	.0	.193	.112	.404	.211
8 NAT S I	.0	.0	.003	.003	.003	.0	.003	.048	.005	.0	.0	.0	.0	.144	.147	.478	.162
9 NAT S II	.0	.002	.002	.002	.0	.0	.0	.002	.090	.002	.002	.0	.002	.086	.117	.288	.388
10 ENGINEER	.0	.002	.002	.002	.0	.002	.0	.0	.002	.016	.0	.0	.0	.106	.041	.771	.057
11 ARCHITEC	.0	.008	.0	.0	.0	.0	.008	.0	.0	.008	.008	.0	.008	.102	.055	.779	.024
12 OTHERS	.0	.0	.0	.0	.0	.0	.018	.0	.0	.0	.0	.0	.0	.140	.474	.316	.053
13 INTERRUP	.0	.0	.0	.0	.0	.0	.0	.0	.0	.0	.0	.0	.0	.0	.0	.0	.0

4.21. TRANSITION MATRIX FOR CHANGE OF SUBJECT W65/S76

	1	2	3	4	5	6	7	8	9	10	11	12	13	14	15	16	17
1 THEOLOGY	.014	.0	.0	.0	.014	.0	.0	.0	.0	.0	.0	.0	.0	.072	.116	.739	.043
2 MEDICINE	.0	.025	.001	.0	.0	.001	.0	.001	.001	.0	.0	.0	.0	.066	.025	.698	.243
3 LAW	.0	.0	.028	.0	.003	.003	.0	.003	.0	.0	.0	.0	.0	.050	.056	.504	.364
4 ECON/BUS	.0	.0	.002	.027	.002	.002	.0	.0	.0	.0	.0	.0	.0	.133	.027	.606	.206
5 SOC SCI	.0	.004	.0	.0	.022	.0	.0	.0	.004	.0	.0	.0	.004	.152	.300	.391	.130
6 HUMANIT	.0	.002	.002	.002	.002	.042	.004	.002	.002	.0	.0	.0	.002	.155	.344	.308	.139
7 HSIT SCI	.006	.006	.0	.0	.0	.006	.037	.0	.0	.0	.0	.0	.0	.193	.112	.410	.230
8 NAT S I	.0	.0	.003	.003	.003	.0	.003	.035	.003	.0	.0	.0	.0	.144	.147	.481	.182
9 NAT S II	.0	.002	.0	.0	.0	.0	.0	.002	.044	.0	.0	.0	.002	.086	.119	.294	.434
10 ENGINEER	.0	.002	.0	.0	.0	.002	.0	.0	.0	.006	.0	.0	.0	.106	.043	.771	.071
11 ARCHITEC	.0	.008	.0	.0	.0	.0	.008	.0	.0	.008	.008	.0	.008	.102	.055	.779	.024
12 OTHERS	.0	.0	.0	.0	.0	.0	.018	.0	.0	.0	.0	.0	.0	.140	.474	.316	.053
13 INTERRUP	.0	.0	.0	.0	.0	.0	.0	.0	.0	.0	.0	.0	.0	.0	.0	.0	.0

APPENDIX B: PROGRAMS

APPENDIX B1: Program for the Computation of the Stocks at the Different Universities in each of the 17 Semesters

```
GET FILE        VERLAUF
RECODE          HSWES65 TO HSWES72,HSWEW65 TO HSWEW72 (1=10)(2=11)(10=1)
                (20=2)(30=3)(40=4)(50=5)(60=6)(70=7)(80=8)(90=9)
RECODE          BRUCHS65 TO BRUCHS73,BRUCHW65 TO BRUCHW72 (1 THRU 9=1)
COMPUTE         HSTOKW65=999
COMPUTE         HSTOKS66=999
COMPUTE         HSTOKW66=999
COMPUTE         HSTOKS67=999
COMPUTE         HSTOKW67=999
COMPUTE         HSTOKS68=999
COMPUTE         HSTOKW68=999
COMPUTE         HSTOKS69=999
COMPUTE         HSTOKW69=999
COMPUTE         HSTOKS70=999
COMPUTE         HSTOKW70=999
COMPUTE         HSTOKS71=999
COMPUTE         HSTOKW71=999
COMPUTE         HSTOKS72=999
COMPUTE         HSTOKW72=999
COMPUTE         HSTOKS73=999
IF              (HSWES65 EQ 999 AND LEAVESEM GT 1 AND LE 17) HSTOKW65=HSIMMAT
IF              (HSWES65 LT 999 AND LEAVESEM GT 1 AND LE 17) HSTOKW65=HSWES65
IF              (HSWEW65 EQ 999 AND LEAVESEM GT 2 AND LE 17) HSTOKS66=HSTOKW65
IF              (HSWEW65 LT 999 AND LEAVESEM GT 2 AND LE 17) HSTOKS66=HSWEW65
IF              (HSWES66 EQ 999 AND LEAVESEM GT 3 AND LE 17) HSTOKW66=HSTOKS66
IF              (HSWES66 LT 999 AND LEAVESEM GT 3 AND LE 17) HSTOKW66=HSWES66
IF              (HSWEW66 EQ 999 AND LEAVESEM GT 4 AND LE 17) HSTOKS67=HSTOKW66
IF              (HSWEW66 LT 999 AND LEAVESEM GT 4 AND LE 17) HSTOKS67=HSWEW66
IF              (HSWES67 EQ 999 AND LEAVESEM GT 5 AND LE 17) HSTOKW67=HSTOKS67
IF              (HSWES67 LT 999 AND LEAVESEM GT 5 AND LE 17) HSTOKW67=HSWES67
IF              (HSWEW67 EQ 999 AND LEAVESEM GT 6 AND LE 17) HSTOKS68=HSTOKW67
IF              (HSWEW67 LT 999 AND LEAVESEM GT 6 AND LE 17) HSTOKS68=HSWEW67
IF              (HSWES68 EQ 999 AND LEAVESEM GT 7 AND LE 17) HSTOKW68=HSTOKS68
IF              (HSWES68 LT 999 AND LEAVESEM GT 7 AND LE 17) HSTOKW68=HSWES68
IF              (HSWEW68 EQ 999 AND LEAVESEM GT 8 AND LE 17) HSTOKS69=HSTOKW68
IF              (HSWEW68 LT 999 AND LEAVESEM GT 8 AND LE 17) HSTOKS69=HSWEW68
IF              (HSWES69 EQ 999 AND LEAVESEM GT 9 AND LE 17) HSTOKW69=HSTOKS69
IF              (HSWES69 LT 999 AND LEAVESEM GT 9 AND LE 17) HSTOKW69=HSWES69
IF              (HSWEW69 EQ 999 AND LEAVESEM GT 10 AND LE 17) HSTOKS70=HSTOKW69
IF              (HSWEW69 LT 999 AND LEAVESEM GT 10 AND LE 17) HSTOKS70=HSWEW69
IF              (HSWES70 EQ 999 AND LEAVESEM GT 11 AND LE 17) HSTOKW70=HSTOKS70
IF              (HSWES70 LT 999 AND LEAVESEM GT 11 AND LE 17) HSTOKW70=HSWES70
IF              (HSWEW70 EQ 999 AND LEAVESEM GT 12 AND LE 17) HSTOKS71=HSTOKW70
IF              (HSWEW70 LT 999 AND LEAVESEM GT 12 AND LE 17) HSTOKS71=HSWEW70
IF              (HSWES71 EQ 999 AND LEAVESEM GT 13 AND LE 17) HSTOKW71=HSTOKS71
IF              (HSWES71 LT 999 AND LEAVESEM GT 13 AND LE 17) HSTOKW71=HSWES71
IF              (HSWEW71 EQ 999 AND LEAVESEM GT 14 AND LE 17) HSTOKS72=HSTOKW71
IF              (HSWEW71 LT 999 AND LEAVESEM GT 14 AND LE 17) HSTOKS72=HSWEW71
IF              (HSWES72 EQ 999 AND LEAVESEM GT 15 AND LE 17) HSTOKW72=HSTOKS72
IF              (HSWES72 LT 999 AND LEAVESEM GT 15 AND LE 17) HSTOKW72=HSWES72
IF              (HSWEW72 EQ 999 AND LEAVESEM EQ 17) HSTOKS73=HSTOKW72
IF              (HSWEW72 LT 999 AND LEAVESEM EQ 17) HSTOKS73=HSWEW72
IF              (BRUCHS65 EQ 1) HSTOKS65=12
IF              (BRUCHW65 EQ 1) HSTOKW65=12
IF              (BRUCHS66 EQ 1) HSTOKS66=12
IF              (BRUCHW66 EQ 1) HSTOKW66=12
IF              (BRUCHS67 EQ 1) HSTOKS67=12
IF              (BRUCHW67 EQ 1) HSTOKW67=12
IF              (BRUCHS68 EQ 1) HSTOKS68=12
IF              (BRUCHW68 EQ 1) HSTOKW68=12
```

```
IF              (BRUCHS69 EQ 1) HSTOKS69=12
IF              (BRUCHW70 EQ 1) HSTOKW70=12
IF              (BRUCHS71 EQ 1) HSTOKS71=12
IF              (BRUCHW71 EQ 1) HSTOKW71=12
IF              (BRUCHS72 EQ 1) HSTOKS72=12
IF              (BRUCHW72 EQ 1) HSTOKW72=12
MISSING VALUES HSTOKW65 TO HSTOKS73(999)
CODEBOOK        HSTOKW65 TO HSTOKS73,BRUCHS65 TO BRUCHS73,BRUCHW65 TO BRUCHW72
```

APPENDIX B2: Program for the Computation of the $n_{ij}/n_{i.}$, given the n_{ij}'s and the $n_{i.}$'s.

```
      DIMENSION RT(13,13,16),C(13,17,16),FM(13,17,16)
      DO 10 K=1,16
   10 READ (5,101) ((FM(I,J,K), J=1,17),I=1,13)
  101 FORMAT(17F4.0)
      DO 14 K=1,16
   14 READ (5,104) ((RT(I,J,K),J=1,13),I=1,13)
  104 FORMAT(13F6.5)
      DO 12 K=1,16
      DO 12 I=1,13
      DO 12 J=1,17
      C(I,J,K)=0.
      DO 12 L=1,13
   12 C(I,J,K)=C(I,J,K) + RT(I,L,K)*FM(L,J,K)
      DO 13 K=1,16
      WRITE(6,1004)
   20 CONTINUE
      DO 21 I=1,13
   21 WRITE(6,102)  (C(I,J,K),J=1,17)
  102 FORMAT(' ',17(F4.3,2X))
   13 WRITE (6,103)
  103 FORMAT('-')
      STOP
      END
```

APPENDIX B3: Adjusting the Transition Matrices for Drop Out Errors

```
      DIMENSION DROP(10,21),UEM(11,16,21),UEMM(11,16,21),DIFF(10,21),
     *SUM(12,21),STOCK(11,11),TM(11,16,21)
      DO 10 K=1,21
   10 READ (5,101) ((UEM(I,J,K),J=1,16),I=1,11)
  101 FORMAT(16F5.3)
      READ (5,120) ((STOCK(I,J),J=1,11),I=1,11)
  120 FORMAT(11F5.0)
      READ (5,110) (DROP(I,1),I=1,10)
  110 FORMAT(10F5.3)
      DO 50 K=1,21
      DO 50 I=1,11
      DO 50 J=1,16
      TM(I,J,K) =0.
      DO 50 L=1,11
   50 TM(I,J,K) = TM(I,J,K) + STOCK(I,L)*UEM(L,J,K)
      DO 60 K=1,21
      DO 60 I=1,12
   60 SUM(I,K) =0.
      DO 70 K=1,21
      DO 70 J=1,12
      DO 70 I=1,11
   70 SUM(J,K) = SUM(J,K) + TM(I,J,K)
      DO 80 K=1,21
      WRITE (6,130) (SUM(I,K),I=1,12)
  130 FORMAT(' ',20X,F6.0)
   80 WRITE (6,131)
  131 FORMAT('-')
      DO 36 K=2,21
      DO 36 I=1,10
   36 DROP(I,K)=DROP(I,1)
      DO 37 K=1,21
      DO 37 I=1,10
   37 UEMM(I,13,K) = DROP(I,K)*UEM(I,13,K)
      DO 38 K=1,21
      DO 38 I=1,10
      DIFF(I,K) = UEMM(I,13,K) - UEM(I,13,K)
   38 UEM(I,13,K)    = UEMM(I,13,K)
      DO 39 K=1,21
      DO 39 I=1,10
      DO 39 J=1,10
      IF(I.NE.J) GO TO 39
      UEM(I,J,K) = UEM(I,J,K) - DIFF(I,K)
   39 CONTINUE
      DO 40 K=1,21
      IF(K/2*2.EQ.K) GO TO 18
      IF(K/2*2.NE.K) GO TO 19
   18 L=65 + K/2
      WRITE (6,1001) L
 1001 FORMAT('-','TRANSITION MATRIX FOR CHANGE OF UNIVERSITY W65/W',I2,'
     *, ADJUSTED FOR DROPOUT ERRORS')
      WRITE (6,1004)
 1004 FORMAT('-','  1      2      3      4      5      6      7      8      9
     * 10      11      12      13      14      15      16')
      GO TO 20
   19 L=65 + (K/2+1)
      WRITE (6,1002) L
 1002 FORMAT('1','TRANSITION MATRIX FOR CHANGE OF UNIVERSITY W65/S',I2,'
     *, ADJUSTED FOR DROPOUT ERRORS')
      WRITE (6,1004)
   20 WRITE (7,1003) ((UEM(I,J,K),J=1,16),I=1,11)
      etc.
```

APPENDIX B4: Calculation of the Mean Number of Semesters Necessary to Obtain a Specific Degree

```
      DIMENSION TRANS(13,17,22),DAUER(22),DISK(13,4,22),SUM(13,4),
     *T(13,4)
      DO 10 K=1,22
   10 READ (5,101)  ((TRANS(I,J,K),J=1,17),I=1,13)
  101 FORMAT(17F4.0)
      READ (5,102) (DAUER(K),K=1,16)
  102 FORMAT(16F5.2)
      READ (5,107) (DAUER(K),K=17,22)
  107 FORMAT(6F5.2)
      DO 20 K=1,22
      DO 20 I=1,13
      DO 20 J=1,4
   20 DISK(I,J,K)=0.
      DO 30 K=1,22
      DO 30 I=1,13
      DO 30 J=14,17
      M=J-13
   30 DISK(I,M,K) = DAUER(K)*TRANS(I,J,K)
      DO 40 I=1,13
      DO 40 J=1,4
   40 SUM(I,J) = 0.
      DO 50 I=1,13
      DO 50 J=1,4
      DO 50 K=1,22
   50 SUM(I,J) = SUM(I,J) + DISK(I,J,K)
      DO 80 I=1,13
      DO 80 J=14,17
      M=J-13
      T(I,M)=0.
      DO 80 K=1,22
   80 T(I,M) = T(I,M) + TRANS(I,J,K)
      DO 70 I=1,13
      DO 70 J=1,4
   70 SUM(I,J)=SUM(I,J)/T(I,J)
      WRITE (6,103)
  103 FORMAT('-','AVERAGE NUMBER OF SEMESTERS TO OBTAIN A SPECIFIC DEGRE
     *E')
      WRITE (6,104)
  104 FORMAT(' ','                          DROP-    LOW      MEDIUM   HIGH')
      WRITE (6,105)
  105 FORMAT(' ','                          OUTS     LEVEL    LEVEL    LEVEL')
      DO 60 I=1,13
   60 WRITE (6,106) (SUM(I,J),J=1,4)
  106 FORMAT(' ',21X,4(F4.1,4X))
      STOP
      END
```

APPENDIX B5: Adjusting the Prediction for Stocks

```
      DIMENSION UEM(13,17,11),STOCK(12,13,22),UEMS(13,17),TS(12,13,22),S
     *JAHR(12,13,11),C(12,17,11),SUM(12,17),SUMS(17),SSTOCK(12,13,22),
     *PROZ(12)
C
C READ INPUT DATA
C UEMS = TRANSITION MATRICES FOR W65 TO FOLLOWING W-SEMESTERS
C UEM  = TRANSITION MATRIX FOR S65/W65
C SSTOCK = STOCK DATA FOR 22 SEMESTERS (S63 TO W73)
C PROZ = PERCENTAGES FOR ADJUSTING FOR STOCK DATA ERRORS
C
      DO 10 I=1,13
   10 READ (5,101) (UEMS(I,J),J=1,17)
      DO 20 K=1,11
   20 READ (5,101) ((UEM(I,J,K),J=1,17),I=1,13)
  101 FORMAT(17F4.3)
      DO 30  K=1,22
   30 READ (5,102) ((SSTOCK(I,J,K),J=1,13),I=1,12)
  102 FORMAT(13F5.0)
      READ (5,109) (PROZ(I),I=1,12)
  109 FORMAT(12F5.3)
C
C ADJUSTING THE STOCK DATA
C
      DO 35 K=1,19
      DO 35 I=1,12
      DO 35 J=1,13
      STOCK(I,J,K)=0.
      IF(I.NE.J) GO TO 35
      STOCK(I,J,K) = PROZ(I) * SSTOCK(I,J,K)
   35 CONTINUE
      DO  36 K=20,22
      DO 36 I=1,12
      DO 36 J=1,13
   36 STOCK(I,J,K)=SSTOCK(I,J,K)
C
C TRANSFORMATION OF SUMMER STOCKS INTO WINTER STOCKS
      DO 40  K=1,21,2
      DO 40  I=1,12
      DO 40  J=1,17
      TS(I,J,K)=0.
      DO 40  L=1,13
   40 TS(I,J,K)=TS(I,J,K) + STOCK(I,L,K)*UEMS(L,J)
C COMPUTATION OF YEARLY STOCKS
      DO 50  K=1,21,2
      DO 50  I=1,12
      DO 50  J=1,17
      L=K+1
      M=L/2
   50 SJAHR(I,J,M)=TS(I,J,K)+STOCK(I,J,L)
C COMPUTATION OF THE NUMBERS OF STUDENTS REMAINING IN W73
      DO 60  K=1,11
      DO 60  I=1,12
      DO 60 J=1,17
      C(I,J,K)=0.
      DO 60 L=1,13
   60 C(I,J,K)=C(I,J,K) + SJAHR(I,L,K)*UEM(L,J,K)
      DO 75 I=1,12
      DO 75  J=1,17
   75 SUM(I,J)=0.
      DO 80  K=1,11
```

```
      DO 80  I=1,12
      DO 80  J=1,17
   80 SUM(I,J)=SUM(I,J) + C(I,J,K)
      WRITE (6,105)
  105 FORMAT('-','TOTAL NUMBER OF STUDENTS AT DIFFERENT SUBJECTS IN W73/
     *74 BY SUBJECT OF ORIGIN')
      DO 90  I=1,12
   90 WRITE (6,106) (SUM(I,J),J=1,17)
  106 FORMAT(' ',17F7.0)
      DO 95 I=1,17
   95 SUMS(I)=0.
      DO 100 J=1,17
      DO 100     I=1,12
  100 SUMS(J)=SUMS(J) + SUM(I,J)
      WRITE (6,107)
  107 FORMAT('-','TOTAL NUMBER OF STUDENTS AT DIFFERENT SUBJECTS IN W73/
     *74')
      WRITE (6,108) (SUMS(I),I=1,17)
  108 FORMAT(' ',17F7.0)
      STOP
      END
```

APPENDIX B6: Program to Predict Transition Probabilities on the Basis of Independent Determinants and Individual Level Data

```
C
C     PROGRAM TO PREDICT TRANSITION PROBABILITIES ON THE BASIS OF INDEPE
C     NDENT VARIABLES
C
C
      DIMENSION COEFF(10,16,37),VAR(37),PROB(10,16),Y(15,10,16),
     *Z(15,10,16),TOTAL(15,10),CONST(10,16)
C
      DO 5 I=1,10
    5 READ (5,100) (CONST(I,J),J=1,16)
  100 FORMAT(10F8.5/6F8.5)
      DO 10 I=1,10
      DO 10 J=1,16
   10 READ (5,101) (COEFF(I,J,K),K=1,37)
  101 FORMAT(10F8.5/10F8.5/10F8.5/7F8.5)
C
      DO 11 I=1,15
      DO 11 J=1,10
      TOTAL(I,J)=0.
      DO 11 K=1,16
   11 Y(I,J,K)=0.
C
   12 READ (8,102,END=14) HSTOK1,HSTOK2,DEGREE,(VAR(K),K=1,37)
  102 FORMAT (28F3.0,12F7.3)
      IF(HSTOK1.GT.10.) GO TO 12
      DO 21 M=1,25
      IF(VAR(M).EQ.999.) GO TO 12
   21 CONTINUE
      I=VAR(1)
      J=HSTOK1
      TOTAL(I,J) = TOTAL(I,J) + 1.
      DO 20 K=1,16
      PROB(J,K) = CONST(J,K)
      DO 13 M=1,37
   13 PROB(J,K) = PROB(J,K) + COEFF(J,K,M) * VAR(M)
   20 Y(I,J,K) = Y(I,J,K) + PROB(J,K)
      GO TO 12
C
   14 DO 15 L=1,15
      DO 15 M=1,10
      DO 15 N=1,16
      IF(TOTAL(L,M).EQ.0.) GO TO 30
      Z(L,M,N) = Y(L,M,N)/TOTAL(L,M)
      IF(Z(L,M,N).LT.0.) Z(L,M,N) = 0.
      IF(Z(L,M,N).GT..999) Z(L,M,N) = .999
      GO TO 15
   30 Z(L,M,N) = 0.
   15 CONTINUE
C
      DO 18 L=1,15
      IF(L/2*2.EQ.L) GO TO 16
      LL = 64 + (L/2 + 1)
      MM = LL + 1
      WRITE (6,103) LL,MM
  103 FORMAT('1','PREDICTED TRANSITION MATRIX FOR CHANGE OF UNIVERSITY
     *W',I2,'/S',I2)
      GO TO 17
   16 LL = 65 + L/2
      WRITE (6,104) LL,LL
  104 FORMAT('-','PREDICTED TRANSITION MATRIX FOR CHANGE OF UNIVERSITY
```

```
     *S',I2,'/W',I2)
  17 WRITE (6,105)
 105 FORMAT(' ','BASEL BERN  ETHL  ETHZ  FRIB   HSG GENEVE LAUSA NEUCH
  18 WRITE (6,106) (Z(L,M,N),N=1,16)
 106 FORMAT(' ',16(F4.3,2X))
     DO 40 L=1,15
  40 WRITE (7,107) ((Z(L,M,N),N=1,16),M=1,10)
 107 FORMAT (16F5.3)
     STOP
     END
```

APPENDIX B7: Program for the Computation of Drop-Outs and Graduates by University of EXIT

```
      DIMENSION AV(21,12),EV(21,12),ATRANS(12,16,21),DSTOCK(12,16,21),
     *SUM(12,16,21),XLEAVE(12,16,21),C(12,16,21)
C
      DO 10 K=1,21
      DO 10 I=1,12
   10 READ (5,101) (ATRANS(I,J,K),J=1,16)
  101 FORMAT(16F5.3)
      READ (5,102) (AV(1,I),I=1,12)
  102 FORMAT(12F4.1)
C
      DO 15 K=1,21
      DO 15 I=1,12
      DO 15 J=13,16
   15 SUM(I,J,K) = 0.
C
C     PROJECTION OF INITIAL STOCKS
C
      DO 40 K=1,20
      M= K+1
      DO 30 I=1,12
      EV(M,I) = 0.
      DO 30 L=1,12
   30 EV(M,I) = EV(M,I) + AV(K,L) * ATRANS(L,I,K)
      DO 35 I=1,12
   35 AV(M,I) = EV(M,I)
   40 CONTINUE
C
C      DIAGONALIZATION OF STOCK-VECTORS
C
      DO 50 K=1,21
      DO 50 I=1,12
      DO 50 J=1,12
      DSTOCK(I,J,K) = 0.
      IF(I.NE.J) GO TO 50
      DSTOCK(I,J,K) = AV(K,I)
   50 CONTINUE
C
C      COMPUTATION OF LEAVERS
C
      DO 60 K=1,21
      DO 60 I=1,12
      DO 60 J=13,16
      C(I,J,K) = 0.
      DO 60 L=1,12
   60 C(I,J,K) = C(I,J,K) + DSTOCK(I,L,K) * ATRANS(L,J,K)
      DO 70 I=1,12
      DO 70 J=13,16
   70 SUM(I,J,1) = C(I,J,1)
      DO 75 K=2,21
      DO 75 I=1,12
      DO 75 J=13,16
      M = K-1
   75 SUM(I,J,K) = SUM(I,J,M) + C(I,J,K)
      DO 78 K=1,21
      DO 78 I=1,10
      DO 78 J=13,16
      XLEAVE(I,J,K) = SUM(I,J,K)/AV(1,I)
   78 CONTINUE
      DO 80 K=1,21
      WRITE (6,103) K

       etc.
```

APPENDIX B8 - A: Program for the Computation of the Transition Matrices of Homogenous Subpopulations

Part I: Creation of a Raw-Output Data File for the Homogenous Subpopulation

```
GET FILE         VERLAUF2
*SELECT IF       (SEX EQ 1)
WRITE CASES      (33F3.0)                P11,P12,P15,P110,P111,P112,P113 TO P116,
                 P22,P28,P210,P211,P212,P213 TO P216,P33,P312,P315,P44,P410,P411,
                 P412,P413 TO P416,P51,P52,P55,P57,P58,P510,P511,P512,P513 TO P51
                 ,P66,P611,P612,P613,P615,P71,P77,P710,P711,P712,P713 TO P715,P82
                 P87,P83,P811,P812,P813 TO P815,P97,P98,P99,P912,P913 TO P915,
                 P102,P105,P107,P108,P1010,P1011,P1012,P1013 TO P1016
FINISH
```

APPENDIX B8 - B: Program for the Computation of the Transition Matrices of Homogenous Subpopulations.

Part II: Computations of the Transition Probabilities

```
      DIMENSION V(81),SUM(15,81),TOTAL(15,10),P(15,81)
C
      DO 5 I=1,15
      DO 5 J=1,10
    5 TOTAL(I,J) = 0.
      DO 6 I=1,15
      DO 6 K=1,81
    6 SUM(I,K) = 0.
   10 READ (8,101,END=15) HSTOK1,SEMNUM,(V(K),K=1,81)
  101 FORMAT(83F3.0)
      IF(HSTOK1.GT.10.) GO TO 10
      I=SEMNUM
      J=HSTOK1
      TOTAL(I,J) = TOTAL(I,J) + 1.
      DO 11 K=1,81
      IF(V(K).NE.1.) GO TO 11
      SUM(I,K) = SUM(I,K) + 1.
      GO TO 10
   11 CONTINUE
      GO TO 10
   15 DO 16 I=1,15
      DO 16 J=1,10
      IF(TOTAL(I,J).EQ.0.) TOTAL(I,J) = 1.
   16 CONTINUE
   20 DO 40 I=1,15
      DO 40 K=1,81
      IF(K.LE.10) GO TO 30
      IF(K.GT.10.AND.K.LE.19) GO TO 31
      IF(K.GE.20.AND.K.LE.22) GO TO 32
      IF(K.GE.23.AND.K.LE.30) GO TO 33
      IF(K.GE.31.AND.K.LE.42) GO TO 34
      IF(K.GE.43.AND.K.LE.47) GO TO 35
      IF(K.GE.48.AND.K.LE.55) GO TO 36
      IF(K.GE.56.AND.K.LE.63) GO TO 37
      IF(K.GE.64.AND.K.LE.70) GO TO 38
      IF(K.GE.71) GO TO 39
   30 P(I,K) = SUM(I,K)/TOTAL(I,1)
      GO TO 40
   31 P(I,K) = SUM(I,K)/TOTAL(I,2)
      GO TO 40
   32 P(I,K) = SUM(I,K)/TOTAL(I,3)
      GO TO 40
   33 P(I,K) = SUM(I,K)/TOTAL(I,4)
      GO TO 40
   34 P(I,K) = SUM(I,K)/TOTAL(I,5)
      GO TO 40
   35 P(I,K) = SUM(I,K)/TOTAL(I,6)
      GO TO 40
   36 P(I,K) = SUM(I,K)/TOTAL(I,7)
      GO TO 40
   37 P(I,K) = SUM(I,K)/TOTAL(I,8)
      GO TO 40
   38 P(I,K) = SUM(I,K)/TOTAL(I,9)
      GO TO 40
   39 P(I,K) = SUM(I,K)/TOTAL(I,10)
   40 CONTINUE
C
      Z = 0.
      DO 50 I=1,15
   50 WRITE (7,106) P(I,1),P(I,2),Z,Z,P(I,3),Z,Z,Z,Z,

      etc.
```

APPENDIX B9: Program to Predict Transition Probabilities on the Basis of Independent Determinants and Semester Means

```
      DIMENSION COEFF(10,16,37),CONST(10,16),AVER(10,37,21),
     *RESULT(10,16,21)
C
      DO 5 I=1,10
    5 READ (8,100) (CONST(I,J),J=1,16)
  100 FORMAT(10F8.5/6F8.5)
      DO 10 I=1,10
      DO 10 J=1,16
   10 READ (8,101) (COEFF(I,J,K),K=1,37)
  101 FORMAT (10F8.5/10F8.5/10F8.5/7F8.5)
      DO 20 I=1,10
      DO 20 K=1,37
   20 READ (9,107) (AVER(I,K,L),L=1,15)
  107 FORMAT(15F5.3)
      DO 21 I=1,10
      DO 21 K=1,37
   21 READ (9,110) (AVER(I,K,L),L=16,21)
  110 FORMAT(6F5.3)
      DO 25 I=1,10
      DO 25 L=10,21
   25 AVER(I,27,L) = 10.* AVER(I,27,L)
      DO 26 I=1,10
   26 AVER(I,37,21) = 10.*AVER(I,37,21)
      DO 27 K=1,21
      DO 27 I=1,10
      DO 27 J=1,10
      COEFF(I,J,16) = 0.
   27 CONTINUE
C
      DO 30 L=1,21
      DO 30 I=1,10
      DO 30 J=1,16
      RESULT(I,J,L) = CONST(I,J)
      DO 30 K=1,37
   30 RESULT(I,J,L) = RESULT(I,J,L) + COEFF(I,J,K) * AVER(I,K,L)
      DO 18 L=1,21
      IF(L/2*2.EQ.L) GO TO 16
      LL = 64 + (L/2 + 1)
      MM = LL + 1
      WRITE (6,103) LL,MM
  103 FORMAT('1','PREDICTED TRANSITION MATRIX FOR CHANGE OF UNIVERSITY
     *W',I2,'/S',I2)
      GO TO 17
   16 LL = 65 + L/2
      WRITE (6,104) LL,LL
  104 FORMAT('-','PREDICTED TRANSITION MATRIX FOR CHANGE OF UNIVERSITY
     *S',I2,'/W',I2)
   17 WRITE (6,105)
  105 FORMAT(' ','BASEL BERN  ETHL  ETHZ  FRIB   HSG GENEVE LAUSA NEUCH
     *ZURIC FORGN INTER DROP   LOW MEDIUM HIGH')
      DO 18 M=1,10
      WRITE (7,120) (RESULT(M,N,L),N=1,16)
  120 FORMAT(16F5.3)
   18 WRITE (6,106) (RESULT(M,N,L),N=1,16)
  106 FORMAT(' ',16(F4.3,2X))
      STOP
      END
```

APPENDIX B10: Program to Predict the Total Number of Students in a Single Semester Using Type 1 Matrices

```
      DIMENSION UEM(12,16,21),STOCK(12,11,22),UEMS(12,16),TS(12,12,22),
     *SJAHR(12,12,11),SSTOCK(12,12,22),PROZ(10),SUM(10),
     *AV(11,12),EV(11,12),AVA(21,12),EVE(21,12),AVO(11,12)
      DO 10 I=1,12
   10 READ (5,101) (UEMS(I,J),J=1,16)
      DO 20 K=1,21
   20 READ (5,101) ((UEM(I,J,K),J=1,16),I=1,12)
  101 FORMAT (16F5.3)
      DO 30 K=1,22
   30 READ (5,102) ((SSTOCK(I,J,K),J=1,11),I=1,10)
  102 FORMAT(11F5.0)
      READ (5,109) (PROZ(I),I=1,10)
  109 FORMAT(10F5.3)
      DO 35 K=1,19
      DO 35 I=1,10
      DO 35 J=1,11
      STOCK(I,J,K) =0.
      IF(I.NE.J) GO TO 35
      STOCK(I,J,K) = PROZ(I) * SSTOCK(I,J,K)
   35 CONTINUE
      DO 36 K=20,22
      DO 36 I=1,10
      DO 36 J=1,11
   36 STOCK(I,J,K) = SSTOCK(I,J,K)
      DO 40 K=1,21,2
      DO 40 I=1,10
      DO 40 J=1,11
      TS(I,J,K) = 0.
      DO 40 L=1,11
   40 TS(I,J,K) = TS(I,J,K) + STOCK(I,L,K) * UEMS(L,J)
      DO 50 K=1,21,2
      DO 50 I=1,10
      DO 50 J=1,11
      L=K+1
      M=L/2
      SJAHR(I,J,M) = TS(I,J,K) + STOCK(I,J,L)
      IF(I.NE.J) GO TO 50
      AVO(M,I) = SJAHR(I,J,M)
   50 CONTINUE
      DO 51 I=1,10
      AV(1,I) = AVO(11,I)
      AV(2,I) =AVO(10,I)
      AV(3,I) = AVO(9,I)
      AV(4,I)=AVO(8,I)
      AV(5,I)=AVO(7,I)
      AV(6,I)=AVO(6,I)
      AV(7,I)=AVO(5,I)
      AV(8,I)=AVO(4,I)
      AV(9,I)=AVO(3,I)
      AV(10,I) = AVO(2,I)
   51 AV(11,I) = AVO(1,I)
      DO 92 K=1,11
   92 WRITE (6,110) (AV(K,I),I=1,10)
  110 FORMAT('-',10F10.0)
      DO 90 K=1,11
      DO 90 I=11,12
   90 AV(K,I) =0.
      DO 89 I=1,11
   89 EV(1,I) = AV(1,I)
      DO 63 N=2,11
```

```
      DO 52 I=1,12
   52 AVA(1,I) = AV(N,I)
      DO 60 K=1,21
      M = K+1
      DO 55 I=1,12
      EVE(M,I) = 0.
      DO 55 L=1,12
   55 EVE(M,I) = EVE(M,I) + AVA(K,L) * UEM(L,I,K)
      DO 58 I=1,12
   58 AVA(M,I) = EVE(M,I)
      J= 2*N-1
      IF(J.EQ.K) GO TO 61
   60 CONTINUE
   61 DO 62 I=1,10
   62 EV(N,I) = EVE(M,I)
   63 CONTINUE
      DO 91 K=1,11
   91 WRITE (6,110) (EV(K,I),I=1,10)
      DO 70 I=1,10
   70 SUM(I) =0.
      DO 80 I=1,10
      DO 80 K=1,11
   80 SUM(I) = SUM(I) + EV(K,I)
      WRITE (6,105) (SUM(I),I=1,10)
  105 FORMAT(10F10.0)
      STOP
      END
```